INTERNATIONAL TECHNOLOGICAL UNIVERSITY
This Book is Donated by:
PROF. SHU-PARK CHAN

Date:

Generalized Networks

Holt, Rinehart and Winston Series in Electrical Engineering, Electronics, and Systems

Other Books in the Series:

Shu-Park Chan, Introductory Topological Analysis of Electrical Networks
George R. Cooper and Clare D. McGillem, Probabilistic Methods of Signal and System Analysis
Woodrow W. Everett, Arlon T. Adams, José Perini, et al., Topics in Intersystem Electromagnetic Compatibility
Mohammed S. Ghausi and John J. Kelly, Introduction to Distributed-Parameter Networks: With Application to Integrated Circuits
D. J. Hamilton, F. A. Lindholm, A. H. Marshak, Principles and Applications of Semiconductor Device Modeling
Roger H. Holmes, Physical Principles of Solid State Devices
Benjamin J. Leon, Lumped Systems
Benjamin J. Leon and Paul A. Wintz, Basic Linear Networks for Electrical and Electronics Engineers
Clare D. McGillem and George R. Cooper, Continuous and Discrete Signal and System Analysis
Samuel Seeley, Electronic Circuits
Amnon Yariv, Introduction to Optical Electronics

Generalized Networks

RICHARD SAEKS

Associate Professor of Electrical Engineering
University of Notre Dame
Notre Dame, Indiana

HOLT, RINEHART AND WINSTON, INC.
New York   Chicago   San Francisco   Atlanta
Dallas   Montreal   Toronto   London   Sydney

To my parents

"I am not concerned that a man does not know of me;
I am concerned that I do not know of him."

— CONFUCIUS, *Analects:* 1 — 16

Copyright © 1972 by Holt, Rinehart, and Winston, Inc.
All rights reserved
Library of Congress Catalog Card Number: 76-162646
ISBN: 0-03-085195-5
Printed in the United States of America
0 1 2 3   038   9 8 7 6 5 4 3 2 1

Preface

Over the past decade a great number of technological and scientific changes have occurred which have significantly altered the role of the engineering theoretician. It is my conviction that this new role necessitates a radical change in the methodology and teaching of the various engineering theories; this book is the result of an endeavor to achieve such a metamorphosis in my own field, network theory.

Possibly the most significant engineering developments of the recent past have been the explosion of solid-state technology with its apparently unending proliferation of new devices and the subsequent application of these devices in the modern digital computer. To the theoretician the former calls for the formulation of ever more general theories, capable of coping with the various devices as they appear, while the latter places in his hands a computational tool of sufficient power to permit such theories to be effectively implemented. These two related developments thus act in concert to call for the establishment of an ever more general theory; the former creating the need and the latter the means.

It is to these needs created by the advances of modern technology that the theory of generalized networks espoused herein is attuned. In order to cope with the ever expanding device technology, the theory is based on fundamental concepts which are independent of the characteristics of the specific devices from which a network might be constructed. Though powerful, such an approach does not lend itself to the simple polynomial coefficient calculation of classical network theory; rather one must invoke the powerful computational capability of the modern computer, also a result of the device explosion. Thus in network theory, as with the other engineering theories, the device explosion has created the need for a generalized theory and simultaneously created the tool needed to implement such a theory.

Although network theory has traditionally been preoccupied with the study of special cases, general theory has not been neglected (though it is often overlooked). The foundations of the generalized theory were laid in the late 1940s by the parallel development of a set of axioms for network theory and the introduction of scattering techniques to the theory. The former yielded definitions for the various network theoretic concepts which were independent of the characteristics of any specific device, while the latter served to define a class of network variables which could be readily identified with physically significant quantities in a wide variety of devices. A significant characteristic of both is that they separated the theory of a network from the theory of its components, thereby paving the way for the "elementless" networks of the generalized theory. In the 1950s and 1960s this trend toward "elementless" networks continued; first, with the formulation of a frequency domain representation for the generalized networks predicated entirely on the network axioms and the characteristics of the scattering variables, and then with the development of a corresponding time domain theory. Finally, in the 1960s, operator theoretic techniques were introduced which served to unify the two representations and allow the formulation of highly general analysis and synthesis procedures. It is the exposition of the resulting theory combined with the great body of applications which have arisen alongside that make up the main substance of this book.

My goal in writing this book is twofold: first, to introduce to the student a methodology in network theory which I believe to be better suited to the needs of modern technology than that presently in vogue; and second, to make available an exposition of one aspect of the theory which has yet to appear in book form. Unfortunately, these two goals are often in conflict; the former demands an intuitive development aimed at the lowest possible level in the student's education, whereas the latter implies a mathematically precise formulation aimed at the mathematically sophisticated. The resultant compromise is possibly best described as "bogus mathematics", this dubious goal being achieved through the use of intuitive discussion and heuristic arguments to motivate the results which are, however, stated formally as definitions and theorems. It is hoped that this approach will serve to motivate the novice network theorist while giving the researcher the skeleton of a rigorous exposition. The proofs are once again a compromise between the fact of exposition and the fiction of motivation. Many of these are mere sketches, these being predicated on the assumption that the expert can fill in the details himself while the novice is uninterested in such.

With this compromise between rigor and motivation, the book may be used for a first graduate course in network theory. In this application the book is self-contained with the necessary results of classical network theory being developed simultaneously with the general theory. In practice, I have used Chapters 1 through 7 and Appendix B for a one-semester course at this level, wherein the majority of students had no previous background in the subject. For those who might prefer to delete the time domain material, Chapters 1 through 3, Sections 4-6 and 5-3, and Chapters 6, 7, and 8 form an essentially self-contained introduction to lumped and distributed networks.

Also, with the compromise between motivation and exposition in mind, I have included about 150 problems, these taking the form of unproven statements (denoted by a star) within the body of the text. My experience is that this serves to give the student an opportunity to familiarize himself with the concepts and verify the results while at the same time trying his hand at the formal method. Moreover, inclusion of this material within the main body of the text allows a considerable expansion of the material covered by the exposition within the limited space available.

The book could readily have been divided into three parts. Chapters 1 through 3, dealing with the basic theory, form the basis for the entire development. Here examples of specific components are used to illustrate the basic ideas, but the concepts themselves are essentially elementless and are applicable to all linear networks. Chapters 4 and 5 are concerned with the analysis and synthesis of generalized networks. Since the techniques are applicable to both the time and frequency domains, an operator theoretic notation is employed which may be translated into either the time or frequency domain. The last four chapters may be viewed as illustrative examples wherein the general theory is brought down to earth for specific applications. These include both constant and time-variable RLC networks, transmission line networks, and multivariable (or "mixed") networks. Finally, there are three appendixes, their purpose being to formalize certain mathematical techniques which are considered only heuristically in the main text. In this way the student may take the time to study distribution theory before reading Chapter 2 wherein it is employed (though only heuristically), or he may skip the Appendix relying on intuition to carry him through the chapter.

For a book of such generalized scope to acknowledge all of those who have contributed ideas and/or techniques would entail a listing of most of the members of the network fraternity, past and present, my teachers through the years, and my students. Of course, such a listing is impossible; I do, however, want to thank those of my colleagues whose comments and suggestions on the preliminary drafts of the manuscript led to the present text. To list but a few: B. D. Anderson, N. DeClaris, B. J. Leon, S. R. Liberty, R-W. Liu, S. K. Mitra, B. R. Myers, R. W. Newcomb, and M. R. Wohlers. Finally I would like to thank my colleague R. Jeffery Leake for the innumerable discussions and conversations wherein were laid the foundations for the viewpoint presented here.

The help and cooperation of the editors and staff at Holt, Rinehart and Winston and the careful typing of the manuscript by Mrs. Carrol Pfender and Mrs. Clarabelle Brown are greatly appreciated.

September 1971   RICHARD SAEKS

Contents

Preface   v

Introduction   **The Network Concept**   1

Chapter 1   **Foundations**   8
- 1-1   Introduction   8
- 1-2   Network Variables   12
- 1-3 n-Ports   18
- 1-4   Networks   27
- 1-5   Fundamental Axioms   34
- 1-6   Passive Networks   47
- 1-7   Discussion   52

Chapter 2   **Time Domain Representation of Linear Networks**   54
- 2-1   Introduction   54
- 2-2   The Scattering Matrix   60
- 2-3   Passive and Lossless Scattering Matrices   68
- 2-4   Reciprocal and Time-Invariant Scattering Matrices   78
- 2-5   Normalized Scattering Matrices   84
- 2-6   General Representations   85
- 2-7   Normality   92
- 2-8   Discussion   95

Chapter 3   **Frequency Domain Representation of Linear Networks**   96
- 3-1   Introduction   96
- 3-2   The System Function   99
- 3-3   Existence of the Scattering Matrix   108
- 3-4   Passive and Lossless Scattering Matrices   112
- 3-5   Network Theorems   123
- 3-6   Discussion   126

Chapter 4   **Topological Analysis**   127
- 4-1   Introduction   127
- 4-2   Graph Matrices and Sources   131
- 4-3   Topological Analysis Via the π Matrix   135

	4-4	Topological Analysis Via the Scattering Matrix 141
	4-5	Analysis with Normalized Scattering Matrices 149
	4-6	Topological Analysis Via the Immittance Matrices 155
	4-7	Cascade Loading 158
	4-8	Analysis of Elementary n-Ports 165
	4-9	Discussion 170

Chapter 5 Realizability Theory 172

5-1 Introduction   172
5-2 Realization of Active π and Scattering Matrices   173
5-3 Synthesis of Passive Networks by Bordering   180
5-4 Equivalent Realizations   187
5-5 Realization of Immittance Matrices   189
5-6 Synthesis of Memoryless n-Ports   193
5-7 Discussion   200

Chapter 6 RLC Networks 202

6-1 Introduction   202
6-2 Properties of RLC Networks   204
6-3 Richards' Theorem   211
6-4 Synthesis of Lossless 1-Ports   213
6-5 Darlington-Type Realizations   220
6-6 Synthesis of Passive n-Ports   231
6-7 Synthesis of Lossless n-Ports   237
6-8 Discussion   241

Chapter 7 Transmission Line Networks 242

7-1 Introduction   242
7-2 Transmission Line Characteristics   244
7-3 The Richards Transformation   249
7-4 Cascade Synthesis   253
7-5 Frequency Dependent Lines   261
7-6 Discussion   264

Chapter 8 Multivariable Networks 265

8-1 Introduction   265
8-2 Multivariable Functions   266
8-3 Multivariable Decomposition   272
8-4 Passive Multivariable Synthesis   284
8-5 Lossless Multivariable Synthesis   287
8-6 Mixed RLC Transmission Line Networks   295
8-7 Variable Parameter Networks   300
8-8 Discussion   303

Chapter 9 Finite Time-Variable Networks; State-Space Techniques 304

9-1 Introduction   304
9-2 Time-Variable RLC Networks   306
9-3 Differential Operators and Dynamical Systems   310
9-4 Separable Impulse Responses   323

9-5	Synthesis Via Reactance Extraction	332
9-6	State-Space Passivity Conditions	342
9-7	Discussion	353

Appendix A   Infinite-Dimensional Vector Spaces and Their Operators   355

- A-1 Introduction 355
- A-2 Vector Spaces 356
- A-3 Operators 359
- A-4 Norms and Inner Products 365
- A-5 Topological Vector Spaces 369
- A-6 Discussion 372

Appendix B   Distributions   373

- B-1 Introduction 373
- B-2 Definitions 375
- B-3 The Kernels Theorem 380
- B-4 The Distributional Fourier Transform 384
- B-5 The Distributional Laplace Transform 387
- B-6 Discussion 389

Appendix C   Graph Theory   390

- C-1 Introduction 390
- C-2 Graph Theory 391
- C-3 Voltage and Current Vectors 397
- C-4 Fundamental Circuits and Cocircuits 402
- C-5 The Graph Matrices 407
- C-6 Matrix Characterization of the Network Variables 411
- C-7 Discussion 414

References   415

Index   427

Generalized Networks

INTRODUCTION

The Network Concept

Before embarking on the formal theory, it is desirable to delineate the scope of the generalized network concept; to determine its extent and to determine how it differs from its neighbors, system theory and control theory.

Classically, network theory dealt with connections of resistors, capacitors, and inductors for which the basic element was the RLC n-port illustrated in Figure I-1. Here it is essential to note that at each of these ports one measures

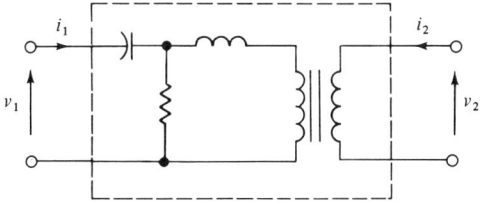

Figure I-1 RLC n-port and variables.

two variables, the port voltage and the port current. Moreover, these variables are not arbitrary; rather they are simultaneously constrained by the two Kirchhoff laws. It is precisely this choice of variables which separates the n-port from the commonly encountered (unilateral) systems.

The use of paired complimentary variables, while constraining the effects

2 INTRODUCTION

of interconnection, also serves to induce feedback into the simplest of structures. Consider, for example, the parallel resistors of Figure I-2. Here if one takes r_1 as the plant with equation

$$v_1 = r_1 i_1 \tag{I-1}$$

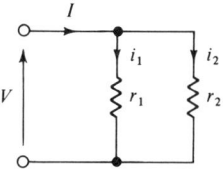

Figure I-2 Parallel resistor feedback structure.

and r_2 as the feedback loop with equation

$$i_2 = \left(\frac{1}{r_2}\right) v_2 = g_2 v_2 \tag{I-2}$$

it is found upon combining Equations (I-1) and (I-2) with the Kirchhoff laws

$$v_1 = V \tag{I-3}$$

$$v_2 = V \tag{I-4}$$

and

$$i_1 + i_2 = I \tag{I-5}$$

that

$$V = \frac{r_1}{1 + r_1 g_2} I \tag{I-6}$$

which is the usual negative feedback equality. Of course, one may equally well let r_2 be the plant and r_1 be the feedback element, in which case

$$V = \frac{r_2}{1 + r_2 g_1} I \tag{I-7}$$

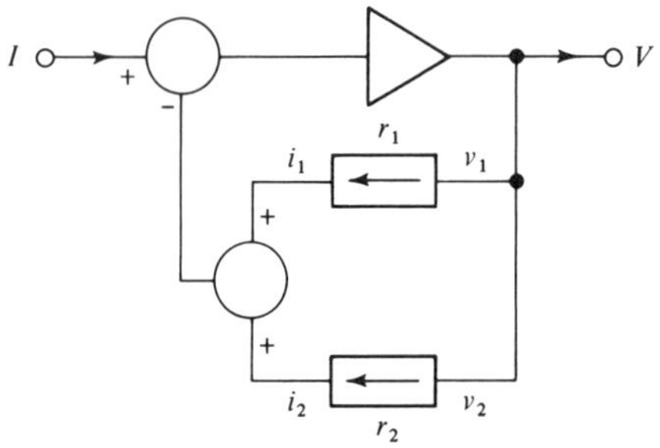

Figure I-3 Block diagram equivalent of parallel resistors.

This simplest of all network configurations thus exhibits a bilateral feedback characteristic which would require the rather complex block diagram of Figure I-3 if the unilateral components of control and system theory were employed rather than the bilateral network components. In this case, five components are needed to represent the five-network equations [Equations (I-1) through (I-5)].

Of course, one need not restrict consideration to electrical devices when studying RLC networks. Any physical system wherein the components are characterized by ordinary differential equations and the connections by Kirchhoff-type conservation laws is equivalent. For example, the fluid flow system of Figure I-4 may be taken as an RLC network. Here, voltage is replaced by

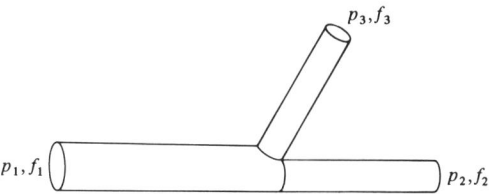

Figure I-4 Fluid network.

pressure and current by mass flow. Once again, however, the essence of the structure is the measurement of two variables at each port, in this case the pipes, which satisfy specified conservation laws. Analogous systems are innumerable; one may measure force and velocity at specified points in a mechanical system, mass flow and mass density in a traffic system, or cost and volume in an economic system. The equations in all cases are similar and they all may be viewed as generalized networks.

Since this book proposes to deal with "all" generalized networks, we should not restrict ourselves to this rather narrow class of networks, all of whose components are characterized by ordinary differential equations. Although such a restriction simplifies the required numerical operations it has no real significance in the theory and is readily dropped. Possibly the most common nondifferential networks are those composed of transmission lines and resistors, as is illustrated by the transmission line n-port of Figure I-5.

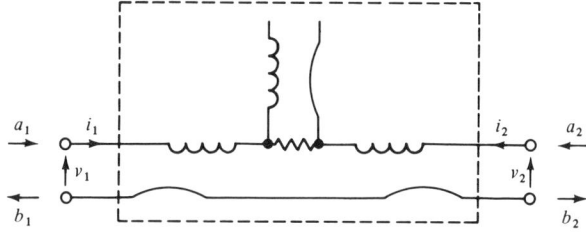

Figure I-5 Transmission line n-port.

In this network, the two variables at each network may once again be taken as the voltage and current (often called the immittance variables). Alternatively, such a network may be described by the reflection coefficients of classical transmission line theory. In this case, one takes as the input to the network an incident wave entering the ports. (This is usually denoted by an a together with an arrow toward the port.) Some of the energy carried by such a wave is absorbed in the network while the remainder is returned to the ports in the form of a reflected wave (which is usually denoted by b together with an arrow away from the port). These waves, known as the scattering variables, are completely equivalent to the immittance variables, but unlike voltage and current which represent two essentially different quantities, both of the scattering variables may be viewed as measures of the same quantity, though traveling in different directions. Although the actual units for the scattering variables differ with the physics of the system at hand, they usually behave in an "energy like" fashion and can be so interpreted. That is, the incident wave may be viewed as a measure of the energy entering a port and the reflected wave as a measure of the energy leaving a port.

Of course, like the immittance variables, the scattering variables also obey certain conservation laws. In particular, if two ports are connected back-to-back, as shown in Figure I-6, the incident wave entering one port must equal

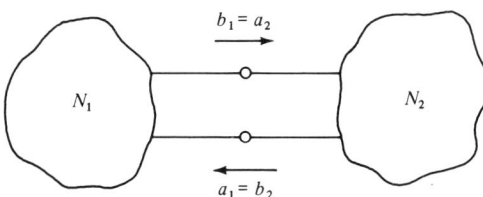

Figure I-6 Back-to-back ports and corresponding connection constraints.

the reflected wave leaving the other (that is, what goes in one side comes out the other). Now that the transmission line case is included within the network

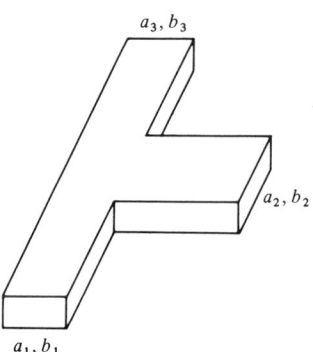

Figure I-7 Waveguide network.

concept, it is a natural step to raise the frequency spectrum under consideration to include the microwave region, replacing the transmission lines with waveguides. In this instance, as illustrated in Figure I-7, one no longer has identifiable pairs of wires between which to measure voltage and current (though the electric and magnetic field modes could serve as such). It is possible, however, to identify a wave of energy propagating down the guide and another returning; hence a network operating in the microwave region does have a physically significant set of scattering variables even though the immittance variables "fail to exist."

Of course, there is no reason to stop our generalizations with the electromagnetic portion of the spectrum; consider the lens of Figure I-8. Here a light

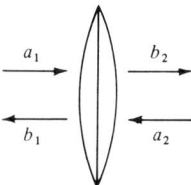

Figure I-8 Optical network.

wave is incident on one side of the lens (taken as a port), some of which is transmitted through the lens, and some of which is reflected. Thus, once again, the scattering approach yields a well-defined set of network variables even though the optical system under consideration does not have measurable immittance variables.

Finally, if one were to replace the incident and reflected waves by beams of particles entering and leaving a medium a completely analogous network struc-

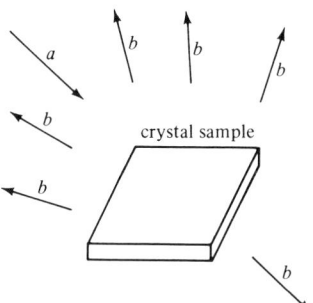

Figure I-9 Electron scattering from a semiconductor.

ture would be obtained. One example of this type is that shown in Figure I-9, wherein a crystal is bombarded by a beam of electrons. Here the beam itself may be taken as the incident "wave," while the reflected wave in this generalized

1-port may be taken as the electrons scattered from the crystal in some specified direction. This configuration, which is commonly employed to discern the characteristics of a semiconductor, is thus one more example of a generalized network.

Clearly, the scope of the incident and reflected wave concept is enormous with heat-energy diagrams, radiation, particle scattering, and information processing readily included. With the possible exception of the RLC example one may always define an n-port as an object possessed of n interfaces with its surroundings, as shown in Figure I-10. Each of these ports is, however, char-

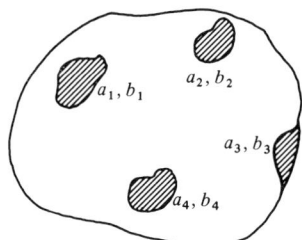

Figure I-10 Generalized n-port.

acterized by two variables, an incident wave entering the port and a reflected wave leaving it, these being constrained by specified conservation laws when one or more n-ports are connected.

Even in the generalized context when the immittance variables have little (if any) physical significance, they may still be identified mathematically, given the scattering variables. Since the incident wave enters the port and the reflected wave leaves it the total quantity flowing through the interface is the difference between that entering and that leaving; hence, the current may be taken as

$$i = a - b \qquad (I-8)$$

Similarly, the voltage or pressure is proportional to the total "activity" at a port, but independent of its direction. It is thus given by the sum of the scattering variables

$$v = a + b \qquad (I-9)$$

Application of Equations (I-8) and (I-9), usually with the addition of a scale factor, allows one to define the immittance variables in networks where only the scattering variables may be physically measured and, conversely, allows the definition of scattering variables in those networks where one may measure only the immittance variables. Thus two well-defined sets of network variables always exist. In particular, the scattering variables may be employed in the RLC case or the immittance variables in the waveguide network. In many respects, the scattering variables are mathematically more convenient to deal with than the immittance variables and are thus emphasized in our theory.

Since they exist in all generalized networks and are physically desirable in some, their use causes no difficulty while often greatly simplifying the mathematics.

If one takes the energy viewpoint of the scattering variables then the variables in certain of our networks are constrained by conservation of energy in addition to the connection constraints. For example, the energy leaving the lens in the reflected wave must not exceed that entering via the incident wave (since the lens may not exhibit an energy gain). Similarly, the number of electrons scattered from a semiconductor sample cannot exceed the number incident on it, while in the heat-energy and information processing networks a similar constraint is imposed by the requirement that the change of entropy be non-negative. Although physically different, these constraints may all be summed up by the requirement that the amplitude of the reflected wave be no greater than that of the incident wave, this being termed "passivity" in the generalized network context. Of course, in the RLC case, passivity corresponds to the requirement that the resistors, capacitors, and inductors all have non-negative values. The choice of physically oriented pairs of variables to describe an n-port thus naturally induces certain restrictions which take on a mathematically similar form even in networks which differ greatly physically. These may thus be defined and studied mathematically in the generalized context without reference to the physics of a specific problem.

In addition to passivity, one may define losslessness by requiring that the amplitude of the incident and the reflected wave be the same; causality by requiring that the incident wave precede the reflected wave; time-invariance by requiring that the characteristics of the n-port be independent of the choice of time scale; and so forth. In all these cases, and more, the fundamental idea is that physical constraints, though differing greatly from network to network, induce similar mathematical restrictions in all cases and hence may be studied in the generalized context. The physical interpretation available for the network variables thus serves to allow the various networks to be classified via physically significant constraints on their behavior, this is turn yielding a powerful structure fundamental to the study of all generalized networks which is not available to system and control theory.

In summary, the generalized networks, though highly varied physically, may always be characterized by pairs of variables, these being simultaneously constrained by the Kirchhoff conservation laws and the physics of the devices. It is precisely these physically motivated restrictions which separate the networks from the systems. Moreover, these constraints serve as the basis of a powerful mathematical structure on which a reasonably complete theory of generalized networks, not applicable to the more general systems, may be predicated.

CHAPTER

1
Foundations

1-1 Introduction

In the introduction the concepts of an n-port, a network, and the network variables were considered from an intuitive standpoint. In this chapter these ideas are formalized. Due to the great generality of the network concept one may not invoke physical and/or geometric considerations in achieving this formalization since the various networks differ greatly in these respects. Rather, it is necessary to adopt an entirely mathematical viewpoint wherein the network variables are taken simply as time functions, independent of whether they represent a flow of electrons, heat, or a quantity of information. Similarly, the geometric variance between different classes of networks precludes the consideration of the specific characteristics of a port in our theory. Here, however, the topological properties of a linear graph [23], [57], [155] may be employed to characterize the connection constraints within a network independently of the port geometry. Thus, whether a port is formed from a pair of wires, a waveguide, or an optical device, its essential properties may be represented by an edge of a linear graph. Even though there is great physical and geometric disparity between the various networks they are therefore unified by two threads, the port variables and the graphical connectivity structure. These two concepts thus form the basis of the entire theory; the port variables alone serve to define an n-port, whereas the graph defines the connection constraints which convert a collection of n-ports into a network.

In essence, an n-port may always be defined as a set of $2n$ time functions, n representing the incident waves at the n ports, and n the reflected waves (or equivalently n port voltages and n port currents). Of course, these $2n$ variables are not independent. Consider, for example, the mirror of Figure 1-1.

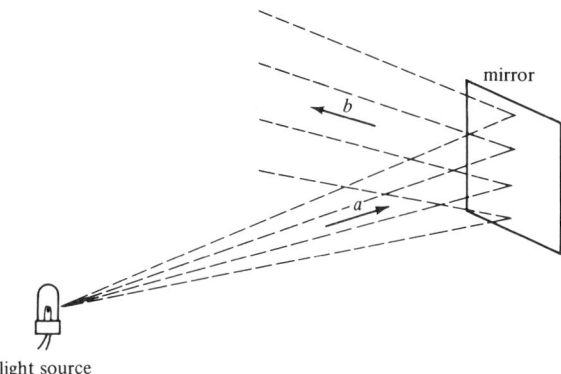

Figure 1-1 Optical 1-port.

This may readily be interpreted as a 1-port with the "quantity" of light impinging upon the surface taken as the incident wave and that leaving the surface as the reflected wave. Now if the mirror is a perfect reflector, all of the incident light is reflected; hence

$$b = a \tag{1-1}$$

is the constraint on this 1-port. Of course, one may invoke Equations (I-8) and (I-9) to convert this constraint to an equivalent one on the immittance variables for the mirror. In this case one finds that

$$v = a + b = 2a \tag{1-2}$$

is arbitrary whereas

$$i = a - b = 0 \tag{1-3}$$

The mirror is therefore a 1-port with arbitrary voltage and zero current; hence it is the optical equivalent of an open circuit. Since one does not desire to introduce physical considerations into our definitions the mirror and the open circuit are defined simply as the 1-port satisfying Equation (1-1), the two being indistinguishable within the scope of the theory. For the sake of tradition, however, the term open circuit is used to describe both. Of course, a real mirror does not reflect all of the light incident upon it; rather it satisfies an equation such as

$$b = Sa \tag{1-4}$$

where S is a positive real constant less than one which characterizes the percentage of the incident light reflected from the mirror. In this case a little manipulation of Equations (I-8) and (I-9) will reveal that the imperfect mirror is the equivalent of a resistor. Here again the optical 1-port is mathematically indistinguishable from a commonly encountered electrical device, and once again both are denoted via the electrical term, resistor. In general, an n-port may be defined simply as a set of admissible signal pairs (of incident and reflected waves), the particular pairs characterizing any given n-port being determined by its physics; this, however, does not enter into our considerations.

In many practical n-ports additional constraints are placed on the allowable signal pairs. For example, one may desire to assume that the incident wave completely determines the reflected wave or that the incident wave precedes the reflected wave in time. While these and other constraints are both physically relevant and useful, they are not needed in the basic definitions wherein n-ports with independent incident and reflected waves, predictors, and so forth, are allowed. Once the basic concept of an n-port has been developed, however, such constraints on the admissible pairs as solvability, causality, passivity, and time-invariance are considered, these playing a fundamental role in the later development.

Like the n-port itself the connection constraints imposed by the interconnection of one or more n-ports may also be formulated in an abstract context, in this case independently of the port geometry. Consider, for example, the network formed by placing a mirror and a lamp in "parallel," as shown in Figure 1-2. Here the mirror satisfies the 1-port constraint of Equation (1-4), whereas the lamp is a "wave source" characterized by

$$b_L = k \tag{1-5}$$

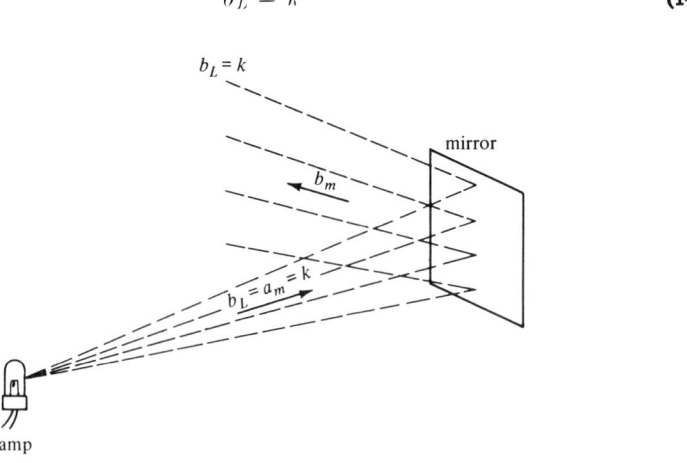

Figure 1-2 Mirror and lamp in parallel.

with k a fixed constant representing the intensity of the lamp. Although neither device bears any physical resemblance to an edge of a graph (that is, an oriented line segment) the constraints imposed on the various network variables by the parallel alignment may be described by identifying an edge with each port (that is, the mirror and the lamp) in the network and paralleling these. More generally, the relevant connection constraints for any interconnected collection of n-ports may be characterized by identifying an edge with each port of each component and joining their vertices so as to correspond to the connections within the network [57], [79]. As with the n-port, the mathematical essence of a network can thus be represented by a simple structure which is independent of the geometry of the n-ports involved.

In summary, the essential properties of a network regardless of its physics or its geometry may be defined by pairs of admissible signals together with a set of connection constraints imposed by a linear graph. The development in the remainder of the chapter thus takes an essentially axiomatic approach which is independent of both physical and geometric considerations. Indeed not only are the network axioms independent of physical and geometric considerations, but they may, in fact, be inconsistent with such. Consider, for example, the simplest network of all — "nothing." In the electrical case this is an open circuit and, as we have seen, its scattering variables are characterized by the equation

$$b = a \qquad (1\text{-}6)$$

In the optical case, however, if a light wave is incident on "nothing" it "keeps on going"; hence no reflected wave is observed and

$$b = 0 \qquad (1\text{-}7)$$

for any a. Thus the same physical structure may induce different mathematical structures upon an appropriate reinterpretation of the network variables. Conversely, the same mathematical structure may be induced by physically distinct devices. In particular, Equation (1-6) represents an open circuit in an electrical network but a mirror in an optical network, whereas Equation (1-7) represents "nothing" in the optical case and a resistor in the electrical case. Consistent with these observations (and warnings) it is proposed to formulate an entirely mathematical theory of networks which includes all of the varied physical and geometrical structures thus far discussed, but is independent of any of them. So as to have a single consistent set of terms and to mollify tradition we do, however, adopt the terminology and notation of classical electric network theory.

The exposition begins with a section on the definition and properties of the network variables, this being predicated on the linear graph connectivity structure. In particular, the function space on which our networks "operate" is established and the space of admissible voltage (current) vectors, defined as a set of functions associated with the edges of a graph which satisfy the Kirchhoff voltage (current) law, is introduced. In this latter case the formal definitions for the voltage and current vectors, along with a review of graph theory, are relegated to Appendix C with an informal review appearing in the present chapter. Finally, the norm and inner product associated with the space of network variables is defined. In the second and third sections the concepts of an n-port and a network are formalized, these being classified as to linearity, time-invariance, passivity, causality, and so forth, in the fourth section. Finally, in the last section the relationships between the various categories of n-port are considered. In particular, the relationship between linearity, passivity, and causality is explored.

Throughout the entire development no consideration of specific circuit elements is necessary. Of course, our familiarity with the common RLC components renders these ideal as examples, for which purpose they are employed.

1-2 Network Variables

Basic to any discussion of network theory is the choice of the network variables. Should discontinuous and/or impulse functions be permitted; should infinite energy inputs be allowed; and so forth? Of course, none of these "pathological" signals can ever be physically achieved, but their inclusion may be of technical value. In fact, the allowed input-output signal space for the networks is taken as D_+.

Definition 1-1

D_+ is the set of real valued functions of a real variable which have infinitely many derivatives (referred to as being smooth) and are together with all their derivatives zero for all t less than some finite T. □

Impulse and step functions are forbidden, but signals having infinite energy are allowed. The general form of a D_+ input is shown in Figure 1-3. Though

Figure 1-3 D_+ signal.

many alternative choices of signal space can be made for special classes of networks [189], [192], [210], [211], D_+ seems to be best suited for the general case [110].

By requiring the input–output signals to be smooth such commonly encountered inputs (at least in theory) as the delta function and its derivatives (see Appendix B on "Distributions") are not permitted. This actually presents no difficulty since these functions can be approximated to any degree of accuracy by members of D_+ [156]. It should be noted that while distributions are not allowed as input-output signals they are needed for the representation of the "impulse response" of a network [6]. Thus, these often troublesome objects have not been completely eliminated.

With our setting for the network variables now established we may proceed to define the sets of admissible voltage and current vectors for a network, these being defined relative to the topology of a network. The formal definition of such are given in Appendix C, in the context of the theory of graphs. Intuitively, however, a graph is a set of vertices V, or points, interconnected by edges E, or lines, such as in Figure 1-4. The vertices are usually denoted by letters, a, b, c,

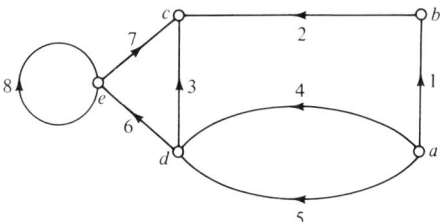

Figure 1-4 A typical graph.

..., and the edges by integers, 1, 2, 3, ..., while the connectivity structure is denoted by a set of ordered triples (j, a, b) where j is an edge name and a and b are vertices. Since there is one such triple per edge we usually identify the edge name and its associated triple using whichever notation is convenient in context. The graph of Figure 1-4 is thus described by the vertex set

$$V = \{a,b,c,d,e\} \tag{1-8}$$

and the edge set

$$E = \{(1,a,b),\ (2,b,c),\ (3,d,c),\ (4,a,d),\ (5,a,d),\ (6,d,e),\ (7,e,c),\ (8,e,e)\} \tag{1-9}$$

We note that the edge of a graph is not uniquely determined by its incident vertices since two edges, in this case 4 and 5, may be connected between the same vertices.

Topologically there is no difference between the edges (j,a,b) and (j,b,a) since they both correspond to an edge connecting vertices a and b [23]. From a network theoretic point of view, however, we desire to associate a voltage function and a current function with each edge of the graph, these in turn requiring the orientation of the edges of our graph [155]. We therefore place an arrowhead on each edge of the graph in the "assumed direction of positive current flow," this being indicated in our connectivity structure by ordering of the vertices a and b. That is, if the arrow on edge j goes from vertex a to vertex b the edge is denoted by (j,a,b), whereas if it goes from b to a the edge is denoted by (j,b,a). With such an orientation one can then associate a voltage and current with each edge of the graph, the current assumed to flow in the direction of the edge orientation and the voltage measured from the tail of the arrow to its head [155], [162].

Of course, the voltages associated with the edges of the graph must also satisfy the Kirchhoff voltage law (KVL) and the currents must satisfy the Kirchhoff current law (KCL). That is, given any circuit in the graph (a formal definition of a circuit and its generalization, the cycle, is given in Appendix C, though the usual intuitive interpretation of this concept suffices for the present purpose), the voltage functions associated with its edges (oriented in a consistent manner around the circuit) must sum to zero. Similarly, the currents entering a node or "supernode" (cocircuit or cocycle in the formal definitions of Appendix C) must sum to zero. Following the notation of the Appendix for each circuit (or cycle) of a graph C, one may define a function e_C which takes the edges of the graph E to the set $\{-1, 0, 1\}$ such that if the jth edge is not in C,

$$e_C(j) = 0 \tag{1-10}$$

while if the jth edge is in C and its orientation coincides with an (arbitrarily defined) orientation around the circuit,

$$e_C(j) = 1 \tag{1-11}$$

And, finally, if the jth edge is in C but its orientation does not coincide with that of C,

$$e_C(j) = -1 \tag{1-12}$$

With the aid of the e_C function we may now define a voltage vector of a graph as a function v from E to D_+ such that for each circuit C

$$\sum_j e_C(j) v(j) = 0 \tag{1-13}$$

That is, a voltage vector is a set of functions in D_+, one associated with each edge of the graph, which sum to zero around each circuit C of the graph. Since $e_C(j)$ is zero if edge j is not in circuit C, Equation (1-13) does not constrain the voltages which are not in circuit C; while if edge j is in circuit C the e_C functions define the sign with which $v(j)$ should appear in the KVL equation. Similarly, for each cocircuit or cocycle S of the graph, an e_S function may be defined which serves the dual purpose in the definition of a current vector as a function i from the edges of a graph E to D_+ such that

$$\sum_j e_S(j) i(j) = 0 \tag{1-14}$$

for each cocircuit S. Although we have defined the voltage and current vectors to be functions from an edge set to D_+ we may, in fact, interpret these as vectors of D_+ functions (in the space D_+^n) where the jth entry in the vector is $v(j)$ or $i(j)$; hence the terminology voltage and current vectors rather than functions. We therefore let v denote a column whose entries are $v(j)$, and v' denote its transpose, a row vector.

For the graph of Figure 1-4, the circuit composed of edges 1, 2, 3, and 5, has the e_C function (for a clockwise circuit orientation)

j	1	2	3	4	5	6	7	8
$e_C(j)$	1	1	-1	0	-1	0	0	0

Hence, the function defined by

j	1	2	3	4	5	6	7	8
$v(j)$	g	g	g	g	0	0	0	0

For some, g in D_+ is not a voltage vector since

$$\sum_j e_C(j)v(j) = g \neq 0 \quad \text{(1-15)}$$

On the other hand, the function defined by

j	1	2	3	4	5	6	7	8
$v(j)$	g	g	g	g	g	g	0	0

satisfies the KVL equation (1-13) for this particular circuit and all other circuits, as the reader may verify for an exercise; hence it is indeed a voltage vector.

Essentially, the voltage vector associates a function from D_+ with each edge of the graph in such a way as to assure that the KVL is satisfied, and similarly for the current vector. In this way, rather than resulting from physical considerations [46], the Kirchhoff laws are true by the fiat of hypotheses. Of course, within the generalized context this is the only practical approach to these concepts since the physics of the various classes of networks differs greatly.

The scattering variables could readily be defined by a technique similar to that of the preceding definitions, that is, as a set of functions from D_+ associated with a graph which satisfies an appropriate set of connection constraints. Alternatively, one may invoke Equations (I-8) and (I-9), defining the scattering variables simply as an appropriate linear combination of the immittance variables. It is this latter approach which we take, though our formal definition differs by a scale factor of one-half from that established heuristically in the Introduction.

Definition 1-2

Let v and i be voltage and current vectors associated with a graph. Then

$$a = v + i$$

and

$$b = v - i$$

are the *incident and reflected waves*, respectively, which are associated with the graph. They are referred to as the *scattering variables*. □

a and b of Definition 1-2 are often called the unnormalized scattering variables, while the more general normalized scattering variables [44], [110], [141], [185], [190], [196] are given by

$$a = R^{-1/2}v + R^{1/2}i \quad \text{(1-16)}$$

and

$$b = R^{-1/2}v - R^{1/2}i \quad \text{(1-17)}$$

16 FOUNDATIONS

Here R is any $n \times n$ diagonal matrix of strictly positive real entries and $R^{1/2}$ is the corresponding matrix of square roots. Although the normalized scattering variables are often of considerable physical interest they seldom yield any conceptual advantage over the unnormalized variables of Definition 1-2; hence they will not generally be employed. In the few cases where they are needed, primarily transmission line theory, their use will be made explicit.

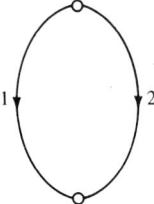

Figure 1-5 Graph with network variables.

In the simple circuit of Figure 1-5, a voltage vector is a pair of functions from D_+, v_1 and v_2, such that

$$v_1 = v_2 \tag{1-18}$$

whereas a current vector is a pair of D_+ functions, i_1 and i_2, satisfying

$$i_1 = -i_2 \tag{1-19}$$

Now combining these equations with the definition of the scattering variables yields

$$a_1 = v_1 + i_1 = v_2 = i_2 = b_2 \tag{1-20}$$

and

$$a_2 = v_2 + i_2 = v_1 - i_1 = b_1 \tag{1-21}$$

which are the corresponding constraints on the scattering variables for this graph.

For the sake of generality only a minimum of structure should be defined on the networks; rather an inner product structure on the signal space D_+ (or more generally the space of n vectors of functions in D_+, denoted by D_+^n) is employed. Since the signal space is independent of the properties of any specific network such a structure can be employed without any loss of generality. The L_2 inner product on the signal space D_+^n and its corresponding norm are defined as follows.

Definition 1-3

Let x and y be in D_+^n. Then the *inner product* of x and y is given by

$$\langle x,y \rangle = \int_{-\infty}^{\infty} x'(u)y(u)du$$

if it exists. Here the prime denotes vector (matrix) transposition and u is an arbitrary variable of integration. □

Definition 1-4

Let x be in $D_+{}^n$. Then the *norm* of x is

$$\|x\| = \langle x,x \rangle^{1/2}$$

if it exists. ☐

It should be noted that neither the inner product nor the norm is assured to exist; hence $D_+{}^n$ is neither an inner product space nor a normed space [22]. If $\|x\|$ exists we say that x is in L_2. In fact, if x and y are both in L_2 the inner product of x and y is assured to exist [22], and hence those functions in $D_+{}^n$ which are L_2 do form an inner product space (as per the definition of Appendix A.) The verification of this fact is left to the reader as an exercise.

In general, the inner product of x and y may be, in some sense, interpreted as a measure of the "angle" between the functions x and y, while the norm of x is, in some sense, a measure of the "size" of x. In the special case where x and y are voltage and current vectors, respectively, the quantity

$$p = i'v = v'i \qquad (1\text{-}22)$$

corresponds to the "instantaneous power" delivered to the network characterized by i and v. The inner product of i and v is thus the total energy delivered to the network by the application of voltage v and current i.

Although energy is most conveniently represented as the inner product of voltage and current, if one invokes the linear relation between the scattering and immittance variables, an equivalent description of the energy dissipated in a network in terms of the scattering variables may be obtained. In fact, this description has the advantage that it may be written purely in terms of the norm rather than the more complex inner product required with the immittance variables.

*Proposition 1-1[1]

Let i and v be current and voltage vectors for a graph and a and b be the corresponding incident and reflected waves; then

$$v'i = i'v = \tfrac{1}{4}(a'a - b'b)$$

and

$$4\langle i,v \rangle = \langle a,a \rangle - \langle b,b \rangle = \|a\|^2 - \|b\|^2 \qquad \square$$

A proof for Proposition 1-1 follows readily from the equalities of Definition 1-2 and (as are those of all starred propositions) is left to the reader as an exercise. In fact, the reader may also verify that the proposition also holds for the case of scattering variables normalized by an arbitrary diagonal R.

---

[1] The star denotes a problem, in the form of an unproven statement, which is left to the reader as an exercise.

In the entire development the fundamental structure underlying the theory is the vector space D_+. In fact, throughout this section we have made the implicit assumption that D_+ and D_+^n are vector spaces whenever we have added the network variables. Of course, as with most interesting vector spaces D_+^n is infinite-dimensional. The verification of the vector space axioms (given in Definition A-1), together with the fact that D_+^n is indeed infinite-dimensional, is left as an exercise.

*Proposition 1-2

D_+^n is an infinite-dimensional vector space. □

1-3 n-Ports

A network is essentially a combination of two structures, one algebraic which characterizes the components, and one topological which characterizes the connection. The basic building block is the n-port. Consistent with the discussion of Section 1-1, one need not attribute any physical significance to an n-port. Rather it is defined simply as a graph of n disconnected edges, the network variables of which are constrained in some predetermined manner. Of course, these edges may be interpreted as waveguides, transmission lines, pipes, or whatever the application at hand might dictate.

Definition 1-5

An *n-port* is a pair, (G_n, N), where G_n is the graph of n disconnected edges and N is a relation on D_+^n which defines a set of *admissible pairs*

$$(a, b) \in N$$

Here a is the incident wave vector of the n-port and b is the reflected wave vector. □

It would also be possible to allow the relation to define admissible voltage–current pairs [106], [110], [192], but since the scattering approach is to be emphasized in this work it is convenient to begin with it.

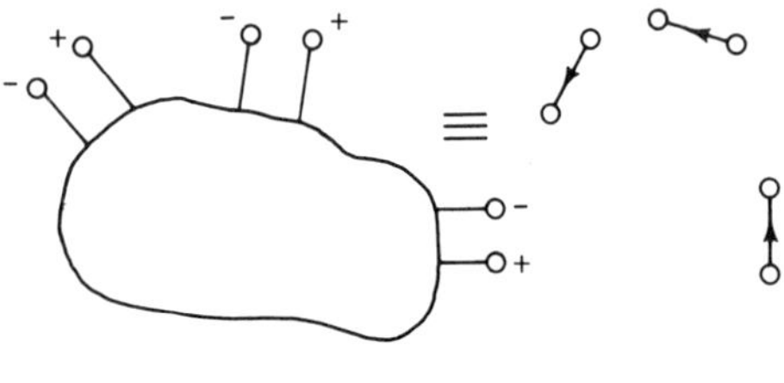

Figure 1-6 Physical and mathematical n-ports.

The relationship between the n-port of Definition 1-1 and the conceptual idea of a black box with $2n$ attached wires is illustrated in Figure 1-6. A vertex is associated with each wire and the terminal pairing is designated by an edge from the "plus" terminal vertex to the "minus" terminal vertex. Since the current leaving an edge at one of its vertices must equal that entering the edge at the other, the current in the paired wires which form a port must be equal and opposite [110]. It is this constraint which distinguishes the n-port from the more general $2n$-terminal device [79]. In particular, an n-port is characterized by $2n$ network variables, while a $2n$-terminal device is characterized by $4n$ network variables [183].

Though the definition of the network is "element independent" it includes all the classical components usually associated with network theory. The definitions of these classical devices along with some less classical, but interesting, devices follow. Both the immittance and scattering formulations are given for the resistor, capacitor, and inductor. All parameters for these components are assumed to be smooth (C^∞) time functions.

Definition 1-6

A *resistor* is a 1-port with admissible pairs

$$(i,v) \in N$$

such that

$$v = r(t)i$$

$r(t)$ is the *resistance*.

A *capacitor* is a 1-port with admissible pairs

$$(i,v) \in N$$

such that

$$i = Dc(t)v$$

$c(t)$ is the *capacitance*. Here D is the time derivative ($D = d/dt$).

An *inductor* is a 1-port with admissible pairs

$$(i,v) \in N$$

such that

$$v = Dl(t)i$$

$l(t)$ is the *inductance*. □

Since the input-output signals i and v for the above definition are in D_+ they are zero for sufficiently negative t. Hence no question of initial condition arises in the definitions for the capacitor and inductor.

Inversion of the defining relation for the incident and reflected waves yields

$$v = \tfrac{1}{2}(a + b) \qquad (1\text{-}23)$$

$$i = \tfrac{1}{2}(a - b) \qquad (1\text{-}24)$$

20　FOUNDATIONS

which upon substitution into the definitions for the components yields equivalent scattering characterizations.

Resistor:
$$b = s(t)a \tag{1-25}$$
such that
$$s(t) = (r(t) - 1)(r(t) + 1)^{-1} \tag{1-26}$$

Capacitor:
$$b = [1 + Dc(t)]^{-1}[1 - Dc(t)]a \tag{1-27}$$

Inductor:
$$b = [Dl(t) + 1]^{-1}[Dl(t) - 1]a \tag{1-28}$$

Here the inverse differential operator of Equation (1-27) is a symbol for the two-sided differential equation
$$[1 + Dc(t)]b = [1 = Dc(t)]a$$
and similarly for that of Equation (1-28).

The basic multiports are the gyrator and the ideal transformer.

Definition 1-7

A *gyrator* is a 2-port with
$$\left(\begin{bmatrix} v_1 \\ v_2 \end{bmatrix}, \begin{bmatrix} i_1 \\ i_2 \end{bmatrix} \right) \in N$$
such that
$$v_1 = \gamma(t)i_2$$
$$v_2 = -\gamma(t)i_1$$

$\gamma(t)$ is the *gyration coefficient*. □

The scattering formulation of the gyrator can be obtained from Equations (1-23) and (1-24) in a manner similar to that of the RLC components of Definition 1-6, the details of which are left to the reader [44]. The one case of special interest, however, is when the gyration coefficient is one. In this case
$$a_2 = v_2 + i_2 = v_1 - i_1 = b_1 \tag{1-29}$$
and
$$b_2 = v_2 - i_2 = -v_1 - i_1 = -a_1 \tag{1-30}$$

Hence the gyrator interchanges (and negates) the scattering variables.

One could readily generalize the gyrator concept by defining a $2n$-port (or an $(n + m)$-port) gyrator by the equations
$$v_1 = \boldsymbol{\gamma}(t)i_2 \tag{1-31}$$
and
$$v_2 = -\boldsymbol{\gamma}(t)'i_1 \tag{1-32}$$

where $\gamma(t)$ is now interpreted as an $n \times n$ matrix (or an $n \times m$ matrix) and the v's and i's are vectors of conformable dimension. This multiport gyrator [93] finds little application in our theory and will be omitted from consideration in the remainder of the development.

Although the gyrator was originally conceived as a theoretical element of electrical circuit theory [174] it has various physical realizations. In the case of an electrical circuit, several transistor realizations [119] are possible, whereas a microwave gyrator may be constructed by the simple process of inserting an appropriate gyroscopic media in a section of waveguide [52]. Of course, as the name indicates, a mechanical gyrator (with force and velocity replacing voltage and current) is a gyroscope [174].

Definition 1-8

An *ideal transformer* is an $(n + m)$-port with

$$\left(\begin{bmatrix} v_1 \\ v_2 \end{bmatrix}, \begin{bmatrix} i_1 \\ i_2 \end{bmatrix} \right) \in N$$

such that

$$v_1 = T(t)' v_2$$

$$i_2 = -T(t) i_1$$

Here v_1 and i_1 are n vectors; v_2 and i_2 are m vectors; and $T(t)$ is an $m \times n$ matrix of time functions, the *turns ratio matrix*. □

The ideal transformer has the effect of shifting voltage up and current down, or the converse. It is significant that the voltage and current in an ideal transformer are decoupled. That is, the voltage at the "1" ports is determined entirely by the voltage at the "2" ports and is independent of the current at either set of ports. Similarly, the current at the "2" ports is determined by that at the "1" ports, but is voltage independent. In this respect the ideal transformer is like its mechanical analog, a gear train, wherein the velocity and force on one gear is determined by the corresponding variables on the other gears, but the two variables are decoupled. Of course, in a gear train the turns ratio matrix corresponds to the gear ratios.

Unlike the immittance case when the ideal transformer is represented by the scattering variables, the two variables are coupled. In fact, one can write the reflected waves at both sets of ports explicitly in terms of the corresponding incident waves. This rather complicated relationship is the subject of the following proposition which the reader may verify as an exercise.

*Proposition 1-3

An ideal transformer is an $(n + m)$-port with

$$\left(\begin{bmatrix} a_1 \\ a_2 \end{bmatrix}, \begin{bmatrix} b_1 \\ b_2 \end{bmatrix}\right) \in N$$

satisfying

$$b_1 = [(I + T'T)^{-1}(T'T - 1)]a_1 + [2(I + T'T)^{-1}T']a_2$$

and

$$b_2 = [2T(I + T'T)^{-1}]a_1 + [(I + TT')^{-1}(I - TT')]a_2$$

Here a_1 and b_1 are n-vectors; a_2 and b_2 are m vectors; and $T(= T(t))$ is the turns ratio matrix. □

Although the equations of Proposition 1-3 are too complex to be of much practical value, they simplify greatly in two cases of special importance. First, if $n = m$ and the turns ratio matrix is orthogonal, that is,

$$T'T = TT' = I \tag{1-33}$$

then

$$(I - T'T) = (I - TT') = 0 \tag{1-34}$$

and the equations of Proposition 1-3 simplify to

$$b_1 = T'(t)a_2 \tag{1-35}$$

and

$$b_2 = T(t)a_1 \tag{1-36}$$

Also, if $n = m = 1$ then the turns ratio matrix is a scaler, k, which represents the actual physical turns ratio. In this case the equations of Proposition 1-3 reduce to

$$b_1 = \frac{k^2 - 1}{k^2 + 1}a_1 + \frac{2k}{k^2 + 1}a_2 \tag{1-37}$$

and

$$b_2 = \frac{2k}{k^2 + 1}a_1 + \frac{1 - k^2}{k^2 + 1}a_2 \tag{1-38}$$

Another special case of interest takes place when $n = m$ and

$$T = R^{1/2} \tag{1-39}$$

where R is a diagonal matrix of strictly positive real numbers. Now

$$a_2 = v_2 + i_2 = R^{-1/2}v_1 - R^{1/2}i_1 = (b_1)_R \tag{1-40}$$

and

$$b_2 = v_2 - i_2 = R^{-1/2}v_1 + R^{1/2}i_1 = (a_1)_R \tag{1-41}$$

Hence the "2" variables are the same as the "1" variables except that they are now normalized to R rather than being unnormalized. The interchanging of a and b, as with the gyrator, is only a mathematical artifice due to the choice of current polarity. The ideal transformer may thus be interpreted as a "normalizer." Of course, if R were replaced by R^{-1} one would obtain a "denormalizer"; and, more generally, a $2n$-port ideal transformer with turns ratio matrix

$$T = R_2^{1/2} R_1^{-1/2} \tag{1-42}$$

converts scattering variables measured at the "1" ports normalized to R_1 to the equivalent variables measured at the "2" ports with the normalization changed to R_2. For this reason one may reduce the problem of studying networks whose scattering variables have arbitrary normalization to an equivalent network with unnormalized variables (that is, $R = I$) by the simple artifice of inserting an ideal transformer into the ports of the given network. Conversely, there are certain applications of the scattering variables (notably Richards' theorem [136], [137]) wherein it is convenient to change the normalization of a network.

Although most of the components we deal with had their origins in classical electric network theory and were only later applied to other classes of networks, the circulator originated as a waveguide element and has only recently been applied to low-frequency electrical networks.

Definition 1-9

A *circulator* is a $3n$-port with

$$\left(\begin{bmatrix} a_1 \\ a_2 \\ a_3 \end{bmatrix}, \begin{bmatrix} b_1 \\ b_2 \\ b_3 \end{bmatrix} \right) \in N$$

such that

$$b_1 = a_3$$
$$b_2 = a_1$$
$$b_3 = a_2$$

Here all of the variables are n vectors. □

In essence, the circulator is a $3n$-port for which the wave entering the "1" ports leaves the "2" ports but is otherwise unperturbed. Similarly, the wave entering the "2" ports leaves the "3" ports and that entering the "3" ports leaves the "1" ports. Of course, the scattering variable constraints which define the circulator could be converted to equivalent constraints on the immittance variables. These, however, prove to be of little value and will not be derived here; the reader may, however, derive them as an exercise.

The final components to be defined are nonclassical, but they are networks in the generalized sense and are useful for examples (and counterexamples). These are the nullator and the norator [49].

Table 1-1 Summary of n-ports (1-ports).

COMPONENT AND SYMBOL	IMMITTANCE CHARACTERIZATION	SCATTERING CHARACTERIZATION
Resistor $r(t)$	$v = r(t)i$	$b = \dfrac{r(t) - 1}{r(t) + 1} a$
Capacitor $c(t)$	$i = D[c(t)v]$	$b = [1 + Dc(t)]^{-1}[1 - Dc(t)]a$
Inductor $l(t)$	$v = D[l(t)i]$	$b = [1 + Dl(t)]^{-1}[1 - Dl(t)]a$
Nullator	$v = i = 0$	$a = b = 0$
Norator	v and i are arbitrary	a and b are arbitrary

Ideal Transformer [$(n + m)$-port] $T(t) : 1$	$v_1 = T'(t)v_2$ $i_2 = -T(t)i_1$	$b_1 = [(1 + T'T)^{-1}(T'T - 1)]a_1$ $+ [2(1 + T'T)^{-1}T']a_2$ $b_2 = [2T(1 + T'T)^{-1}]a_1$ $+ [(1 + TT')^{-1}(1 - TT')]a_2$
Ideal Transformer [$2n$-port, $T(t)$ orthogonal] $T(t) : 1$	$v_1 = T'(t)v_2$ $i_2 = -T(t)i_1$	$b_1 = T'(t)a_2$ $b_2 = T(t)a_1$
Ideal Transformer [2-port] $k : 1$	$v_1 = kv_2$ $i_2 = -ki_1$	$b_1 = \dfrac{k^2 - 1}{k^2 + 1} a_1 + \dfrac{2k}{k^2 + 1} a_2$ $b_2 = \dfrac{2k}{k^2 + 1} a_1 + \dfrac{1 - k^2}{k^2 + 1} a_2$

Element	Immittance	Scattering
Gyrator [2-port] $\gamma(t)$	$v_1 = \gamma(t) i_2$ $v_2 = -\gamma(t) i_1$	$b_1 = \dfrac{\gamma^2-1}{\gamma^2+1} a_1 + \dfrac{2\gamma}{\gamma^2+1} a_2$ $b_2 = \dfrac{-2\gamma}{\gamma^2+1} a_2 + \dfrac{\gamma^2-1}{\gamma^2+1} a_2$
Gyrator [2-port, $\gamma = 1$]	$v_1 = i_2$ $v_2 = -i_1$	$b_1 = a_2$ $b_2 = -a_1$
Gyrator [$(n+m)$ - port] $\gamma(t)$	$v_1 = \gamma(t) i_2$ $v_2 = -\gamma'(t) i_1$	—
Circulator [$3n$ - port]	—	$b_1 = a_3$ $b_2 = a_1$ $b_3 = a_2$

Definition 1-10

A *nullator* is a 1-port with

$$(a,b) \in N$$

such that

$$a = b = 0$$

A *norator* is a 1-port with

$$(a,b) \in N$$

where a and b are arbitrary. □

The nullator and norator are two components which have the same characteristics in either the immittance or the scattering variables. For the nullator a and b are always zero, and hence their linear combinations $v = \tfrac{1}{2}(a + b)$ and $i = \tfrac{1}{2}(a - b)$ are also zero; whereas for the norator both a and b are arbitrary, and hence so is their sum and difference, i and v.

The various n-ports which we have thus far defined are tabulated in Table 1-1. Here both the immittance and scattering characterizations are given together with the symbol used to represent these components in schematic network diagrams. In these diagrams we adopt the policy of denoting sets of similar ports by a single port together with a ⟩ symbol. For instance, an $(n + m)$-port ideal transformer may be represented by the same symbol as used for a 2-port ideal transformer with the addition of the ⟩ symbol which denotes the grouping of the "1" and "2" ports, respectively.

The various n-ports can be classified in a number of ways by such characteristics as linearity, passivity, time-invariance, and so forth. Fundamental to many of these classifications is the ability of an n-port to act as an amplifier. For this purpose it is necessary that, at least for some input, the energy carried by the reflected wave exceed that delivered to the network by the incident wave. One measure of this ability to amplify is the norm of an n-port which is a (least upper) bound on the ratio of the energy delivered by an n-port to its surroundings to that delivered to the n-port by the surroundings.

Definition 1-11

Let (G_n,N) be an n-port. Then the *norm* of (G_n,N), denoted by $\|N\|$, is

$$\|N\| = \sup_{\substack{(a,b)\epsilon N \\ 0<\|a\|<\infty}} \left\{ \frac{1}{\|a\|} \|b\| \right\} \qquad \square$$

It is noteworthy that only nonzero inputs with finite norm are considered in the definition for the norm of N. Although this is clearly necessary if the mathematics of Definition 1-11 are to be well defined, it is also reasonable on physical grounds, for if the input energy to an n-port were either infinite or undefined, the concept of an energy gain would be meaningless. We do, however, allow for the possibility of an infinite-energy reflected wave, in which case the norm of N is defined to be infinite.

For an open circuit

$$b = a \qquad (1\text{-}43)$$

for all admissible pairs (a,b) in N; hence

$$\|N\| = \sup_{\substack{(a,b)\epsilon N \\ 0<\|a\|<\infty}} \left\{ \frac{1}{\|a\|} \|b\| \right\} = \sup_{\substack{(a,b)\epsilon N \\ 0<\|a\|<\infty}} \left\{ \frac{1}{\|a\|} \|a\| \right\} \qquad (1\text{-}44)$$

$$\sup_{\substack{(a,b)\epsilon N \\ 0<\|a\|<\infty}} \{1\} = 1$$

As expected, the open circuit is not capable of amplification since it is never capable of delivering more (or less) energy to its surroundings than it receives from them. The norator, on the other hand, is capable of amplification since its input-output pairs are totally unrelated, thus allowing the ratio

$$\frac{1}{\|a\|} \|b\| \qquad (1\text{-}45)$$

to be arbitrarily large. The norm of the norator is thus infinite. The norms of various n-ports listed in Table 1-1 are characterized by the following proposition, the proof of which is left to the reader as an exercise.

*Proposition 1-4

The transformer, gyrator, and circulator have a norm of one. The resistor with scattering parameter $s(t)$ has norm

$$\sup_t |s(t)|$$

whereas the norator has an infinite norm. ☐

Rigorously speaking, the nullator has no norm since no inputs with $a \neq 0$ are allowed (thus rendering Definition 1-11 meaningless). It is, however, convenient to take the norm of such to be zero, this being done in the sequel.

1-4 Networks

In the previous section we considered the basic building block for the network, the n-port. Here several n-ports (with n variable) are combined by identifying the vertices of their associated graphs, G_n, to form a network. The main problem is to combine the constraints imposed by the various n-ports with the constraints imposed by the connection so as to discern the characteristics of the admissible network variables. Usually one inserts additional ports at certain predetermined points in the network and is asked to find the characteristics of the n-port thus realized by the network. For example, consider the inductor-loaded gyrator of Figure 1-7. Here a 1-port (between points x and y) is realized by the network. The gyrator is characterized by the equations

$$v_1 = i_2 \tag{1-46}$$

and

$$v_2 = -i_1 \tag{1-47}$$

whereas the inductor is described by

$$v_L = D[l(t)i_L] \tag{1-48}$$

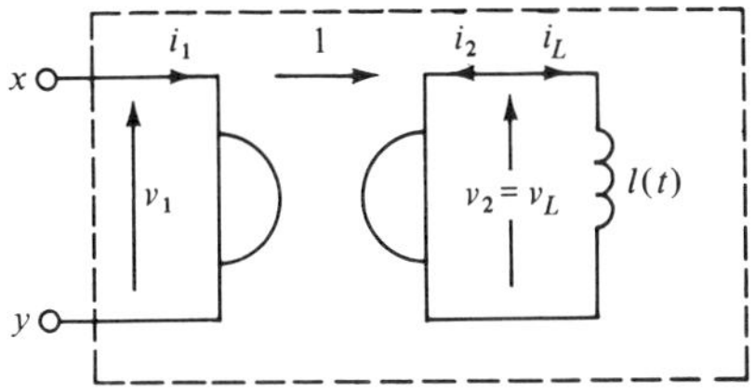

Figure 1-7 Inductor-loaded gyrator.

Combining Equations (1-46) through (1-48) with the connection constraints

$$i_2 = -i_L \qquad (1\text{-}49)$$

and

$$v_2 = v_L \qquad (1\text{-}50)$$

yields

$$i_1 = -v_2 = -v_L = -D[l(t)i_L] = D[l(t)i_2] = D[l(t)v_1] \qquad (1\text{-}51)$$

which is precisely the behavior of a capacitor. Hence the network of Figure 1-7, composed of a gyrator and an inductor, realizes a capacitor. More generally, if one replaces the inductor of Figure 1-7 with an arbitrary 1-port load characterized by the equation

$$v_L = F(i_L) \qquad (1\text{-}52)$$

where $F(\cdot)$ is any operator (possibly including derivatives, delays, and so forth), then the network realizes a 1-port characterized by

$$i_1 = F(v_1) \qquad (1\text{-}53)$$

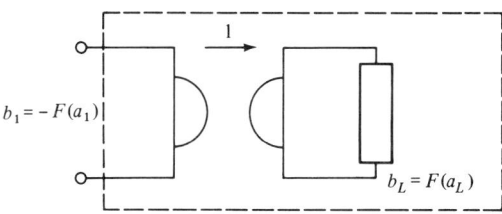

Figure 1-8 Arbitrarily loaded gyrator.

The configuration of Figure 1-7 may thus be viewed as a device which interchanges the role of voltage and current between the load and the port. When a similar configuration is characterized by its scattering variables, as shown in Figure 1-8, it acts as a "negative scattering converter." That is, if the load is characterized by the operator equation

$$b_L = F(a_L) \qquad (1\text{-}54)$$

then the port scattering variables are characterized by the equation

$$b_1 = -F(a_1) \qquad (1\text{-}55)$$

This result may be obtained by combining the gyrator characteristics

$$a_2 = b_1 \qquad (1\text{-}56)$$

and

$$a_1 = -b_2 \qquad (1\text{-}57)$$

of Equations (1-29) and (1-30) with the connection constraints

$$a_1 = b_L \qquad (1\text{-}58)$$

and

$$a_L = b_1 \qquad (1\text{-}59)$$

and the load characteristics of Equation (1-54). The details are left to the reader as an exercise.

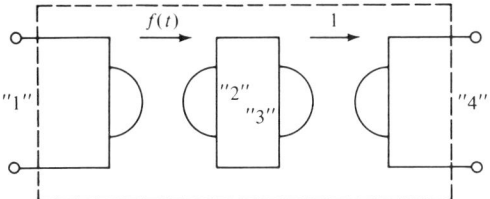

Figure 1-9 Realization of a transformer via cascaded gyrators.

Another example is the 2-port of Figure 1-9, which is constructed by cascading two gyrators. In this case one has the equalities

$$v_1 = f(t)i_2 \qquad (1\text{-}60)$$

$$v_2 = -f(t)i_1 \qquad (1\text{-}61)$$

characterizing the first gyrator and

$$v_3 = i_4 \qquad (1\text{-}62)$$

$$v_4 = -i_3 \qquad (1\text{-}63)$$

characterizing the second gyrator, which when combined with the connection constraints

$$v_2 = v_3 \qquad (1\text{-}64)$$

and

$$i_2 = -i_3 \qquad (1\text{-}65)$$

yields

$$v_1 = f(t)i_2 = -f(t)i_3 = f(t)v_4 \qquad (1\text{-}66)$$

and

$$i_4 = v_3 = v_2 = -f(t)i_1 \qquad (1\text{-}67)$$

which are the equations of a 2-port ideal transformer with turns ratio $f(t)$. Two gyrators thus combine to allow the realization of an ideal transformer.

Consistent with the results of the preceding examples, a network may be said to be a collection of n-ports for which the vertices of their graphs have been identified in some appropriate manner, whereas the network variables are the incident and reflected waves (or voltage and current vectors) which simultaneously satisfy the connection constraints imposed by the graph and the component relations imposed by the various n-ports with which the network is constructed. More formally, we have the following.

Definition 1-12
Let

$$(G_{n_i}, N_i) \qquad i = 1, 2, \cdots, m$$

be n-ports. Then a *network* is an $(m + 1)$-tuple

$$(G, N_1, N_2, \cdots, N_m)$$

where G is a graph formed from the G_{n_i} by identifying their vertices in some arbitrary manner. □

As an illustration of this definition consider the network composed of two cascaded gyrators, shown in Figure 1-9. Here the components are two 2-ports, the gyrators, with admissible pairs

$$N_1 = \left(\begin{bmatrix} v_1 = f(t)i_2 \\ v_2 = -f(t)i_1 \end{bmatrix}, \begin{bmatrix} i_1 \\ i_2 \end{bmatrix} \right) \quad (1\text{-}68)$$

and

$$N_2 = \left(\begin{bmatrix} v_3 = i_4 \\ v_4 = -i_3 \end{bmatrix}, \begin{bmatrix} i_3 \\ i_4 \end{bmatrix} \right) \quad (1\text{-}69)$$

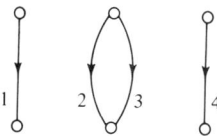

Figure 1-10 Formation of a graph for the network of Figure 1-9.

while the graph G is that shown in Figure 1-10. According to Definition 1-12, the network is represented by the 3-tuple

$$(G, N_1, N_2)$$

Although physically a network does, in fact, occur as an interconnection of several independent components, one can, with some convenience, view any network as containing only a single component, a composite n-port. That is, the two 2-ports forming the network of Figure 1-9 may be viewed as a single 4-port

$$\hat{N} = N_1 \oplus N_2 = \left(\begin{bmatrix} v_1 = f(t)i_2 \\ v_2 = -f(t)i_1 \\ v_3 = i_4 \\ v_4 = -i_3 \end{bmatrix}, \begin{bmatrix} i_1 \\ i_2 \\ i_3 \\ i_4 \end{bmatrix} \right) \quad (1\text{-}70)$$

where the 1 and 2 variables are decoupled from the 3 and 4 variables. Similarly, in the network of Figure 1-7, the 2-port gyrator and 1-port inductor may be viewed as a composite 3-port satisfying

$$\hat{N} = N_1 \oplus N_2 = \left(\begin{bmatrix} v_1 = i_2 \\ v_2 = -i_1 \\ v_3 = D[l(t)i_3] \end{bmatrix}, \begin{bmatrix} i_1 \\ i_2 \\ i_3 \end{bmatrix} \right) \quad (1\text{-}71)$$

The process of combining m separate n_i-ports into a single composite (Σn_i)-port is termed "taking the direct product of the N_i" and, in general, if

$$N_i = \{a_i, b_i\} \qquad i = 1, 2, \cdots, m \tag{1-72}$$

(or similarly for the immittance variables), then the direct product of the N_i is the (Σn_i)-port given by

$$\hat{N} = \Sigma \oplus N_i = \left\{ \begin{bmatrix} a_1 \\ a_2 \\ \cdot \\ \cdot \\ \cdot \\ a_m \end{bmatrix}, \begin{bmatrix} b_1 \\ b_2 \\ \cdot \\ \cdot \\ \cdot \\ b_m \end{bmatrix} \right\} \tag{1-73}$$

Clearly, if the original components were uncoupled then there is no coupling between variables of different indices in $\hat{N}$. In the very common case where the b_i can be obtained from the a_i by a matrix equation of the type

$$b_i = S_i(t) a_i \qquad i = 1, 2, \cdots, m \tag{1-74}$$

then the direct product of the relations N_i coincides with the direct product of the matrices $S_i(t)$. That is,

$$b = \begin{bmatrix} b_1 \\ b_2 \\ \cdot \\ \cdot \\ \cdot \\ b_m \end{bmatrix} = \begin{bmatrix} S_1(t) & & & 0 \\ & S_2(t) & & \\ & & \cdot & \\ & & & \cdot \\ 0 & & & S_m(t) \end{bmatrix} \begin{bmatrix} a_1 \\ a_2 \\ \cdot \\ \cdot \\ \cdot \\ a_m \end{bmatrix} = \hat{S} a \tag{1-75}$$

In essence, the direct product, either for arbitrary component relations, as in Equation (1-73), or for matrix equations, is a convenient formality which allows one to write several sets of simultaneous equations as one large set of equations, this without in any way changing the constraints imposed by the equations. For the present purposes the direct product is simply a mathematical artifice which allows one to view any network as being composed of only a single component rather than m separate ones. Of course, such a change in viewpoint in no way changes the network.

Definition 1-13

Let $(G, N_1, N_2, \cdots, N_m)$ be a network. Then its associated *composite* (Σn_i)-*port* is

$$\hat{N} = \Sigma \oplus N_i$$

and the network is denoted by $(G, \hat{N})$. □

The variables admissible to a network are incident and reflected waves (or voltages and currents) which simultaneously satisfy the connection constraint imposed by the graph G and the component constraint imposed by the n-ports, N_i, or equivalently by the composite (Σn_i)-port, $\hat{N}$. If one takes the viewpoint that $\hat{N}$ is a set of simultaneously admissible component variables, the network variables are then the intersection of this set with the set of simultaneously admissible variables for the graph. Denoting this latter set by $\hat{G}$ we have the following.

Definition 1-14

Let a network $(G, N_1, N_2, \cdots, N_m)$ have composite (Σn_i)-port $\hat{N}$ and graph variables constrained by the relation $\hat{G}$. Then the *admissible variables to the network* are given by the set of admissible pairs

$$\hat{N} \cap \hat{G}$$ □

The configuration of Figure 1-11 is composed of an $(m + n)$-port ideal transformer loaded in an n-port constrained by the equation

$$v_3 = Z(t)i_3 \tag{1-76}$$

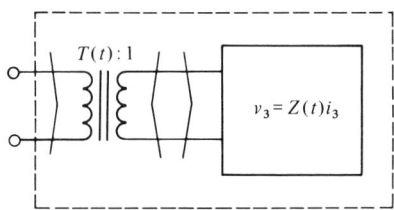

Figure 1-11 $(m + n)$-port ideal transformer loaded in n-port impedance.

The corresponding network thus has two components characterized by the equations

$$N_1 = \left(\begin{bmatrix} v_1 \\ v_2 \end{bmatrix} = \begin{bmatrix} T'(t)v_2 \\ v_2 \end{bmatrix}, \begin{bmatrix} i_1 \\ i_2 \end{bmatrix} = \begin{bmatrix} i_1 \\ -T(t)i_1 \end{bmatrix} \right) \tag{1-77}$$

and

$$N_2 = (v_3] = Z(t)i_3], i_3]) \tag{1-78}$$

where v_1 and i_1 are m vectors representing the variables measured at the m input ports, and v_2, v_3, i_2, and i_3 are n vectors representing the internally measured variables. Combining Equations (1-77) and (1-78) yields the composite $(m + 2n)$-port for the network characterized by

$$N = \left(\begin{bmatrix} v_1 \\ v_2 \\ v_3 \end{bmatrix} = \begin{bmatrix} T(t)'v_2 \\ v_2 \\ Z(t)i_3 \end{bmatrix}, \begin{bmatrix} i_1 \\ i_2 \\ i_3 \end{bmatrix} = \begin{bmatrix} i_1 \\ -T(t)i_1 \\ i_3 \end{bmatrix} \right) \tag{1-79}$$

which completely describes the component constraints for the network. The graph for the network, shown in Figure 1-12, induces connection constraints characterized by the relation

$$\hat{G} = \left(\begin{bmatrix} v_1 \\ v_2 \\ v_3 \end{bmatrix} = \begin{bmatrix} v_1 \\ v_3 \\ v_2 \end{bmatrix}, \begin{bmatrix} i_1 \\ i_2 \\ i_3 \end{bmatrix} = \begin{bmatrix} i_1 \\ -i_3 \\ i_3 \end{bmatrix} \right) \tag{1-80}$$

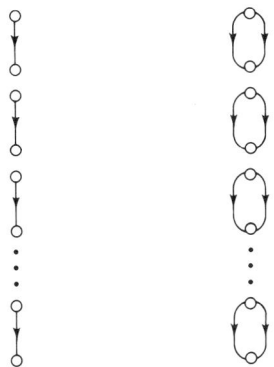

Figure 1-12 Graph for the transformer network of Figure 1-11.

Taking the intersection of the relations of Equations (1-79) and (1-80) by combining the constraints imposed by the separate relations yields for the admissible network variables

$$N \cap \hat{G} = \left(\begin{bmatrix} v_1 \\ v_2 \\ v_3 \end{bmatrix} = \begin{bmatrix} T(t)'Z(t)T(t)i_1 \\ Z(t)T(t)i_1 \\ Z(t)T(t)i_1 \end{bmatrix}, \begin{bmatrix} i_1 \\ i_2 \\ i_3 \end{bmatrix} = \begin{bmatrix} i_1 \\ -T(t)i_1 \\ T(t)i_1 \end{bmatrix} \right) \tag{1-81}$$

where for convenience we have taken i_1 as the "independent" variable with all other variables written in terms of it. Of course, in practice one is seldom interested in characterizing all network variables; rather one would like to have a constraint only on the variables which are accessible at the ports of an n-port realized by the network. In this example these are the "1" variables which are constrained via

$$N_1 = (v_1] = T(t)'Z(t)T(t)i_1], i_1]) \tag{1-82}$$

From Equation (1-82) one observes that the effect of the ideal transformer is to implement a congruence transformation [132] on the $Z(t)$ matrix. In the special case where $n = m = 1$ with the load corresponding to a resistor

$$Z(t) = r \tag{1-83}$$

and the transformer having a turns ratio of k

$$T(t) = k \tag{1-84}$$

the effect of the transformer is to scale the impedance by a factor of k^2

$$v_1 = k^2 r i_1 \qquad (1\text{-}85)$$

which is the commonly encountered result for a 2-port ideal transformer [78]; that is, the impedance is scaled up by the square of the turns ratio.

The most interesting case of an ideal transformer is when $m = n$ and $T(t)$ is an orthogonal matrix (or a scaler). In this case the effect of an ideal transformer on a load, as illustrated in Figure 1-13, will always be that of a congruence

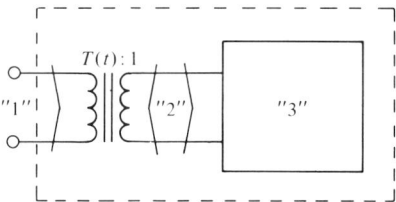

Figure 1-13 $(n + n)$-port orthogonal ideal transformer loaded in an n-port.

transformation (which is not the case for general $T(t)$ when scattering variables are employed). The reader is invited to confirm the following proposition to this effect.

*Proposition 1-5

Let a network have the form shown in Figure 1-13 where the turns ratio matrix $T(t)$ is orthogonal (that is, $T'(t)T(t) = T(t)T'(t) = 1$). Then the characteristics of the n-port realized by this network (at the "1" ports) are

$$\begin{aligned} v_1 &= T(t)'Z(t)T(t)i_1 & \text{if } v_3 &= Z(t)i_3 \\ i_1 &= T(t)'Y(t)T(t)v_1 & \text{if } i_3 &= Y(t)v_3 \\ b_1 &= T(t)'S(t)T(t)a_1 & \text{if } b_3 &= S(t)a_3 \end{aligned}$$
□

Although Proposition 1-5 is stated for the special case where the load is characterized by a simple "matrix algebraic" input-output relation, the result is in fact quite general. In particular, the same result holds for a load whose input-output relation is defined by a distributional kernel or a system function, these to be defined in the following chapters.

1-5 Fundamental Axioms

In the previous sections the n-port and network were defined in an abstract manner, including essentially any phenomena "observable" at a discrete set of points. In this section the classification of networks into various categories for which mathematical tools are available will be considered. In some cases, linearity or time-invariance, the required properties are never achieved physically but the mathematical techniques applicable to them are so powerful as to justify their use. To the contrary, other axioms are satisfied by all real devices, while some, passivity or losslessness, may or may not be physically realizable.

The tools applicable to linear networks are sufficiently powerful that it is advantageous to approximate real devices by linear ones.

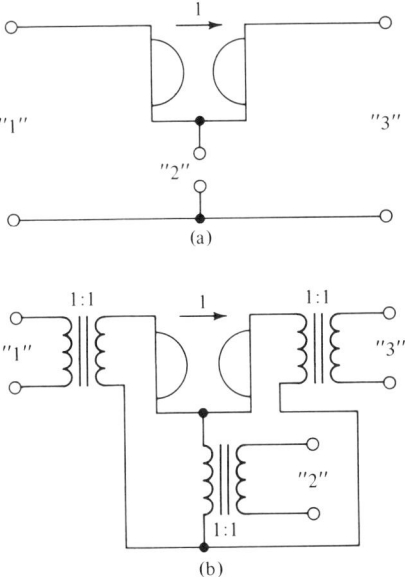

Figure 1-14 Circulator realization of a 3-port circulator.

Definition 1-15

An n-port is *linear* if whenever (a,b) and $(\boldsymbol{a},\boldsymbol{b})$ are in N. Then for all real c and d

$$(ca + d\boldsymbol{a}, cb + d\boldsymbol{b})$$

is also in N. □

This is the usual definition for a linear function [94], but as stated here also includes relations.

All of the circuit elements which have thus far been considered are linear. In the case of n-ports whose input-output relations are characterized by integrals and/or derivatives, this results from the fact that the sum of two integrals is the integral of the sum and that a constant multiplier commutes with the integral sign; and similarly for the derivative. Moreover, all components whose characteristics are determined by a matrix relation are linear, for if

$$M_1(t)b = M_2(t)a \tag{1-86}$$

and

$$M_1(t)\boldsymbol{b} = M_2(t)\boldsymbol{a} \tag{1-87}$$

then clearly

$$M_1(t)(cb + d\boldsymbol{b}) = M_2(t)(ca + d\boldsymbol{a}) \tag{1-88}$$

Even components which cannot readily be represented by a matrix relation may be linear. For example, the norator is linear since all possible (a,b) pairs are admissible; hence certainly $(ca + d\boldsymbol{a}, cb + d\boldsymbol{b})$ is admissible for any c,d and admissible pairs (a,b) and $(\boldsymbol{a},\boldsymbol{b})$. Also, the nullator is linear, for if (a,b) and $(\boldsymbol{a},\boldsymbol{b})$ are admissible they must be zero, in which case $(ca + d\boldsymbol{a}, db + d\boldsymbol{a})$ is also zero, hence admissable.

Although linearity is defined in terms of the scattering variables, the corresponding relationship with i and v replacing a and b yields precisely the same characteristics, since the relationship between these two sets of variables

$$\begin{bmatrix} a \\ b \end{bmatrix} = \begin{bmatrix} 1 & | & 1 \\ \hline 1 & | & 1 \end{bmatrix} \begin{bmatrix} v \\ i \end{bmatrix} \tag{1-89}$$

and its inverse

$$\begin{bmatrix} v \\ i \end{bmatrix} = \frac{1}{2}\begin{bmatrix} 1 & | & 1 \\ \hline 1 & | & 1 \end{bmatrix} \begin{bmatrix} a \\ b \end{bmatrix} \tag{1-90}$$

is linear. Linearity of the immittance variables is thus transformed to an equivalent linearity for the scattering variables, and conversely.

One class of n-ports for which linearity may be illustrated graphically are the "so-called" memoryless n-ports. These are characterized by the property that the relationship constraining their network variables is a function of only the instantaneous values of the network variables and independent of past or future values, derivatives, or the means by which the present values of the network variables are reached. Of course, the constraint itself may be a function of time. Formally, we have the following.

Definition 1-16

An n-port (G_n, N) is *memoryless* if there exists a function

$$f(\cdot,\cdot,t): R^n \times R^n \times R \to R^m$$

(m arbitrary) such that for every t

$$f(a(t), b(t), t) = 0$$

if and only if (a,b) is an admissible pair in N. □

It is important to note that although the network variables, a and b, are in the function space D_+^n the function f of the definition operates on R^n rather than D_+^n. As such it "sees" only the instantaneous values of the network variables, but not the function which brought the network variables to these values.

We leave it to the reader to verify that all of the n-ports thus far defined, except for the inductor and capacitor, are memoryless. For instance, one may verify that the norator is memoryless by letting f be the zero function, in which case the equality of Definition 1-16 is satisfied for all a and b.

In the case of the memoryless networks where the relation between i and v (or a and b) is a function only of the instantaneous values of the variables, and not their past or present values, one may conveniently represent a 1-port by its i-v (or a-b) curve (for a fixed time) in the plane. Now, a 1-port is linear if whenever two points on its curve are admissible then the hyperplane [94] connecting those two points with the origin also represents admissible pairs of the network. For this reason the resistor, whose i-v curve is a line through the origin, is clearly linear, while the norator is linear since its i-v curve is the whole plane. On the other hand, the nullator is linear since its i-v curve only has one point, the origin. These curves are illustrated in Figure 1-15. A common ex-

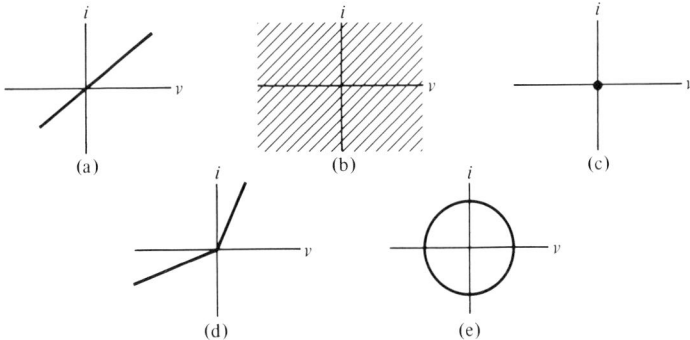

Figure 1-15 i-v curves for a 1-port. (a) Resistor. (b) Norator. (c) Nullator. (d) Diode. (e) 1-port with the unit as its i-v curve.

ample of a nonlinear 1-port is the diode whose i-v curve is shown in Figure 1-15(d). This is composed of one line emanating from the origin in the first quadrant and another in the third. Another nonlinear 1-port is that shown in Figure 1-15(e), whose i-v curve is a circle.

A network characteristic related to, but weaker than, linearity is that of a bilateral n-port. An n-port is said to be bilateral if whenever (a,b) is admissible then so is $(-a,-b)$. Clearly, negating the scattering variables is equivalent to negating the immittance variables; thus one may say that an n-port is bilateral if whenever (v,i) is admissible so is $(-v,-i)$. In an electrical network, reversing the two wires which form a port is equivalent to negating both variables (since it reverses the polarity of the port); hence one may characterize a bilateral n-port as one whose behavior is unchanged by reversing the polarity of each port.

If a network is linear and has an admissible pair (a,b) then the pair

$$(-a + 0a, -b + 0b) = (-a,-b) \tag{1-91}$$

where $(\boldsymbol{a},\boldsymbol{b}) = (a,b)$, $c = -1$, and $d = 0$, is also admissible. Thus every linear network is bilateral. Bilaterality is, however, not sufficient to assure linearity.

In fact, a memoryless 1-port is bilateral if and only if its i-v (a-b) curve is symmetric about the origin. The 1-port with the circle as its i-v curve is thus bilateral, though it is not linear. On the other hand, the diode is both nonlinear and nonbilateral.

Causality is a concept which has both physical and philosophical significance [37], [152]. Basically, it implies that a network is not capable of predicting the future.

Definition 1-17

An n-port is *causal* if whenever (a,b) and $(\boldsymbol{a},\boldsymbol{b})$ are in N and

$$a(t) = \boldsymbol{a}(t) \qquad \text{for } t < T$$

then

$$b(t) = \boldsymbol{b}(t) \qquad \text{for } t < T \qquad \square$$

An alternate approach to causality [131] requires that if

$$a(t) = 0 \qquad \text{for } t < T \qquad (1\text{-}92)$$

then

$$b(t) = 0 \qquad \text{for } t < T \qquad (1\text{-}93)$$

Though this definition is equivalent to the above in the linear case (see Proposition 1-20), it is weaker in the general case [152].

Consider the 1-port characterized by the admissible pairs

$$N = (a(t), b(t) = a(t + d)) \qquad (1\text{-}94)$$

for which the output coincides with the input except for a shift of the time scale by d units. If d is negative the output follows the input, in which case the device is a delay, whereas if d is positive the output precedes the input, the device thereby serving as a predictor. Now in its delay mode (d negative), if two incident waves coincide on some initial segment

$$a(t) = \boldsymbol{a}(t) \qquad \text{for } t < T \qquad (1\text{-}95)$$

then for t less than $T - d$

$$b(t) = a(t + d) = \boldsymbol{a}(t + d) = \boldsymbol{b}(t) \qquad (1\text{-}96)$$

since the middle equality is valid whenever the argument of a, that is, $t + d$, is less than T. In particular, if d is negative, Equation (1-96) holds for all t less than T, which is sufficient to assure the causality of the delayer. If, however, the 1-port is acting as a predictor with d positive, it is not causal, as is illustrated by the input-output pairs shown in Figure 1-16. Here one pair has zero input for all time and hence also zero output for all time, while the second pair has zero input for t less than zero and nonzero input thereafter. The output due to

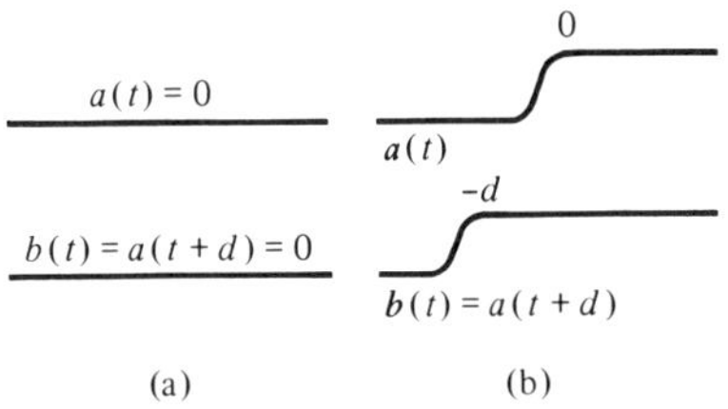

Figure 1-16 Input-output pairs illustrating the noncausality of the ideal prediction.

this latter input is zero prior to $t = -d$ and nonzero thereafter. The two inputs thus coincide prior to $t = 0$, but their outputs do not. The predictor is therefore noncausal.

Of course, the commonly encountered circuit elements are causal, as might be expected of such elements which have real-world origins. The characterization of the causal circuit elements is considered in Proposition 1-8 at the end of this section.

Passivity and losslessness are possibly the most important network axioms. They are essentially energy concepts which determine whether or not a network may deliver energy to its surroundings. These are, of course, external concepts for which a network containing a trivially connected source may be passive. The intuitive idea that passivity implies that a network contains no sources thus may not hold true, though it is possible to realize a passive network without sources.

In all of these energy-related concepts one is interested in power or its integral, energy, but not in the particular variables employed to represent the n-port. One may thus use either the immittance or scattering variables and, in fact, may interchange them at will via the "power equality"

$$i'v = v'i = \tfrac{1}{4}(a'a - b'b) \tag{1-97}$$

derived in Proposition 1-1. In practice one is often more interested in whether the power or energy is positive or negative, finite or infinite, than in knowing its actual value. Since these characteristics are unaffected by the constant multiplier, $\tfrac{1}{4}$, it is generally deleted from the scattering formulation.

Definition 1-18

An n-port is *passive* if for every (a,b) in N and all real t

$$\int_{-\infty}^{t} (a'a - b'b)du \geq 0$$

A network that is not passive is *active*. □

Note that since the network variables are in D_+ the integrals used in the definition of passivity are assured to exist [210]. An alternate definition of passivity [143], [145], [186], [192] requires only that

$$\int_{-\infty}^{\infty} (a'a - b'b)du = \langle a'a \rangle - \langle b'b \rangle \geq 0 \qquad (1\text{-}98)$$

In this case, however, it is necessary to restrict consideration to L_2 functions, for which the integral is finite. Moreover, the fundamental relationship between passivity and causality (obtained in the following section) would not result from the weaker definition of Equation (1-98). In particular, the predictor is active via our definition but passive via the weaker definition. These two characterizations of passivity are thus not, in general, equivalent.

Although the definition of passivity requires that the energy absorbed by the network be non-negative for all time it does not require the instantaneous power to be non-negative. In fact, it is possible for a passive network to absorb energy, store it, and then return it to the outside world at a later time. Of course, during the period in which energy is being returned, the instantaneous power is negative.

For a resistor the power dissipated is

$$i'v = i'r(t)i = i^2 r(t) \qquad (1\text{-}99)$$

which is assured to be greater than zero for all possible input currents i if and only if $r(t)$ is non-negative for all t. Of course, a similar argument is applicable to

$$\int_{-\infty}^{t} i'v\, du = \int_{-\infty}^{t} r(u)i^2 du \qquad (1\text{-}100)$$

Hence, the resistor is passive if and only if $r(t)$ is non-negative for all t. Now, the corresponding scattering characterization for the passivity of the resistor may be obtained by substitution into Equation (1-26). If $r(t)$ is non-negative then

$$|s(t)| = |(r(t) + 1)^{-1}(r(t) - 1)| \leq 1 \qquad (1\text{-}101)$$

Alternatively, this characterization of the passivity of a resistor in terms of its scattering representation may be taken directly from the power dissipation characteristics of the resistor. In this case

$$\begin{aligned} a'a - b'b &= a'a - [s(t)a]'[s(t)a] \\ &= a^2 - s^2(t)a^2 = a^2[1 - s^2(t)] \end{aligned} \qquad (1\text{-}102)$$

which is non-negative if and only if the absolute value of $s(t)$ is less than or equal to 1, this being the same characterization of passivity as was obtained before.

The energy absorbed by a capacitor is given by the integral

$$W(t) = \int_{-\infty}^{t} i'v\, du = \int_{-\infty}^{t} v(c(u)v)du \qquad (1\text{-}103)$$

which upon integration by parts and an application of the fact that $v(-\infty) = 0$ (since v is in D_+) becomes

$$W(t) = \frac{v^2(t)c(t)}{2} + \frac{1}{2}\int_{-\infty}^{t} v^2(u)\dot{c}(u)du \qquad (1\text{-}104)$$

Now, if $c(t)$ is negative for any t, one may choose a $v(t)$ from D_+ which renders the first term of Equation (1-104) both dominant and negative; whereas if $\dot{c}(t)$ is negative, a $v(t)$ may be found which renders the second term dominant over the first and negative [168]. On the other hand, if both $c(t)$ and its derivative $\dot{c}(t)$ are non-negative for all t, the energy $W(t)$ is clearly positive; hence a capacitor is passive if and only if $c(t)$ and $\dot{c}(t)$ are non-negative for all time. Although this result may be at odds with our intuition obtained from the constant capacitor (for which $\dot{c}(t) = 0$ is indeed non-negative), it is physically valid and, in fact, accounts for the ability of a parametric amplifier to exhibit gain [171]. In that case the sinusoidally varying capacitance is always positive, but its derivative is periodically negative, thereby rendering the device active.

In any physical device passivity is determined not only by the physics of the device, but also by the location and number of ports. For example, as analyzed above, the capacitor is a 1-port which may be active. If, however, one were to view a capacitor as a 2-port, one the usual electrical port and the other a port where the mechanical or "pump" energy to change the capacitance is supplied (a small boy with a crank), the capacitor would then be passive. Similarly, a tunnel diode itself is passive, but if the bias needed to render its negative resistance usable is included within the device, it becomes active [171]. The concept of passivity is thus not only dependent on the physics of a device, but also on where and how one chooses to define its boundaries.

Definition 1-19

An n-port is *lossless* if it is passive and for every (a,b) in N where a is L_2 (that is, a is L_2 if $\langle a,a \rangle$ is finite) then

$$\langle a,a \rangle - \langle b,b \rangle = 0 \qquad \square$$

Losslessness can be viewed as the boundary between passivity and activity, for though it permits the network to receive energy in finite time, this energy must be stored and returned—it may not be dissipated.

The energy absorbed by a resistor, as given by Equation (1-100) is

$$\int_{-\infty}^{t} i^2 r(u) du \qquad (1\text{-}105)$$

where $r(t)$ is non-negative for the resistor to be passive. Now, if the resistor is also to be lossless,

$$\int_{-\infty}^{\infty} i^2 r(u) du \qquad (1\text{-}106)$$

must be zero whenever

$$a = r(t)i + i = [r(t) + 1]i \qquad (1\text{-}107)$$

is in L_2. For this to be assured, the integrand itself must be zero for all such i; hence $r(t)$ must be identically zero. As expected, then, the only lossless resistor is the trivial one.

Referring back to Equation (1-104) for the energy dissipated by a capacitor, if

$$a = \dot{v} + (c(t)v)\dot{} \qquad (1\text{-}108)$$

is L_2, $v(t)$ must go to zero at infinity; hence $W(\infty)$ is given by

$$2W(\infty) = \int_{-\infty}^{\infty} v^2(u)\dot{c}(u)du \qquad (1\text{-}109)$$

Now the argument employed above for the resistor will suffice to assure that the capacitor is lossless if and only if $\dot{c}(t)$ is zero; hence the only lossless capacitor is the constant one [168].

If we calculate the power delivered to a transformer we have

$$P(t) = i'v = i'_1 v_1 + i'_2 v_2 = i'_1[T'(t)v_2] + [-i'_1 T(t)']v_2 = i'_1 T'(t)v_2 - i'_1 T(t)'v_2 = 0 \qquad (1\text{-}110)$$

The power is, therefore, zero for all time; hence so is its integral, the energy. The transformer is thus lossless and, in fact, is instantaneously lossless. That is, at no time does the transformer either receive or generate energy.

Definition 1-20

An n-port is *instantaneously lossless* if for every (a,b) in N

$$P(t) = a(t)'a(t) - b(t)'b(t) = 0$$

for all t. ☐

Along with the transformer, the circulator, gyrator, nullator, and several other elements are also instantaneously lossless. The reader is invited to determine all of the instantaneously lossless circuit elements (Proposition 1-13).

Although passivity is the most important of the energy-based n-port classifications, it is sometimes useful to separate the active devices into those capable of exhibiting only a finite amount of energy gain and those which may serve as infinite sources of energy. Of course, the latter do not occur in the real world (being frowned upon by the power industry).

Definition 1-21

An n-port is *nonenergetic* if there exists a finite real number M such that for all (a,b) in N and all real t

$$\int_{-\infty}^{t} (Ma'a - b'b)du \geq 0$$

Otherwise, the n-port is *energetic*. ☐

Clearly, every passive n-port is nonenergetic if one chooses $M = 1$. On the other hand, a 1-port with the scattering representation defined by the admissible pairs

$$(a,b = Da) \in N \qquad (1\text{-}111)$$

is energetic. To see this it suffices to observe that the input-output pair illustrated in Figure 1-17 causes the 1-port to deliver T units of energy to its surroundings given an input energy of one unit. Clearly, it suffices to choose

$$T > M \qquad (1\text{-}112)$$

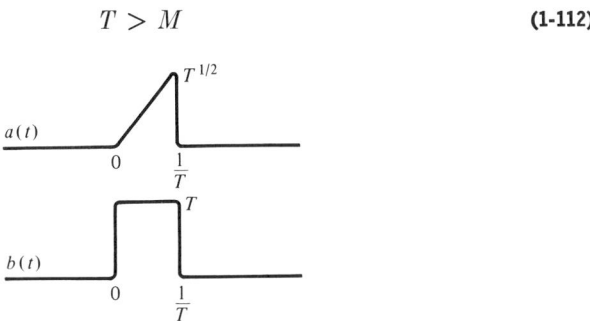

Figure 1-17 Input-output pair for a network characterized by $b = Da$.

to find a signal pair which causes the 1-port to exhibit an energy gain greater than any given M; hence no M satisfying the conditions of Definition 1-21 exists, and the 1-port defined by Equation (1-111) is indeed energetic.

Although the definitions for passive, lossless, and energetic networks are predicated on an intuitive interpretation of the integral of $a'a - b'b$ as energy, they prove to be physically relevant in classes of networks for which the concept of energy is essentially meaningless. For example, in a communications network one may define the entropy of the incident wave as

$$\int_{-\infty}^{t} -|a|' \ln |a| \, du \qquad (1\text{-}113)$$

and that of the reflected wave as

$$\int_{-\infty}^{t} -|b|' \ln |b| \, du \qquad (1\text{-}114)$$

The reader is now invited to determine the relationship between the entropy function and the energy function.

Time-invariance, like linearity, is another property which is probably never achieved in the real world. In this case, however, many real devices are "almost" time-invariant, and, as with linearity, an assumption of time-invariance allows a number of powerful mathematical tools to be invoked.

Definition 1-22

An n-port is *time-invariant* if for every (a,b) in N and every real T, $(a_T, b_T) = (a(t-T), b(t-T))$ is also in N. An n-port which is not time-invariant is said to be *time-variable*. □

In essence, time-invariance implies that if the time scale of the input is shifted, then the output is unchanged except for a corresponding time shift.

For a resistor, one has the equation

$$b(t) = s(t)a(t) \tag{1-115}$$

If the time scale of the input is shifted, say to

$$a_T(t) = a(t-T) \tag{1-116}$$

the corresponding output is

$$b_T(t) = s(t)a_T(t) = s(t)a(t-T) \tag{1-117}$$

Now, this is the required shifted version of the original output if and only if $s(t)$ is not a function of t (since a time parameter in $s(t)$ is not shifted when the input is shifted). As might be expected, the resistor is time-invariant if and only if $s(t)$, or equivalently its resistance, is independent of time. Of course, a similar argument holds for the other RLC circuit elements, which are also time-invariant if and only if they are constant. Although this relationship between time-invariance and constancy holds for most n-ports, it is possible to construct time-invariant n-ports which have variable parameters. These pathological n-ports are considered in more detail in the following chapter.

The concept of time-invariance is closely related to that of periodicity, which corresponds to a weakened form of time-invariance wherein one deals with a single specified T rather than all real T. That is, an n-port, N, is periodic in T if whenever (a,b) is in N, $(a_T, b_T) = (a(t-T), b(t-T))$ is also in N. As the name implies, this corresponds to the class of networks whose parameters are periodic functions in T. Note that the definition of periodicity does not require that the input and output signals for the network be periodic [that is, $a(t-T) = a(t)$], but only that the component coefficients be periodic. Clearly, a time-invariant n-port is periodic in every T. In fact, the reader may verify that a network is time-invariant if and only if it is periodic in every T.

In the general definition of an n-port, N is a relation, but not necessarily a function. In the special case where N is a function, the network is said to be *solvable*.

Definition 1-23

An n-port is *solvable* if for every f in D_+^n there exists a unique (a,b) in N such that $a = f$; and, moreover, if a is L_2 so is b. □

Unlike most of the other definitions, solvability is dependent on the choice of the network variables. In particular, a network may be a function when represented by its immittance variables (for instance, a -1-ohm resistor), but not when represented by scattering variables, or conversely (for instance, an ideal transformer). On the other hand, the nullator or norator may never be represented as functions [9], [49], whereas the 1-ohm resistor has a functional representation in both the immittance and scattering variables. So as to eliminate any ambiguity, the term solvable will be reserved for the case when the admissible pairs for the network take the form of a function with the incident wave as the input and the reflected wave as the output. The requirement that an L_2 a be mapped into an L_2 b is simply a technicality required to guarantee the existence of certain integrals encountered in the sequel, and has no deep interpretation.

The last property to be considered is reciprocity.

Definition 1-24

An n-port is *reciprocal* if for any two ports, k and l, and any two admissible pairs in N having the form

$$(a,b) = \begin{bmatrix} 0 \\ 0 \\ 0 \\ a_k \\ 0 \\ \cdot \\ \cdot \\ 0 \end{bmatrix}, \begin{bmatrix} b_1 \\ b_2 \\ \cdot \\ \cdot \\ \cdot \\ \cdot \\ b_n \end{bmatrix} \qquad (\boldsymbol{a},\boldsymbol{b}) = \begin{bmatrix} 0 \\ 0 \\ 0 \\ 0 \\ \boldsymbol{a}_l \\ \cdot \\ \cdot \\ 0 \end{bmatrix}, \begin{bmatrix} \boldsymbol{b}_1 \\ \boldsymbol{b}_2 \\ \cdot \\ \cdot \\ \cdot \\ \cdot \\ \boldsymbol{b}_n \end{bmatrix}$$

where $a_k = \boldsymbol{a}_l$, then $b_l = \boldsymbol{b}_k$. □

This definition, which corresponds to the intuitive idea of a symmetric matrix representation, coincides with the classical Lorentz definition [110] only for a restricted class of networks. Lorentz reciprocity requires that

$$\int_{-\infty}^{t} a(t-u)'\boldsymbol{b}(u)\,du = \int_{-\infty}^{t} \boldsymbol{a}(t-u)'b(u)\,du \qquad (1\text{-}118)$$

for all admissible pairs (a,b), $(\boldsymbol{a},\boldsymbol{b})$ and real t. The relationship between the two forms of reciprocity will be considered in more detail later.

One final network axiom is closure. Unlike most of our previous definitions, closure has no real physical significance, but rather is a mathematical technicality which will be needed later.

Definition 1-25

An n-port is *closed* if whenever (a_i,b_i) is a sequence of admissible pairs in N which is weakly convergent (in the sense of Definition A-27) to a pair (a,b) then (a,b) is also in N. □

Definition 1-25 is the usual topological concept of closure [22], [134] relative to the topology of weak convergence, and hence might be better termed "weak closure." Since we do not consider closure relative to the topology of strong convergence, the term "closed" suffices.

The definition of a closed n-port completes the list of fundamental network axioms. Although we have already considered the implications of the axioms on many of the circuit elements defined in Section 1-3, the reader is invited to complete the process, the results being summarized by the following propositions.

*Proposition 1-7

All of the circuit elements (listed in Table 1-1) are linear. ☐

*Proposition 1-8

All of the circuit elements except the norator are causal. ☐

*Proposition 1-9

A resistor is passive if and only if $r(t)$ is non-negative for all t. An inductor (capacitor) is passive if and only if $l(t)$ and $\dot{l}(t)$ [$c(t)$ and $\dot{c}(t)$] are non-negative for all t. The gyrator, ideal transformer, circulator, and nullator are always passive. The norator is active. ☐

*Proposition 1-10

The circuit elements are time invariant if and only if their parameters are constant. ☐

*Proposition 1-11

All of the circuit elements except the gyrator, circulator, and norator are reciprocal. ☐

*Proposition 1-12

The lossless circuit elements are constant inductors and capacitors, the ideal transformer, gyrator, circulator, and nullator. ☐

Two additional circuit elements which satisfy all of the axioms are the open and short circuit.

*Definition 1-26

An *open circuit* is a 1-port with $b = a$. A *short circuit* is a 1-port with $b = -a$. ☐

*Proposition 1-13

The open and short circuits are lossless, time invariant, solvable, causal, and reciprocal. □

*Proposition 1-14

The circuit elements which are instantaneously lossless are the transformer, gyrator, nullator, circulator, open circuit, and short circuit. □

*Proposition 1-15

All of the circuit elements except the norator are nonenergetic. □

1-6 Passive Networks

Of the various network axioms passivity (together with linearity) is the one which proves most useful. This is primarily due to the fact that general networks can often be reduced to passive ones. In this section the relation between passivity and the other axioms is considered.

Though the alternate definition of passivity (given in the previous section) is not, in general, an equivalent definition, if one assumes causality and solvability, then the two definitions are equivalent [145], [152].

Proposition 1-16

A causal, solvable n-port is passive if and only if for every (a,b) in N with a in L_2

$$\langle a,a \rangle - \langle b,b \rangle \geq 0$$

Proof

If the network is passive then for each (a,b) and t

$$\int_{-\infty}^{t} (a'a - b'b)\, du = \int_{-\infty}^{t} a'a\, du - \int_{-\infty}^{t} b'b\, du \geq 0 \qquad \text{(1-119)}$$

Upon letting t go to infinity this yields

$$\langle a,a \rangle - \langle b,b \rangle = \lim_{t \to \infty} \int_{-\infty}^{t} (a'a - b'b)\, ds \geq 0 \qquad \text{(1-120)}$$

where the limit exists if a is in L_2 (in which case solvability implies that b is also in L_2). Conversely, if the n-port is causal, solvable, and

$$\langle a,a \rangle - \langle b,b \rangle \geq 0 \qquad \text{(1-121)}$$

for all (a,b) in N for which a is L_2, then letting

$$\boldsymbol{a} = a \qquad \text{for } t < T \qquad \text{(1-122)}$$

and

$$\boldsymbol{a} = 0 \qquad \text{for } t > T + \epsilon \qquad \text{(1-123)}$$

and choosing (a,b) to be the unique member of N with a as its first entry, we have

$$0 \le \langle a,a \rangle - \langle b,b \rangle = \int_{-\infty}^{\infty} (a'a - b'b) \, du$$

$$= \int_{-\infty}^{T} (a'a - b'b) \, du + \int_{T}^{T+\epsilon} (a'a - b'b) \, du + \int_{T+\epsilon}^{\infty} (a'a - b'b) \, du \quad \text{(1-124)}$$

Here a is in L_2 since a is in D_+ and $a = 0$ for large t, while b is in L_2 via solvability. Now, if ϵ is chosen sufficiently small the middle term on the right side of Equation (1-124) may be neglected, while the last term is negative since

$$a = 0 \quad \text{for } t > T + \epsilon \quad \text{(1-125)}$$

Thus

$$0 \le \int_{-\infty}^{T} (a'a - b'b) \, du \quad \text{(1-126)}$$

For $t < T$, $a = a$, and since the network is causal, $b = b$ for $t < T$. Thus Equation (1-126) becomes

$$0 \le \int_{-\infty}^{T} (a'a - b'b) \, du \quad \text{(1-127)}$$

Since this argument holds for all T and any (a,b) in N (not necessarily L_2), the network is passive. □

Causality is needed in the proof of the above proposition to assure that changing a after T does not affect b before T. Consider, for example, the ideal predictor for which

$$b(t) = a(t + d) \quad \text{(1-128)}$$

where d is positive. Now, for an L_2 a

$$\langle b,b \rangle = \int_{-\infty}^{\infty} b(t)'b(t) \, dt$$

$$= \int_{-\infty}^{\infty} a(t+d)'a(t+d) \, dt \quad \text{(1-129)}$$

$$= \int_{-\infty}^{\infty} a(u)'a(u) \, du = \langle a,a \rangle$$

since the time scale shift induced by d only changes the "dummy" variable of integration but not the value of the integral (when the limits of integration are infinite). Hence

$$\langle a,a \rangle - \langle b,b \rangle = \langle a,a \rangle - \langle a,a \rangle = 0 \quad \text{(1-130)}$$

showing that the condition of the theorem may be satisfied for a solvable active network if it is not causal.

The condition of Proposition 1-16 that

$$\langle a,a \rangle - \langle b,b \rangle \ge 0 \quad \text{(1-131)}$$

for all (a,b) in N (a in L_2) may be reformulated in terms of the norm of the n-port [149], for if

$$\langle a,a \rangle - \langle b,b \rangle = \|a\|^2 - \|b\|^2 \geq 0 \tag{1-132}$$

then

$$1 \geq \frac{\|b\|}{\|a\|} \tag{1-133}$$

implying that

$$\|N\| = \sup_{0 < \|a\| < \infty} \left\{ \frac{\|b\|}{\|a\|} \right\} \leq 1 \tag{1-134}$$

The reader may verify that the converse is also true, which then completes the proof of the following proposition.

*Proposition 1-17

For an n-port,

$$\|N\| \leq 1$$

if and only if

$$\langle a,a \rangle - \langle b,b \rangle \geq 0$$

for all admissible pairs (a,b) in N with a in L_2. □

If the passivity condition of Proposition 1-16 is replaced by the weaker condition that the n-port merely be nonenergetic, analogous results are obtained with the positivity of the inner product replaced by a simple boundedness condition [150]. The reader may verify the following.

*Proposition 1-18

A causal, solvable n-port is nonenergetic if and only if there exists a finite real M such that for every (a,b) in N with a in L_2

$$M \langle a,a \rangle - \langle b,b \rangle \geq 0 \quad \square$$

As before, the inner product condition can be replaced by an equivalent condition on the norm.

*Proposition 1-19

For an n-port,

$$\|N\| < \infty$$

if and only if there exists a finite real M such that

$$M \langle a,a \rangle - \langle b,b \rangle \geq 0$$

for all (a,b) in N with a in L_2. □

In the above proposition, one can, in fact, show that the norm of N is bounded by $|M|$; thus upon setting M equal to one, Proposition 1-17 is obtained as a special case.

The combined result of Propositions 1-18 and 1-19, that every causal solvable nonenergetic n-port has finite norm, is possibly the most important reason for employing the scattering variables rather than the immittance variables, for which no analogous result is available (a capacitor is causal, solvable, and lossless but has infinite norm when represented with voltage as the input and current as the output). Since all physically realizable n-ports satisfy these conditions, the powerful mathematical tools applicable only to n-ports with finite norm may be invoked when one employs the scattering variables, but not when the immittance variables are employed.

In the preceding development we have already encountered a relationship between causality and passivity. This relationship is, in fact, quite close and of fundamental importance [150], [152], [192]. It is most easily investigated in the linear case where a simplified characterization of causality is available. This is formulated with the aid of the following lemma.

Lemma 1-1

Let N be the relation representing a linear n-port. Then (0,0) is in N.

Proof

Let (a,b) be an arbitrary pair in N. Then, by linearity,

$$(0a, 0b) = (0,0)$$

is also in N. □

With the aid of this lemma one may show that the simplified causality condition of Equations (1-92) and (1-93) is equivalent to the general causality condition for linear n-ports [152].

Proposition 1-20

A linear n-port is causal if and only if for every admissible pair (a,b) in N with $a(t) = 0$ for $t < T$, then $b(t) = 0$ for $t < T$.

Proof

If an n-port is not causal then there exists an (a,b) in N and an $(\boldsymbol{a},\boldsymbol{b})$ in N with

$$a(t) = \boldsymbol{a}(t) \quad \text{for } t < T \tag{1-135}$$

but

$$b(t) \neq \boldsymbol{b}(t) \quad \text{for some } t < T \tag{1-136}$$

By linearity, $(a - \boldsymbol{a}, b - \boldsymbol{b})$ is also in N and

$$(a - \boldsymbol{a})(t) = 0 \quad \text{for } t < T \tag{1-137}$$

and

$$(b - \boldsymbol{b})(t) \neq 0 \quad \text{for some } t < T \tag{1-138}$$

Thus, if a linear n-port is not causal the weaker characterization also fails to hold. Conversely, if there exists an admissible pair (a,b) in N with

$$a(t) = 0 \quad \text{for } t < T \tag{1-139}$$

but with

$$b(t) \neq 0 \quad \text{for some } t < T \tag{1-140}$$

then (a,b) and $(0,0)$ are both in N and fail to satisfy the condition for causality. Thus, the failure of the weaker characterization also implies the failure of the stronger; hence the two characterizations are equivalent for linear networks. □

Theorem 1-1

Every linear nonenergetic (hence passive) n-port is causal.

Proof

If N is not causal then Proposition 1-20 implies that there exists an admissible pair (a,b) in N with

$$a(t) = 0 \quad \text{for } t < T \tag{1-141}$$

but

$$b(t) \neq 0 \quad \text{for some } t < T \tag{1-142}$$

Thus

$$M \int_{-\infty}^{T} a'a \, du - \int_{-\infty}^{T} b'b \, du = -\int_{-\infty}^{T} b'b \, du < 0 \tag{1-143}$$

showing that a noncausal network is of necessity energetic. □

Theorem 1-1 yields an intuitively palatable result since it shows that if the output of an n-port precedes its input, then the n-port delivers energy to its surroundings which has not previously been delivered to it. Intuition, however, does not indicate the need for linearity in the hypothesis, though no proof (or counterexample) is known in the general case.

A comparison of the propositions at the end of the last section, concerning the passive and causal circuit elements, reveals that all of the passive elements are indeed causal, as required by Theorem 1-1. Moreover, the predictor, which might upon a cursory inspection appear to be passive since the energy in its output is the same as that delivered by its input, is, in fact, active, as is implied by the theorem (from its noncausality).

In the preceding, the relationship between passive and nonenergetic networks and their norms has been considered. The reader is invited to consider the corresponding results for lossless n-ports. In particular, the following may be verified.

*Proposition 1-21

The norm of every lossless network is one. □

Proposition 1-21 is not sufficient, one counterexample being the 2-port of Figure 1-18, a fact which the reader may verify. The 2-port of Figure 1-18 is,

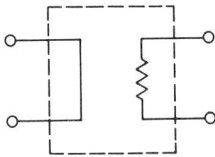

Figure 1-18 2-port with norm one.

in fact, causal and solvable; hence one may not expect to render Proposition 1-21 sufficient by adding these conditions to the hypotheses, as was done in the passive and nonenergetic cases.

1-7 Discussion

The purpose of this chapter has been the formalization of the intuitive idea of a network and the classification of such. Basic to the definitions is the linear graph, which serves both as the medium by which connection can take place and as a basis by which "nonphysical" definitions of voltage and current can be made.

The scattering variables, although defined as linear combinations of voltage and current, are taken as basic. This is justified, in part, by the convenient formulation of passivity and causality achieved, but, moreover, by the dominance of the scattering approach in the sequel. This is, in turn, due primarily to the powerful existence theorems (of Chapters 2 and 3) for networks characterized by the scattering variables which have no immittance counterpart.

The various network properties were defined externally in that they were not based on the properties of any particular circuit elements. However, they did coincide with the intuitive concept of these properties when applied to the "classical" RLC components.

In the final section, the relationship between the various network axioms and properties was considered. This was done most successfully in the linear case, wherein a simplified characterization of causality was found. Due to this fact and also to the number of tools applicable only to linear networks (in particular, the kernels theorem discussed in Appendix B) the sequel will be restricted to the linear case. Though this is indeed a restriction of some significance, a theory of generalized linear networks exists, whereas the results for nonlinear networks are only applicable to scattered special cases.

CHAPTER

2
Time Domain Representation of Linear Networks

2-1 Introduction

In the preceding chapter an n-port was defined as a relation on D_+^n. Though such a definition allows a great deal of generality it is not amenable to convenient analysis. In this chapter only linear networks are considered, the restriction thereby permitting such networks to be characterized by linear operators on the space D_+^n. For example, a linear solvable n-port may be characterized by the scattering operator

$$b = Sa \qquad (2\text{-}1)$$

which defines the unique reflected wave engendered by any incident wave (from D_+^n) applied to the network. Similarly, if each voltage vector induces a unique current vector, the network may be symbolized by the admittance operator

$$i = Yv \qquad (2\text{-}2)$$

or an impedance operator may be used to symbolize a correspondence between current input and voltage output vectors,

$$v = Zi \qquad (2\text{-}3)$$

In dealing with such operators one must remember that they are only symbols which indicate that some correspondence between input and output exists; the symbol, however, gives no clue whatsoever as to the form of that correspondence or its properties. Although a surprisingly large part of our theory can be (and is in Chapters 4 and 5) obtained entirely by manipulating such symbols, it is often desirable to represent an operator by some type of equation which allows one to calculate the network output when given the input. For various special classes of networks such representations abound. These include ordinary and partial differential equations [207], the frequency domain characterization of classical electric network theory [186], [192], state equations [207], and so forth. For the purposes of the present chapter it is the integral operator

$$\int_{-\infty}^{\infty} H(t,q)(\cdot)dq \qquad (2\text{-}4)$$

which proves to be most suitable. Here $H(t,q)$ is a distributional weighting function (see Appendix B on "Distributions") in two variables which, in some sense, corresponds to the output measured at time t, due to an impulsive input at time q. Of course, we do not allow our networks to have impulsive inputs so that this physical interpretation has no real meaning. The terminology associated with impulse responses is, however, convenient and is adopted.

In Appendix A the underlying properties of vector spaces and their associated operator spaces are studied. In particular, it is shown that the space of operators on D_+^n (that is, $\text{Hom}(D_+^n, D_+^n)$) is an algebra under operator addition and composition (multiplication). For this purpose, the sum of two operators, K and L, is defined by the formula

$$(K + L)x \equiv Kx + Lx \qquad (2\text{-}5)$$

That is, the operator $K + L$ takes the vector x to the sum of the vectors Kx and Lx (that is, the output of the sum is the sum of the outputs). Similarly, operator composition or multiplication is defined by

$$(KL)x = K(Lx) \qquad (2\text{-}6)$$

Hence, the output of KL operating on x is the output of K operating on Lx (the output of L operating in x). With these operations (together with scaler multiplication), the operators on the space D_+^n form an algebra [94], the properties of which are essentially the same as those of the "matrix algebra." The latter is, in fact, just the algebra of operators on the space of real n vectors. It is also of some to importance consider the relation between an operator and the inner product $\langle , \rangle$. This is best achieved through the concept of an operator adjoint, which is defined by the equality

$$\langle x, Ky \rangle = \langle K^*s, y \rangle \qquad (2\text{-}7)$$

for all x and y in D_+.

In all of the above, no specific consideration is given to the actual correspondence indicated by the various operators. Rather, one merely says that if the correspondence symbolized by K and that symbolized by L are known, then the new correspondences symbolized by $K + L$, KL and K^*, are known. At no point in the development, however, does one assume a particular correspondence. In essence, one is forming new symbols from the old without ever studying particular symbols. Operator theory is thus "a game" of symbol manipulation; the particular meaning of any symbol, however, is outside the rules of the game. If one desires, as we do, to go beyond this level, it is necessary to represent an operator by some form of equation. This equation, which for the purpose of this chapter is the integral operator of Equation (2-4), then allows one to consider the particular properties of specified n-ports by studying the distributional weighting function $H(t,q)$ which is used to represent the n-port (and allows one to calculate the output of an n-port induced by a specified input).

As a preliminary to formulating the integral representation of a network, let us consider the "simpler" problem of calculating the output of the n-port at a specific time, say t, due to a given input. (Note that since a network may contain integrators, delays, predictors, and so forth, it is, in general, necessary to know the entire input, both before and after time t, if one is to calculate the output at t.) Assume that an n-port is characterized by an operator on the space D_+, say

$$x = Hy \qquad \text{(2-8)}$$

and let H_t be the induced operator, which associates with an input y the real number $x(t)$. That is,

$$x(t) = H_t y \qquad \text{(2-9)}$$

Now if H is linear, so is H_t. Hence, this operator of Equation (2-9) is a functional [22] taking the space D_+ to the reals. By definition, such a functional is just a distribution and can therefore always be represented by the integral operator

$$x(t) = \int_{-\infty}^{\infty} H_t(q) y(q) dq \qquad \text{(2-10)}$$

where $H_t(q)$ is a distributional weighting function in the (dummy) variable q. (Rigorously speaking, a distribution is a functional on the space D, which is zero both before and after some T_1 and T_2. Fortunately, an extension from the space D to D_+, as discussed in Appendix B, is possible, thereby allowing one to apply the above argument with inputs in D_+ rather than D.) Of course, a representation such as that in Equation (2-10) can be obtained for any time t. Hence, one may calculate the entire output, a function in D_+, by considering a parameterized by t family of such integral representations. Taking this viewpoint, and applying the appropriate representation theorem [156], [157], [179], as discussed in Appendix B, one can represent the operator H taking the space D_+ to itself by an integral operator of the form

$$\int_{-\infty}^{\infty} H(t,q)(\cdot)dq \qquad (2\text{-}11)$$

where q is the variable of integration and t is a parameter representing the time at which the output is observed. Of course, if one allows $H(t,q)$ to be a matrix of distributional weighting functions, the resultant operator may be used to represent operators on the space D_+^n. The details inherent in justifying this form of integral representation are nontrivial. The fact that it can be done at all and, in particular, that the parameter t can be "promoted" to a full variable is, however, assured by Schwartz' "kernels theorem" [156], [157], [179].

In dealing with the representations of Equation (2-11) it is important to remember that the distributional weighting functions are not functions in the usual sense, nor is the integral the usual functional integral in the classical sense. Fortunately, the distributions are so defined that they can be manipulated by exactly the same rules normally used for functions and integrals; hence, in the sequel no distinction is made between distributional weighting functions and "functions." It should be noted, however, that this holds true only so long as the distributional weighting function remains under the integral sign. In fact, those distributional weighting functions which are not derived from locally integrable functions (see Appendix B) are not even defined when taken out from under the integral sign.

In Appendix B it is shown that the identity operator is represented by the kernel (or impulse response)

$$x(t) = y(t) = \int_{-\infty}^{\infty} \delta(t - q)y(q)dq \qquad (2\text{-}12)$$

whereas the delay (or predictor) is represented by

$$x(t) = y(t + T) = \int_{-\infty}^{\infty} \delta(t + T - q)y(q)dq \qquad (2\text{-}13)$$

and the differentiator by

$$x(t) = \dot{y}(t) = \int_{-\infty}^{\infty} \dot{\delta}(t - q)y(q)dq \qquad (2\text{-}14)$$

Finally, consider the operator whose weighting function is the unit step (at $t - q$). Now

$$\int_{-\infty}^{\infty} U(t - q)y(q)dq = \int_{-\infty}^{t} y(q)dq = y(q)^{(-1)}\Big|_{-\infty}^{t} \qquad (2\text{-}15)$$

where $y(q)^{(-1)}$ is the indefinite integral of $y(q)$. Since y is assumed to be in D_+, so is $y^{(-1)}$, thus

$$y(-\infty)^{(-1)} = 0 \qquad (2\text{-}16)$$

and
$$\int_{-\infty}^{\infty} U(t-q)y(q)dq = y(q)^{(-1)}\Big|_{-\infty}^{t} = y(t)^{(-1)} \quad \text{(2-17)}$$

$U(t-q)$ is thus the weighting function for the integral operator D^{-1}.

Having formulated an integral representation for the "abstract" operators on D_+, it would be desirable to discern the effect of the algebraic operations (of addition and composition) on the integral representation. In particular, if the weighting functions $K(t,q)$ and $L(t,q)$ are known for operators K and L, then what are the weighting functions for the operators KL, $K+L$ and K^*? In essence, we desire to find the relationship between the symbolic formulas for manipulating operators and the integrals required to calculate the actual output from a given input. Let us, therefore, assume that the operator K is characterized by the equation

$$x(t) = \int_{-\infty}^{\infty} K(t,q)y(q)dq \quad \text{(2-18)}$$

and L is characterized by

$$z(t) = \int_{-\infty}^{\infty} L(t,q)y(q)dq \quad \text{(2-19)}$$

Now, operator addition is symbolized by

$$(K+L)y = Ky + Ly = \int_{-\infty}^{\infty} K(t,q)y(q)dq + \int_{-\infty}^{\infty} L(t,q)y(q)dq$$
$$= \int_{-\infty}^{\infty} [K(t,q) + L(t,q)]y(q)dq = x(t) + z(t) \quad \text{(2-20)}$$

The weighting function of a sum is, therefore, just the sum of the weighting functions. Similarly, the weighting function of the composition of K and L may be obtained by substitution into the definition of KL. Here it is convenient to use a different dummy variable of integration in representing the two operators; hence let K be represented as

$$x(t) = \int_{-\infty}^{\infty} K(t,u)y(u)du \quad \text{(2-21)}$$

and L as

$$y(u) = \int_{-\infty}^{\infty} L(u,q)z(q)dq \quad \text{(2-22)}$$

Now, application of the formula for operator composition gives

$$KLz = K(Lz) = \int_{-\infty}^{\infty} K(t,u)\left[\int_{-\infty}^{\infty} L(u,q)z(q)dq\right]du \quad \text{(2-23)}$$

which, upon commuting the kernel $K(t,u)$ with the integral sign (with respect to q) and changing the order of integration, becomes

$$KLz = \int_{-\infty}^{\infty} \left[\int_{-\infty}^{\infty} K(t,u)L(u,q)du\right]z(q)dq = x(t) \quad \text{(2-24)}$$

The weighting function for the composite operator KL is thus given by

$$\int_{-\infty}^{\infty} K(t,u)L(u,q)du \equiv K(t,q)\mathrm{o}L(t,q) \tag{2-25}$$

and is denoted by the "o" operation. It should be noted here that even though

$$K(t,q)\cdot L(t,q) = L(t,q)\cdot K(t,q) \tag{2-26}$$

$$K(t,q)\mathrm{o}L(t,q) \neq L(t,q)\mathrm{o}K(t,q) \tag{2-27}$$

This results from the fact that in Equation (2-25) the t and q variables would be interchanged if the order of the operators were reversed. We should also note that in Equation (2-25) u is a (dummy) variable of integration, and thus the product

$$K(t,q)\mathrm{o}L(t,q) \tag{2-28}$$

is a function of the t and q variables only.

Finally, we would like to find the weighting function for K^* given that of K. In this case, if K is represented by $K(t,q)$, K^* will be represented by $K(q,t)'$. That is, the weighting function for the adjoint is obtained from $K(t,q)$ by interchanging the two variables and transposing the matrix. The reader may verify this by confirming the equality

$$\begin{aligned}\langle x, Ky \rangle &= \int_{-\infty}^{\infty} x'(t)\left[\int_{-\infty}^{\infty} K(t,q)y(q)dq\right]dt \\ &= \int_{-\infty}^{\infty} \left[\int_{-\infty}^{\infty} K(q,t)'x(q)dq\right]' y(t)dt = \langle K^*x, y \rangle \end{aligned} \tag{2-29}$$

Consistent with the preceding development, one may characterize an n-port via the essentially symbolic operator algebra, or alternatively, via the "hard" integral representation with its corresponding algebra. The primary purpose of this chapter is to formulate the various operators which prove to be useful in the characterization of n-ports and networks, and to study the properties of their corresponding integral representations.

Three different network characterizations are considered. These are the π and Q matrices, applicable to essentially all linear networks, and the less general but more powerful scattering matrix S. Additional consideration is given to the more classical admittance and impedance matrices (referred to jointly as immittance matrices) and the augmented network matrices. These, however, fail to exist in many interesting situations and are thus relegated to a secondary position, the theorems relating to their properties appearing primarily as exercises.

Considerable effort is expended on a study of the concept of normality, this being the property which distinguishes those networks which are "physically" realizable, or are idealizations of such, from those which are essentially mathematical constructs. The non-normal networks (primarily nullators, norators, and devices constructed therefrom), though often maligned, yield a rich supply of examples and counterexamples. In fact, the nullator might have been more appropriately termed "the universal counterexample," a role in which it serves supremely.

Two sections of the chapter are devoted to the translation of the concepts of causality, passivity, losslessness, reciprocity, and time invariance into constraints on the scattering matrix of the network.

Finally, the characteristics of the normalized scattering matrix and their relationship to the various network axioms are considered. Since these results follow from the corresponding properties for the unnormalized scattering matrix, their derivation is, for the most part, left as an exercise.

2-2 The Scattering Matrix

The class of networks characterized by scattering matrices is less general than those to which the π and Q matrices of Section 2-6 apply, but unlike these, the S matrix yields an explicit input-output relationship. This, together with the fact that the scattering matrix exists for a significant class of networks, justifies its predominance in our development.

Though the admittance and impedance descriptions of a network are essentially electrical in nature, the scattering concept, in one form or another, transcends the boundary between the various disciplines. Scattering techniques are variously employed in electromagnetic field theory, microwave studies, and particle physics, as well as in network theory [25], [107], [109], [122], [133]. In network theory, the scattering matrix was formulated as the n-port extension of the reflection parameters long employed in the study of transmission lines. The generalization is due primarily to the endeavors of Belevitch [28], [30]–[32] and Carlin [50].

The most common physical interpretation of the incident and reflected waves is as measures of the power delivered to the network and the power extracted from the network, respectively. Thus, the scattering matrix represents the ratio of the power incident on the network to that which is reflected. Of course, the reflected power may be perturbed in some way (phase shift, delay, and so forth), the characteristics of which are also inherent in the scattering matrix.

The 1-ohm resistor, which absorbs all power delivered to it, has a zero scattering matrix, while an open circuit which reflects all power (without modification) has a scattering matrix equal to one. Upon inspection of the definition for the (normalized) scattering variables

$$a = R^{-1/2}v + R^{1/2}i \tag{2-30}$$

$$b = R^{-1/2}v - R^{1/2}i \tag{2-31}$$

it is observed that they actually have the units of "square root of power"; hence the intuitive basis for the equation

$$P = v'i = \tfrac{1}{4}(a'a - b'b) \tag{2-32}$$

derived in Chapter 1. The power dissipated or stored in the network is thus the difference between the power delivered to the network by a and that reflected in b. The formal definition for the scattering matrix follows.

Definition 2-1

A *scattering matrix* for an n-port is an $n \times n$ matrix of distributions in two variables, $S(t,q)$, such that for each (a,b) in N

$$b(t) = \int_{-\infty}^{\infty} S(t,q)a(q)dq$$

$\square$

As has been mentioned previously, the power of the scattering matrix in network theory is due primarily to the class of networks which possess a scattering matrix description. The first of our existence theorems, this one due to Newcomb and Anderson [6], [110], [113], follows.

Theorem 2-1

Every linear, solvable, nonenergetic n-port has a scattering matrix.

Proof

The proof requires essentially three steps. First, we employ the kernels theorem [156], [179] to construct a scattering matrix which is valid for inputs in D^n. Then the implications of causality (since the n-port is linear and nonenergetic) are delineated, and finally, these are used to show that the scattering matrix constructed via the kernels theorem for inputs in D^n is, in fact, well defined for all inputs in D_+^n.

Since the network is linear and solvable the correspondence between inputs a and outputs b clearly defines a linear operator S such that

$$b = Sa \qquad (2\text{-}33)$$

which takes D_+^n to itself. Now, if $\{a_i\}$ is a sequence of vectors in D^n converging (uniformly) to zero, that is,

$$\lim_{i \to \infty} \|a_i\| = 0 \qquad (2\text{-}34)$$

where the norms exist since all vectors in D^n are L_2, we have, via solvability, that the corresponding outputs

$$b_i = Sa_i \qquad (2\text{-}35)$$

are also in L_2 (though not necessarily in D^n); and, since the n-port is nonenergetic, there exists a real (positive) M such that

$$M\|a_i\|^2 - \|b_i\|^2 = M\langle a_i,a_i \rangle - \langle b_i,b_i \rangle \geq 0 \qquad (2\text{-}36)$$

Rearranging Equation (2-36) now yields

$$\sqrt{M}\|a_i\| \geq \|b_i\| \geq 0 \tag{2-37}$$

which, in combination with Equation (2-34) implies that

$$\lim_{i \to \infty} \|b_i\| = 0 \tag{2-38}$$

Hence, if a sequence of vectors $\{a_i\}$ in D^n converge (uniformly) to zero then the induced sequence of outputs (in $D_+{}^n \cap L_2$) also converges (uniformly) to zero. Moreover, if the sequence $\{b_i\}$ converges (uniformly) to zero it also converges (weakly) to zero, for any given ϕ in D^n the Schwartz inequality [22] yields

$$\left| \int_{-\infty}^{\infty} b_i(q) \phi(q) dq \right| \leq \|b_i\| \, \|\phi\| \tag{2-39}$$

which clearly converges to zero. The restriction of the operator S to inputs in D^n is therefore a linear continuous operator taking D^n to D'^n (the outputs are actually in $D_+{}^n$ but this may be viewed as a subset of D'^n) where continuity is defined to mean that a (uniformly) convergent sequence in D^n is taken to a (weakly) convergent sequence in D'^n. This is, however, precisely the hypothesis of the kernels theorem ([156], [179], Theorem B-1). Hence there exists a unique matrix of distribution in two variables, $S(t,q)$, such that for any incident wave a in D^n the resultant reflected wave b is given by

$$b(t) = \int_{-\infty}^{\infty} S(t,q) a(q) dq \tag{2-40}$$

$S(t,q)$ is thus the required scattering matrix when the incident wave is in D^n. It remains to be shown that it is, in fact, well defined for all inputs in $D_+{}^n$ and yields the required output for these inputs. To this end we have the following lemma.

Lemma 2-1

$$S(t,q) = 0 \quad t < q$$

Proof

Let $m(t,q)$ be a smooth function in two variables such that

$$m(t,q) = \begin{cases} 1, & q \leq t \\ 0, & q > t + \epsilon \end{cases} \tag{2-41}$$

for some specified $\epsilon > 0$, and consider the scattering matrix $S(t,q)m(t,q)$. Upon fixing t at some arbitrary value and invoking the formula for the product of a distribution and a smooth function (Definition B-4), we have, for any a in D^n, that

$$\int_{-\infty}^{\infty} [S(t,q)m(t,q)]a(q)dq = \int_{-\infty}^{\infty} S(t,q)[m(t,q)a(q)]dq \qquad (2\text{-}42)$$

Now,

$$m(t,q)a(q) = a(q) \qquad q \leq t \qquad (2\text{-}43)$$

Hence, since the n-port is causal (being linear and nonenergetic), its response at t to the input $m(t,q)a(q)$ (in D^n) is the same as to the input $a(q)$ (in D^n); thus the right side of Equation (2-42) is equal to

$$\int_{-\infty}^{\infty} S(t,q)a(q)dq \qquad (2\text{-}44)$$

or

$$\int_{-\infty}^{\infty} [S(t,q)m(t,q)]a(q) = \int_{-\infty}^{\infty} S(t,q)a(q)dq \qquad (2\text{-}45)$$

for all t and $a(q)$ in D^n. Since the kernel representation for the operator S, restricted to D^n, is unique, Equation (2-45) implies that

$$S(t,q)m(t,q) = S(t,q) \qquad (2\text{-}46)$$

for all t and q. Finally, since $m(t,q) = 0$ for $q > t + \epsilon$, so is $S(t,q)$; and, moreover, since this is true for all $\epsilon > 0$ we have

$$S(t,q) = 0 \qquad t < q \qquad (2\text{-}47)$$

as was to be shown. □

Finally, with this lemma in our repertory, we may complete the proof of Theorem 2-1.

Proof of Theorem 2-1 (cont.)

Let a be in D_+^n and let $m(t,q)$ be a function of the type used in the proof of the lemma. Now, the response of the n-port at t to the incident wave a is, by causality, the same as to the input $m(t,q)a(q)$, which is in D^n. Hence, if $b = Sa$ we have

$$b(t) = \int_{-\infty}^{\infty} S(t,q)[m(t,q)a(q)] = \int_{-\infty}^{\infty} [S(t,q)m(t,q)]a(q)dq = \int_{-\infty}^{\infty} S(t,q)a(q)dq \quad (2\text{-}48)$$

where the last equality results from the substitution of Equation (2-46). Since Equation (2-48) is valid for all a in D_+^n, $S(t,q)$ is, by Definition 2-1, the required scattering matrix which is well defined for all a in D_+^n. □

Since all passive n-ports are nonenergetic ($M = 1$), the theorem clearly holds in this special case as well.

A resistor is characterized by the equality

$$b(t) = s(t)a(t) \tag{2-49}$$

Hence

$$b(t) = s(t) \int_{-\infty}^{\infty} \delta(t-q)a(q)dq = \int_{-\infty}^{\infty} s(t)\delta(t-q)a(q)dq \tag{2-50}$$

so its scattering matrix is

$$S(t,q) = s(t)\delta(t-q) \tag{2-51}$$

A little algebra will reveal that an equivalent form of this scattering matrix is also given by

$$S(t,q) = s(q)\delta(t-q) \tag{2-52}$$

The details of the derivation, which, of course, apply to the distribution $\delta(t-q)$, whether or not it represents a scattering matrix, are left to the reader as an exercise.

Though the scattering matrix has many advantages over the more classical immittance matrices, the measurement of the scattering parameters of an n-port is not as easily carried out as that of the immittance parameters. This difficulty is by-passed through the use of augmented networks which allow the scattering parameters of a given network to be calculated from the (easily measured) immittance parameters of the augmented networks. Given an n-port, the series augmented network is formed by placing a 1-ohm resistor in series with each port, while the shunt augmented network is formed by placing a 1-ohm resistor in parallel with each port. These are shown in Figure 2-1. If the im-

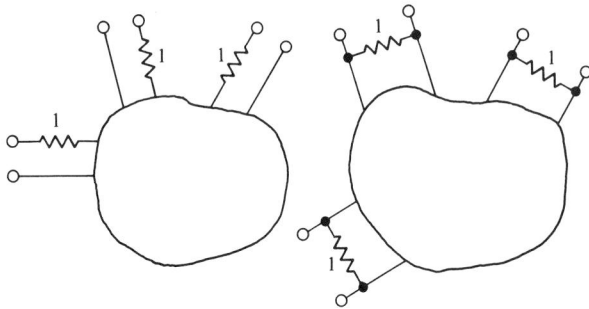

Figure 2-1 Series and shunt augmented networks.

mittance description for the admissible pairs of a network is employed, the admissible pairs for the augmented networks may be calculated therefrom, by adding the voltage drop in the series resistors to the voltage of N, and similarly for the shunt augmented networks.

*Definition 2-2

Let an n-port be described by admissible pairs

$$(i,v) \quad \epsilon\, N$$

Then the *series augmented network* is the n-port with admissible pairs

$$(i, i+v) \quad \epsilon\, N^a$$

and the *shunt augmented network* is the n-port with admissible pairs

$$(i+v, v) \quad \epsilon\, N_a \qquad \square$$

Though the terms admittance and impedance matrix have been employed rather freely (assuming the reader to be familiar with the concept), it is convenient to formalize these ideas prior to examining the relationship between the augmented networks and the scattering matrix.

*Definition 2-3

An *admittance matrix* for an n-port is an $n \times n$ matrix of distributions in two variables, $Y(t,q)$, such that for each

$$(a,b) = (v+i, v-i) \quad \epsilon\, N$$

$$i(t) = \int_{-\infty}^{\infty} Y(t,q) v(q)\, dq \qquad \square$$

*Definition 2-4

An *impedance matrix* for an n-port is an $n \times n$ matrix of distributions in two variables, $Z(t,q)$, such that for each

$$(a,b) = (v+i, v-i) \quad \epsilon\, N$$

$$v(t) = \int_{-\infty}^{\infty} Z(t,q) i(q)\, dq \qquad \square$$

The difference between the immittance matrices and the scattering matrix for a network is simply the variable on which they operate. Unfortunately, if the immittance variables are used, passivity does not assure finiteness for the norm of the network (defined in terms of the immittance variables). Hence, an existence theorem similar to Theorem 2-1 is not applicable (even if solvability is assumed for voltage and current inputs).

Since the derivative operator is represented by

$$\dot{\delta}(t-q) \tag{2-53}$$

while the integral is represented by

$$U(t-q) \tag{2-54}$$

the capacitor, whose immittance variables are constrained by

$$i = (c(t)v)\dot{} = D(c(t)v) \tag{2-55}$$

has admittance and impedance matrices

$$Y(t,q) = \dot{\delta}(t-q)c(q) \tag{2-56}$$

and

$$Z(t,q) = U(t-q)\frac{1}{c(t)} \tag{2-57}$$

respectively.

Notice that in forming these matrices if one wants $c(t)$ to be differentiated, it must be a function of q rather than the parameter t, and similarly for the integration of $1/c(t)$. In particular, the admittance

$$Y(t,q) = \dot{\delta}(t-q)c(t) \tag{2-58}$$

corresponds to the admissible pairs of the form

$$i = c(t)\dot{v} \tag{2-59}$$

which does not, in general, represent a capacitor. A similar argument for the inductor yields the immittance matrices

$$Y(t,q) = U(t-q)\frac{1}{l(t)} \tag{2-60}$$

and

$$Z(t,q) = \dot{\delta}(t-q)l(q) \tag{2-61}$$

Of course, a memoryless n-port which is characterized by a matrix equation

$$v = Z(t)i \tag{2-62}$$

or

$$i = Y(t)v \tag{2-63}$$

can be converted to an immittance matrix in the present sense via

$$Z(t,q) = \delta(t-q)Z(t) = \delta(t-q)Z(q) \tag{2-64}$$

and

$$Y(t,q) = \delta(t-q)Y(t) = \delta(t-q)Y(q) \tag{2-65}$$

A gyrator may, therefore, be characterized by the matrix

$$Z(t,q) = \begin{bmatrix} 0 & -\delta(t-q) \\ \delta(t-q) & 0 \end{bmatrix} \tag{2-66}$$

with $\gamma(t)$ taken here as one.

It is now possible to consider the relationship between the scattering matrix of a network and the immittance matrices for the augmented networks.

*Proposition 2-1

Consider an n-port and let $Y^a(t,q)$ be the admittance matrix of the series augmented n-port. Then the scattering matrix of the given n-port (not the augmented network) is

$$S(t,q) = \delta(t-q) - 2Y^a(t,q) \qquad \square$$

*Proposition 2-2

Consider an n-port and let $Z_a(t,q)$ be the impedance matrix of the shunt augmented n-port. Then the scattering matrix of the given n-port (not the augmented network) is

$$S(t,q) = 2Z_a(t,q) - \delta(t-q) \qquad \square$$

Consistent with the propositions, the problem of calculating the scattering matrix of a network is reduced to that of calculating the immittance matrix for an augmented network. Although immittance matrices are not assured to exist, the augmenting resistors guarantee that the augmented networks are well behaved; hence the following proposition.

*Proposition 2-3

Let $S(t,q)$, $Z_a(t,q)$, and $Y^a(t,q)$ be the scattering, augmented impedance, and augmented admittance matrices for an n-port. Then either they all exist or none of them exist. $\qquad \square$

The augmented matrices yield the most convenient means for calculating the scattering matrix of an inductor or capacitor. The reader may check that these are given by the following proposition.

*Proposition 2-4

The scattering matrices for the inductor and capacitor are

$$S(t,q) = \delta(t-q) - 2U(t-q)\frac{1}{l(t)} \exp\left(-\int_q^t \frac{1}{l(u)} du\right)$$

and

$$S(t,q) = -\delta(t-q) + 2U(t-q)\frac{1}{c(t)} \exp\left(-\int_q^t \frac{1}{c(u)} du\right) \qquad \square$$

In working with the various network matrices it is often convenient to adopt the operator notation

$$L \leftrightarrow \int_{-\infty}^{\infty} L(t,q)(\cdot) dq$$

$$g = Lf \leftrightarrow \int_{-\infty}^{\infty} L(t,q)f(q) dq$$

and

$$KL \leftrightarrow \int_{-\infty}^{\infty} K(t,u)L(u,q) du$$

where the integral sign and explicit "t,q" dependence is suppressed. In general, we will establish the policy of using the operator notation for the "essentially" algebraic results that hold in the frequency domain (to be defined in Chapter 3), as well as for the time domain representation of this chapter. The explicit form of the operator will be given for the results that apply to only one domain of representation. For future reference, we also summarize the admittance, impedance, and scattering matrices of the common circuit elements in Table 2-1.

2-3 Passive and Lossless Scattering Matrices

In the preceding section, the admissible pairs description of a linear network was reformulated as an integral equation. Though this is clearly advantageous with respect to calculating input-output pairs, its full potential can only be realized if the constraints imposed on the admissible pairs by the various network axioms (passivity, causality, and so forth) can be translated into constraints on the integral operators. Fortunately, in the case of the scattering matrix, necessary and sufficient conditions for it to represent a passive, lossless, time invariant, causal, and/or reciprocal network are possible. These in turn permit the properties of a network to be determined by inspection (?) of its scattering matrix.

In this section, passivity, losslessness, and causality are considered, with the characteristics of reciprocity and time invariance reserved for the next section.

In this chapter, only linear networks are considered; hence Theorem 1-1 requires that passive networks be causal. It is thus convenient to study causality as a necessary condition for passivity prior to passivity itself.

Theorem 2-2

Let $S(t,q)$ be the scattering matrix of an n-port. Then

$$S(t,q) = 0 \quad \text{for } t < q$$

if and only if the n-port is causal.

Proof

If

$$S(t,q) = 0 \quad \text{for } t < q \qquad \textbf{(2-67)}$$

then for any (a,b) in N with

$$a(t) = 0 \quad \text{for } t < T \qquad \textbf{(2-68)}$$

$b(t)$ is given by

$$b(t) = \int_{-\infty}^{\infty} S(t,q)a(q)dq \qquad \textbf{(2-69)}$$

Table 2-1 IMMITTANCE AND SCATTERING MATRICES OF THE CIRCUIT ELEMENTS

CIRCUIT ELEMENT	$Z(t,q)$	$Y(t,q)$	$S(t,q)$
RESISTOR	$\delta(t-q)r(t)$	$\delta(t-q)\dfrac{1}{r(t)}$	$\dfrac{r(t)-1}{r(t)+1}\delta(t-q)$
CAPACITOR	$\dfrac{1}{c(t)}U(t-q)$	$\dot{\delta}(t-q)c(q)$	$-\delta(t-q)+2U(t-q)\dfrac{1}{c(t)}\exp\left(-\int_q^t \dfrac{1}{c(u)}du\right)$
INDUCTOR	$\dot{\delta}(t-q)l(q)$	$\dfrac{1}{l(t)}U(t-q)$	$\delta(t-q)-2U(t-q)\dfrac{1}{l(t)}\exp\left(-\int_q^t \dfrac{1}{l(u)}du\right)$
TRANSFORMER $[T(t)\equiv T]$	—	—	$\begin{bmatrix} [1+T'T']^{-1}[T'T-1] & 2[1+T'T']^{-1}T' \\ 2T[1+T'T']^{-1} & [1+TT']^{-1}[1-TT'] \end{bmatrix}\delta(t-q)$
TRANSFORMER $[T(t)\text{ ORTHOGONAL}]$	—	—	$\begin{bmatrix} 0 & T'(t) \\ T(t) & 0 \end{bmatrix}\delta(t-q)$
GYRATOR	$\begin{bmatrix} 0 & -\gamma(t) \\ \gamma(t) & 0 \end{bmatrix}\delta(t-q)$	$\begin{bmatrix} 0 & \dfrac{1}{\gamma(t)} \\ -\dfrac{1}{\gamma(t)} & 0 \end{bmatrix}\delta(t-q)$	$\begin{bmatrix} \dfrac{\gamma(t)^2-1}{\gamma(t)^2+1} & \dfrac{2\gamma(t)}{\gamma(t)^2+1} \\ \dfrac{-2\gamma(t)}{\gamma(t)^2+1} & \dfrac{\gamma(t)^2-1}{\gamma(t)^2+1} \end{bmatrix}\delta(t-q)$
CIRCULATOR	—	—	$\begin{bmatrix} 0 & 0 & 1 \\ 1 & 0 & 0 \\ 0 & 1 & 0 \end{bmatrix}\delta(t-q)$
NULLATOR	—	—	—
NORATOR	—	—	—

Upon combining Equations (2-67) and (2-68), the integrand is seen to be zero for

$$t < q < T \tag{2-70}$$

Hence, the response $b(t)$ is zero for $t < T$. This, together with Proposition 1-18, then assures causality.

The converse of the theorem results from essentially the same argument applied in the proof of Lemma 2-1. That is,

$$S(t,q) = S(t,q)m(t,q) \tag{2-71}$$

where $m(t,q)$ is a smooth function such that

$$m(t,q) = \begin{cases} 1, & q \leq t \\ 0, & q > t + \epsilon \end{cases} \tag{2-72}$$

and ϵ is any strictly positive constant. Since Equation (2-71) is valid for any ϵ, the required result follows. □

The circuit elements, the scattering matrices of which are tabulated in Table 2-1, all satisfy the causality condition of the theorem and, hence, as we already know, they are causal. More generally, any $S(t,q)$ which is composed of linear combinations of the distributions defined by $U(t - q)$, $\delta(t - q)$, and $\delta^{(n)}(t - q)$, satisfies the condition

$$S(t,q) = 0 \quad t < q \tag{2-73}$$

and is thus causal. A scattering matrix which may not be causal is that of the predictor-delay, for which

$$S(t,q) = \delta(t + T - q) \tag{2-74}$$

Now, this is zero except when

$$t = q - T \tag{2-75}$$

showing, via the theorem, that the network is causal if and only if T is less than or equal to zero. That is, the network is operating in its delay mode.

Upon combining Theorem 2-2 with Theorem 1-1, a necessary condition for $S(t,q)$ to represent a passive network is obtained.

Corollary 2-1

For a linear passive network

$$S(t,q) = 0 \quad \text{for } t < q \qquad \square$$

In the preceding chapter it was shown that

$$\|N\| \leq 1 \tag{2-76}$$

for a passive network. Translating this statement into the integral operator notation we write

$$\|S(t,q)\| \triangleq \|N\| \leq 1 \tag{2-77}$$

which we take as the definition of the norm of $S(t,q)$. Equation (2-77) is therefore still another necessary condition on the scattering matrix of a passive n-port. A third necessary condition follows from Theorem 2-1 to the effect that a linear, passive, solvable n-port has a scattering matrix. Finally, if an n-port is solvable it maps L_2 inputs to L_2 outputs. On the surface this latter condition might appear to result from the norm condition of Equation (2-77), but, in fact, the validity of that condition is dependent on an *a priori* assumption that N maps L_2 and L_2 (since otherwise the integrals used in the derivation of Equation (2-77) may not exist); hence, the L_2 mapping condition is, indeed, an independent necessary constraint on a passive n-port. The combination of these four necessary conditions is, in fact, also sufficient.

Theorem 2-3

An n-port is linear, solvable, and passive if and only if the following four conditions are satisfied:
  1. the n-port is characterized by a scattering matrix, $S(t,q)$
  2. $S(t,q) = 0 \quad t < q$
  3. $\|S(t,q)\| \leq 1$
  4. L_2 incident waves are mapped to L_2 reflected waves. □

Proof

The necessity of the four conditions follows from the aformationed theorems. Conversely, if the four conditions are satisfied, the network is represented by a scattering matrix; hence, it is linear and every incident wave is mapped to a unique reflected wave. Since L_2 is mapped to L_2 the existence of $S(t,q)$ also implies that the n-port is solvable. Now, condition 3, together with Theorem 2-2, implies that the n-port is causal; hence Propositions 1-16 and 1-17 imply that the n-port is passive if and only if

$$\|S(t,q)\| = \|N\| \leq 1 \qquad (2\text{-}78)$$

but this is precisely condition 3; hence the n-port is passive, as required, and our proof is complete. □

Clearly, a similar set of conditions holds for nonenergetic n-ports with the norm condition weakened to a finiteness constraint. In this case we have the following result, the details of which are left to the reader.

Theorem 2-4

An n-port is linear, solvable, and nonenergetic if and only if the following four conditions are satisfied:
  1. the n-port is characterized by a scattering matrix, $S(t,q)$
  2. $S(t,q) = 0 \quad t < q$
  3. $\|S(t,q)\| < \infty$
  4. L_2 incident waves are mapped to L_2 reflected waves. □

The predictor-delay has the scattering matrix

$$S(t,q) = \delta(t + T - q) \tag{2-79}$$

Hence, condition 1 of the theorem is satisfied, whereas it was shown previously that the causality condition 2 is satisfied when the 1-port is in the delay mode, $T \leq 0$. Finally, the output of this 1-port is a duplication of the input, except for the time scale shift; hence

$$\|S(t,q)\| = \|\delta(t + T - q)\| = 1 \tag{2-80}$$

and L_2 inputs are mapped to L_2 outputs. Invoking the theorem, it is thus seen that the predictor-delay is passive only in its delay mode and, in fact, it is energetic in its predictor mode. The circuit elements of Table 2-1 all have scattering matrices (except for the nullator and norator) satisfying the causality condition. They are thus passive if and only if

$$\|S(t,q)\| \leq 1 \tag{2-81}$$

and the L_2 mapping condition is satisfied. We leave it as an exercise for the reader to verify that the passivity conditions for these elements obtained via Theorem 2-3 and Equation (2-81) do indeed coincide with those already obtained for the circuit elements in Chapter 1. A scattering matrix satisfying the conditions of Theorem 2-3 is variously termed a passive scattering matrix, a bounded scattering matrix, or a bounded real scattering (BR).

It is often convenient to replace condition 3 of the theorem by

$$\delta(t - q) - S(t,q)^* \circ S(t,q) \geq 0 \tag{2-82}$$

where $S(t,q)^*$ is the adjoint of $S(t,q)$

$$S(t,q)^* = S(q,t)' \tag{2-83}$$

and the positivity for the operator of Equation (2-82) is that defined in Appendix A (Definition A-22).

The equivalence of Equation (2-82) to condition 3 can be shown by replacing

$$\|N\| \leq 1 \tag{2-84}$$

with its equivalent

$$0 \leq \langle a,a \rangle - \langle b,b \rangle \tag{2-85}$$

Now,

$$b(t) = \int_{-\infty}^{\infty} S(t,q)a(q)dq = Sa \tag{2-86}$$

where Sa is the operator notation for the integral. Substituting Equation (2-86) into (2-85) yields for all L_2 a in $D_+{}^3$

$$0 \leq \langle a,a \rangle - \langle b,b \rangle = \langle a,a \rangle - \langle Sa,Sa \rangle = \langle a,a \rangle - \langle a,S^*Sa \rangle = \langle a,(1 - S^*S)a \rangle \tag{2-87}$$

Now, Equation (2-87) is just the definition of positivity for the operator

$$(1 - S^*S) = \delta(t - q) - S(t,q)^* \circ S(t,q) \quad (2\text{-}88)$$

Replacing condition 3 in the theorem by Equation (2-88) yields the following Corollary.

Corollary 2-2

An n-port is linear solvable and passive if and only if the following four conditions are satisfied:
  1. the n-port is characterized by a scattering matrix, $S(t,q)$
  2. $S(t,q) = 0$ for $t < q$
  3'. $0 \leq \delta(t - q) - S(t,q)^* \circ S(t,q)$
  4. L_2 incident waves are mapped into L_2 reflected waves. □

The operator

$$1 - S^*S \quad (2\text{-}89)$$

is called the resistivity matrix of the network. It arises whenever the "power" characteristics of a scattering matrix are considered.

For the resistor

$$S(t,q) = s(t)\delta(t - q) \quad (2\text{-}90)$$

hence

$$\delta(t - q) - S(t,q)^* \circ S(t,q) = \delta(t - q) - s(q)\delta(q - t) \circ s(t)\delta(t - q) \quad (2\text{-}91)$$
$$= [1 - s(t)^2]\delta(t - q)$$

which is positive if and only if

$$|s(t)| \leq 1 \quad (2\text{-}92)$$

Now, Equation (2-92), combined with the "obvious" existence and causality of the resistor, yields the desired passivity conditions for the resistor.

Losslessness requires that a network be passive and, in addition, that

$$\langle a,a \rangle - \langle b,b \rangle = 0 \quad (2\text{-}93)$$

whenever

$$\langle a,a \rangle < \infty \quad (2\text{-}94)$$

Upon repeating the argument used in Corollary 2-2, with the inequality replaced by an equality, Theorem 2-5 is obtained.

Theorem 2-5

An n-port is linear solvable and lossless if and only if the following four conditions are satisfied:
  1. the n-port is characterized by a scattering matrix, $S(t,q)$
  2. $S(t,q) = 0$ for $t < q$
  3L. $0 = \delta(t - q) - S(t,q)^* \circ S(t,q)$
  4. L_2 incident waves are mapped into L_2 reflected waves. □

Condition 3L is a strengthening of condition 3′; thus it might be expected that condition 3 can also be strengthened to allow the characterization of losslessness. In fact, this is not the case, for

$$\|S(t,q)\| = 1 \tag{2-95}$$

is a necessary but not sufficient condition for losslessness. An example of a strictly passive network (passive and not lossless) with norm one is given in

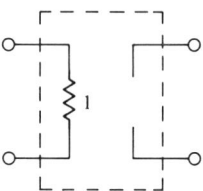

Figure 2-2 A strictly passive network with norm one.

Figure 2-2. This network is composed of two uncoupled 1-ports, an open circuit and a 1-ohm resistor, the admissible pairs for which are given by

$$\left(\begin{bmatrix} a_1 \\ a_2 \end{bmatrix}, \begin{bmatrix} b_1 \\ b_2 \end{bmatrix} \equiv \begin{bmatrix} a_1 \\ 0 \end{bmatrix}\right) \tag{2-96}$$

and its scattering matrix is

$$S(t,q) = \begin{bmatrix} \delta(t-q) & 0 \\ 0 & 0 \end{bmatrix} \tag{2-97}$$

The network is strictly passive since the 1-ohm resistor is strictly passive; thus

$$\|S(t,q)\| \leq 1 \tag{2-98}$$

For the input $\begin{bmatrix} a_1 \\ 0 \end{bmatrix}$ the output is also $\begin{bmatrix} a_1 \\ 0 \end{bmatrix}$; hence

$$1 \geq \|S(t,q)\| = \sup_{\substack{(a,b) \text{ in } N \\ 0 < \|a\| < \infty}} \left\{\frac{\|b\|}{\|a\|}\right\} \geq \frac{\left\|\begin{bmatrix} a_1 \\ 0 \end{bmatrix}\right\|}{\left\|\begin{bmatrix} a_1 \\ 0 \end{bmatrix}\right\|} = 1 \tag{2-99}$$

Now, the two inequalities of Equation (2-99) are simultaneously satisfied only when

$$\|S(t,q)\| = 1 \tag{2-100}$$

which was to be shown.

Theorem 2-5 yields a straightforward means of confirming the losslessness of the transformer and circulator. In fact, any memoryless n-port characterized by a scattering matrix of the form

$$S(t,q) = O(t)\delta(t-q) \tag{2-101}$$

where the matrix $O(t)$ is orthogonal, that is,

$$O(t)'O(t) = 1 \qquad (2\text{-}102)$$

is (instantaneously) lossless. Clearly, conditions 1 and 2 of Theorem 2-5 are satisfied while

$$\begin{aligned}
\delta(t - q) - S(t,q)* \text{ o } S(t,q) &= \delta(t - q) - O(q)'\delta(q - t) \text{ o } O(t)\delta(t - q) \\
&= \delta(t - q) - \int_{-\infty}^{\infty} O(u)'\delta(u - t)O(u)\delta(u - q)du \\
&= \delta(t - q) - \int_{-\infty}^{\infty} \delta(u - t)\delta(u - q)du \\
&= \delta(t - q) - \delta(t - q) = 0
\end{aligned} \qquad (2\text{-}103)$$

Hence, the theorem assures that such an n-port is lossless. We leave it to the reader to confirm that $O(t)$ is, in fact, orthogonal for the transformer, gyrator, and circulator.

An interesting and often useful corollary to the preceding development is the following.

*Proposition 2-5

Let S_1 and S_2 be scattering matrices for passive n-ports. Then

$$S_1 S_2$$

is also the scattering matrix of a passive n-port. □

Proposition 2-5 is indicative of the possibility of realizing complex scattering matrices as a product of simple factors. To achieve this potential, however, it is necessary to connect realizations of S_1 and S_2 so as to realize their product. This can be accomplished by the use of a circulator, a realization for which is shown in Figure 2-3. Here the product of S_1 and S_2 is realized at one set of

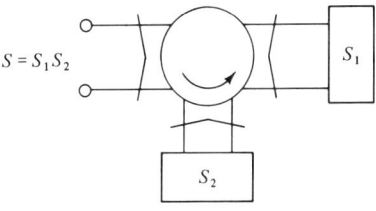

Figure 2-3 Circulator realization of $S_1 S_2$.

circulator ports when the others are loaded with realizations of S_1 and S_2. The symbol) in the figure is used to denote n-ports, all similarly connected.

*Proposition 2-6

The network of Figure 2-3 realizes $S_1 S_2$ at ports "1."

Repeated application of Proposition 2-6 yields a realization for the product

$$S = \prod_{i=1}^{n} S_i \tag{2-104}$$

which is passive if all of the factors are passive. The string of circulators required to achieve the realization is shown in Figure 2-4. The structure of Figure 2-4 is one of the most powerful tools of network synthesis. In many applications,

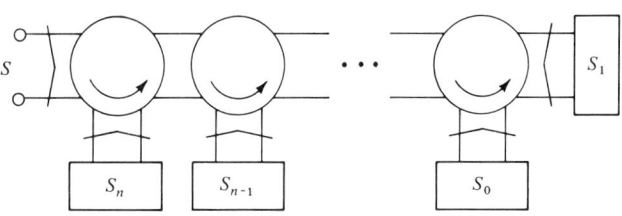

Figure 2-4 Circulator realization of $S = \prod_{i=1}^{n} S_i$.

the circulators of Figure 2-4 are replaced by their gyrator-transformer equivalent (obtained in Chapter 1).

Consider the 2-port with the scattering matrix

$$S(t,q) = \begin{bmatrix} 0 & (\tfrac{1}{2})\delta(t-q) \\ (\tfrac{1}{2})\delta(t-q) & 0 \end{bmatrix} \tag{2-105}$$

Clearly, this 2-port is not itself one of the circuit elements which have thus far been discussed. It, however, may be factored as

$$S(t,q) = S_1(t,q) \circ S_2(t,q)$$

$$= \begin{bmatrix} (\tfrac{1}{2})\delta(t-q) & 0 \\ 0 & (\tfrac{1}{2})\delta(t-q) \end{bmatrix} \circ \begin{bmatrix} 0 & \delta(t-q) \\ \delta(t-q) & 0 \end{bmatrix} \tag{2-106}$$

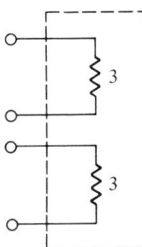

Figure 2-5 Realization of $S_1(t,q)$.

where $S_1(t,q)$ represents the 2-port of Figure 2-5, composed of two uncoupled 3-ohm resistors, and $S_2(t,q)$ corresponds to an ideal transformer with unity turns ratio (see Table 2-1). Now, invoking Proposition 2-6 allows one to realize $S(t,q)$ via a 6-port circulator loaded with the network of Figure 2-5 [which realizes $S_1(t,q)$] at one set of ports and an ideal transformer at the other. This realization is illustrated in Figure 2-6. Of course, there are many other possible

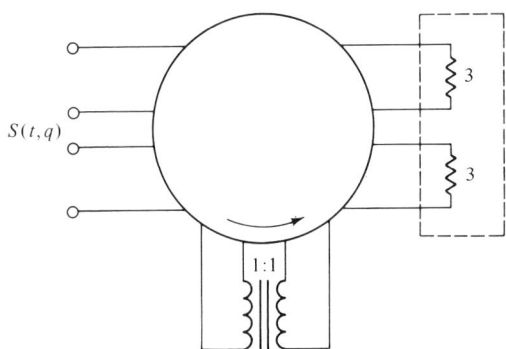

Figure 2-6 Realization of a 2-port, $S(t,q)$, by scattering matrix factorization.

realizations for $S(t,q)$ and, in fact, that of Figure 2-6 is by no means the "best." The technique by which this realization was obtained is, however, illustrative of a very powerful network synthesis procedure, via the factorization of scattering matrices [27], which will be encountered on numerous occasions in the remaining chapters.

The passivity and losslessness conditions for admittance and impedance matrices are not as readily delineated as those for the scattering matrix. We therefore restrict consideration to the memoryless case; yet it is still necessary to hypothesize the existence of $Y(t,q)$ and $Z(t,q)$, whereas, in the scattering case, it is implied.

*Proposition 2-7

A linear memoryless n-port characterized by an immittance matrix is passive if and only if the following two conditions are satisfied:

1. $Y(t,q) = Y(t)\delta(t-q)$
2. $0 \leq Y(t,q) + Y(t,q)^* \; (0 \leq Y(t) + Y(t)')$

and similarly for $Z(t,q)$. □

Consider the gyrator with impedance matrix.

$$Z(t,q) = Z(t)\delta(t-q) = \begin{bmatrix} 0 & \gamma(t)\delta(t-q) \\ -\gamma(t)\delta(t-q) & 0 \end{bmatrix} \quad (2\text{-}107)$$

Here $Z(t,q)$ exists and clearly satisfies condition 1. Moreover,

$$Z(t,q)^* = Z(t)'\delta(t-q) = \begin{bmatrix} 0 & -\gamma(q)\delta(q-t) \\ \gamma(q)\delta(q-t) & 0 \end{bmatrix}$$

$$= \begin{bmatrix} 0 & -\gamma(t)\delta(t-q) \\ \gamma(t)\delta(t-q) & 0 \end{bmatrix}$$

(2-108)

from which it is clear that

$$Z(t,q) + Z(t,q)^* = 0 \geq 0 \qquad (2\text{-}109)$$

verifying the fact that the gyrator is indeed passive.

*Proposition 2-8

A linear memoryless n-port characterized by an immittance matrix is lossless if and only if the following two conditions are satisfied:

1. $Y(t,q) = Y(t)\delta(t-q)$
2. $0 = Y(t,q) + Y(t,q)^* \ (0 = Y(t) + Y(t)')$

and similarly for $Z(t,q)$. □

In Equation (2-109) we had, in fact, equality rather than inequality; hence the gyrator is, as expected, lossless.

2-4 Reciprocal and Time-Invariant Scattering Matrices

The scattering matrix interpretations for reciprocity and time-invariance are so straightforward and readily tested that they are often taken as the definitions of these properties.

Theorem 2-6

Let an n-port be characterized by a scattering matrix $S(t,q)$. Then the n-port is reciprocal if and only if $S(t,q)$ is a symmetric matrix.

Proof

For reciprocity it is required that for inputs

$$\begin{bmatrix} 0 \\ 0 \\ 0 \\ \cdot \\ \cdot \\ a_k \\ \cdot \\ \cdot \\ 0 \end{bmatrix} \quad \text{and} \quad \begin{bmatrix} 0 \\ 0 \\ \cdot \\ a_l \\ \cdot \\ \cdot \\ \cdot \\ 0 \end{bmatrix} \qquad (2\text{-}110)$$

where
$$a_k = a_l \tag{2-111}$$

that the output at port l due to the first input is equal to the output at port k due to the second input. Since these inputs are zero except at the significant port, the outputs of interest are given by

$$b_l(t) = \int_{-\infty}^{\infty} S_{lk}(t,q) a_k(q) dq \tag{2-112}$$

and

$$b_k(t) = \int_{-\infty}^{\infty} S_{kl}(t,q) a_l(q) dq \tag{2-113}$$

now, we require that

$$b_l = b_k \tag{2-114}$$

whenever

$$a_k = a_l \tag{2-115}$$

which is the case if and only if

$$S_{kl}(t,q) = S_{lk}(t,q) \tag{2-116}$$

Since the argument holds for all k and l the n-port is reciprocal if and only if

$$S(t,q)' = S(t,q) \tag{2-117}$$

which was to be shown. □

The above argument does not hold if the Lorentz definition of reciprocity [Equation (1-79)] is used. In that case a linear network is reciprocal if and only if it is time-invariant and its scattering matrix is symmetric. This unexpected pathology in the classical reciprocity definition was discovered by Anderson and Newcomb [4], and is the reason for the nonstandard definition of reciprocity used in this work. The implications of the Lorentz definition will be considered again later after the properties of time-invariant scattering matrices have been derived.

Since any 1×1 matrix is symmetric, it follows immediately from the theorem that any 1-port characterized by a scattering matrix is reciprocal. The theorem, however, presupposes the existence of a scattering matrix and thus does not imply that all 1-ports are reciprocal (if they do not have scattering matrices). In fact, the norator is a nonreciprocal 1-port.

Of the network elements thus far defined, only the norator, circulator, and gyrator are nonreciprocal. Now, the norator has no scattering matrix [44] and is thus not characterized by the theorem. Of the latter two, the circulator is defined by the admissible pairs

$$\left(\begin{bmatrix} a_1 \\ a_2 \\ a_3 \end{bmatrix}, \begin{bmatrix} b_1 \\ b_2 \\ b_3 \end{bmatrix} = \begin{bmatrix} a_3 \\ a_1 \\ a_2 \end{bmatrix} \right) \tag{2-118}$$

Hence, its scattering matrix is

$$S(t,q) = \begin{bmatrix} 0 & 0 & \delta(t-q) \\ \delta(t-q) & 0 & 0 \\ 0 & \delta(t-q) & 0 \end{bmatrix} \quad \text{(2-119)}$$

which is clearly nonsymmetric. The nonreciprocity of the gyrator is most clearly illustrated by its admittance matrix,

$$Y(t,q) = \begin{bmatrix} 0 & \delta(t-q) \\ -\delta(t-q) & 0 \end{bmatrix} \quad \text{(2-120)}$$

This satisfies

$$Y(t,q)' = -Y(t,q) \quad \text{(2-121)}$$

which is the definition of an antireciprocal network.

The scattering matrix interpretation of time-invariance is the subject of Theorem 2-7.

Theorem 2-7

Let an n-port be characterized by a scattering matrix $S(t,q)$. Then it is time-invariant if and only if

$$S(t,q) = S(t-q)$$

Proof
If

$$S(t,q) = S(t-q) \quad \text{(2-122)}$$

then an input $a(t)$ yields an output

$$b(t) = \int_{-\infty}^{\infty} S(t-q)a(q)dq \quad \text{(2-123)}$$

whereas an input $a(t-T)$ yields

$$\int_{-\infty}^{\infty} S(t-q)a(q-T)dq = \int_{-\infty}^{\infty} S(t-(u+T))a(u)du$$

$$= \int_{-\infty}^{\infty} S((t-T)-u)a(u)du = b(t-T) \quad \text{(2-124)}$$

Here

$$u = t - T \quad \text{(2-125)}$$

and the last equality is true since the integral is over the variable u with the parameter $(t-T)$ of Equation (2-124) replacing the parameter t of Equation (2-123). Since the argument holds for any input, the network is time-invariant if

$$S(t,q) = S(t-q) \quad \text{(2-126)}$$

Conversely, if the n-port is time-invariant, $(a(t - T), b(t - T))$ is an admissible pair whenever $(a(t), b(t))$ is itself admissible. Hence,

$$b(t) = \int_{-\infty}^{\infty} S(t,q)a(q)dq \tag{2-127}$$

and

$$b(t - T) = \int_{-\infty}^{\infty} S(t,q)a(q - T)dq \tag{2-128}$$

But, upon changing the variable t to $t - T$ and the dummy variable to u, Equation (2-127) becomes

$$b(t - T) = \int_{-\infty}^{\infty} S(t - T, u)a(u)du \tag{2-129}$$

whereas changing the variable $q - T$ to u in Equation (2-128) yields

$$b(t - T) = \int_{-\infty}^{\infty} S(t, u + T)a(u)du \tag{2-130}$$

Now, since Equations (2-129) and (2-130) hold for all $a(t)$, equating them yields

$$S(t - T, u) = S(t, u + T) \tag{2-131}$$

which is valid for all t, u, and T. In particular, if one lets

$$t = q \tag{2-132}$$

then $u = 0$ and Equation (2-131) becomes

$$S(t - q, 0) = S(t, q) \tag{2-133}$$

which implies that $S(t,q)$ does, indeed, have the required form. ☐

In essence, Theorem 2-7 implies that the scattering matrix of a time-invariant network is a function of only one variable,

$$v = t - q \tag{2-134}$$

Definition 2-5

Let $S(t,q)$ be a scattering matrix. Then the *age variable* is

$$v = t - q \quad ☐$$

This is consistent with the intuitive interpretation of the variables t and q as the time at which the output is observed and the time at which the input is applied, respectively. If a network is time-invariant, then the response is independent of when the input is applied, since the characteristics of the network do not change with time. Rather, the output is a function only of the time lapse between observation of the output and application of the input. A time-invariant scattering matrix is often denoted by $S(v)$ rather than $S(t,q)$ as an indication of its dependence on only a single variable.

The scattering matrix for a resistor is

$$S(t,q) = s(t)\delta(t - q) = s(t)\delta(v) \tag{2-135}$$

which is a function of v only, if and only if the resistor is constant. Similar arguments can be given to show that all of the circuit elements are time-invariant if and only if they are constant. This is not the case with more general circuit elements. Consider, for example, the (linearized) solenoid of Figure 2-7. When

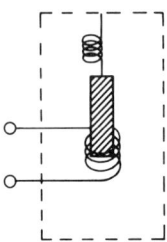

Figure 2-7 Nonconstant time-invariant device.

the input is applied the core is drawn into the coil, thereby increasing the inductance. The network is still time-invariant, however, since the change of inductance is a function of the age variable rather than absolute time (that is, if the input is delayed, then so is the motion of the core and the change of inductance). Upon making a small signal approximation, the impedance matrix for the solenoid is

$$Z(t,q) = L(t - q)\dot{\delta}(t - q) = L(v)\dot{\delta}(v) \tag{2-136}$$

which is, by Theorem 2-7, time-invariant. By contrast, the impedance matrix of an inductor is

$$Z(t,q) = L(q)\dot{\delta}(t - q) \tag{2-137}$$

which does not coincide with that of the solenoid and is, in general, time-variable.

It is sometimes convenient to use the age variable even for time-variable networks (see Chapter 3). In this case the substitution

$$S(t,q) = S(t, t - v) \equiv \mathbf{S}(t,v) \tag{2-138}$$

is made.

An alternative characterization of time-invariance is given by the following proposition.

*Proposition 2-9

A linear n-port is time-invariant if and only if (a,b) is in N implies that $(\dot{a},\dot{b})$ is also in N. ☐

The characterization of time-invariance is essentially the same for all of the network matrices.

*Proposition 2-10

An n-port characterized by a Y, Z, Y^a, or Z_a matrix is time-invariant if and only if the matrix representations are functions only of the age variable. □

Similarly, the various immittance matrices represent reciprocal networks if and only if they are symmetric.

*Proposition 2-11

The matrices $Y(t,q)$, $Z(t,q)$, $Y^a(t,q)$, and $Z_a(t,q)$ represent reciprocal networks if and only if they are symmetric. □

Having established the characteristics of the time-invariant scattering matrices, it is now possible to verify our previous claim that a necessary condition for a network to be Lorentz reciprocal is that it be time-invariant [4]. (The necessary and sufficient condition for Lorentz reciprocity that the network be time-invariant and reciprocal, in our sense, is derived in Chapter 3.) We begin with the "convolution" characterization of Lorentz reciprocity of Equation (1-118) which requires that

$$\int_{-\infty}^{t} a(u)'\boldsymbol{b}(t-u)du = \int_{-\infty}^{t} \boldsymbol{a}(u)'b(t-u)du \qquad (2\text{-}139)$$

for all admissible pairs (a,b) and $(\boldsymbol{a},\boldsymbol{b})$. Now, if a scattering matrix exists one may write

$$\boldsymbol{b}(t-u) = \int_{-\infty}^{\infty} S(t-u,q)\boldsymbol{a}(q)dq \qquad (2\text{-}140)$$

and

$$b(t-u) = \int_{-\infty}^{\infty} S(t-u,q)a(q)dq \qquad (2\text{-}141)$$

Substituting Equations (2-140) and (2-141) into Equation (2-139), and rearranging the order of integration, leaves

$$\int_{-\infty}^{t}\int_{-\infty}^{\infty} a(u)'S(t-u,q)\boldsymbol{a}(q)dqdu = \int_{-\infty}^{t}\int_{-\infty}^{\infty} \boldsymbol{a}(u)'S(t-u,q)a(q)dqdu \qquad (2\text{-}142)$$

Since the integrand on the right side of Equation (2-142) is a 1×1 matrix, it may be transposed without changing its value. Moreover, both u and q are dummy variables of integration; hence, they may be interchanged without affecting the value of the integral, a function of t. Carrying out both of these operations on the right side of Equation (2-142) results in the equality

$$\int_{-\infty}^{t}\int_{-\infty}^{\infty} a(u)'S(t-u,q)\boldsymbol{a}(q)dqdu = \int_{-\infty}^{t}\int_{-\infty}^{\infty} a(u)'S'(t-q,u)\boldsymbol{a}(q)dqdu \qquad (2\text{-}143)$$

Now, if Equation (2-143) is to hold for all possible a and $\boldsymbol{a}$, then it follows that

$$S(t-u,q) = S(t-q,u)' \qquad (2\text{-}144)$$

for all t,q, and u. This being the case, let us consider the implication of Equation (2-143) in the special case where
$$u = 0 \qquad (2\text{-}145)$$
for which
$$S(t,q) = S(t - 0, q) = S(t - q, 0)' \qquad (2\text{-}146)$$
Now, Equation (2-146) implies that $S(t,q)$ is a function only of the variable $t - q$, and it is, by Theorem 2-7, therefore, time-invariant. We have thus verified the contention that every Lorentz reciprocal n-port characterized by a scattering matrix is time-invariant. It should, however, be observed that the argument presupposes the existence of a scattering matrix, and, therefore, is not assured to be valid for networks which do not have scattering matrices. The reader is invited to investigate this case and determine whether or not those Lorentz reciprocal n-ports which do not have scattering matrices are, in fact, time-invariant.

2-5 Normalized Scattering Matrices

In Chapter 1, the normalized scattering variables were defined by
$$a + R^{-1/2}v + R^{1/2}i \qquad (2\text{-}147)$$
and
$$b = R^{-1/2}v - R^{1/2}i \qquad (2\text{-}148)$$

Though these variables have little conceptual advantage over the usual (unnormalized) variables, on occasion they permit certain operational simplifications. In particular, the characteristic impedance for a transmission line is most easily defined as the normalization constant for its scattering matrix [48], [186].

The normalized variables are most easily examined by viewing them as the (unnormalized) scattering variables of a new network characterized by normalized voltage and current variables.

Definition 2-6

Let v and i be the voltage and current vectors associated with a graph. Then for any diagonal matrix R of strictly positive entries
$$\boldsymbol{v} = R^{-1/2}v$$
and
$$\boldsymbol{i} = R^{1/2}i$$
are the *normalized voltage and current vectors*, while
$$a = \boldsymbol{v} + \boldsymbol{i} = R^{-1/2}v + R^{1/2}i$$
and
$$b = \boldsymbol{v} - \boldsymbol{i} = R^{-1/2}v - R^{1/2}i$$

are the *normalized scattering variables*. Here $R^{1/2}$ represents the matrix whose entries are the square roots of those in R, and $R^{-1/2}$ is its inverse. □

Definition 2-7

The *normalized scattering matrix* $S(t,q)_R$, *admittance matrix* $Y(t,q)_R$, *impedance matrix* $Z(t,q)_R$; *augmented admittance matrix* $Y^a(t,q)_R$, and *augmented impedance matrix* $Z_a(t,q)_R$ are defined exactly as their unnormalized antecedents, but relating the normalized variables, v, i, a, and b. □

The normalized matrices thus correspond to the unnormalized matrices except for a voltage and current scaling.

*Proposition 2-12

$$Y(t,q)_R = R^{1/2} Y(t,q) R^{1/2}$$

$$Z(t,q)_R = R^{-1/2} Z(t,q) R^{-1/2}$$

$$S(t,q)_R = [Z(t,q) + R\delta(t-q)]^{-1}[Z(t,q) - R\delta(t-q)]$$

$$= [R\delta(t-q) + Y(t,q)]^{-1}[R\delta(t-q) - Y(t,q)]$$ □

Of course, the relationship between the various normalized matrices satisfies the same algebraic constraints as do the unnormalized matrices; hence, these relationships also hold in the unnormalized case $(R = 1)$. Interestingly, the normalized augmented matrices can be interpreted as the unnormalized immittance matrices for augmented networks, wherein the resistor in the ith port has resistance R_{ii} rather than 1. As in the unnormalized case, this serves as a means of measuring the normalized scattering parameters of a network.

The scaling of variables clearly has no effect on the t,q constraints which characterize causality and time-invariance, nor does it effect the energy integral of a network; thus, the following proposition.

*Proposition 2-13

The passivity, losslessness, causality, reciprocity, and time-invariance conditions for a scattering matrix are an invariant of the normalization (that is, the necessary and sufficient conditions for an unnormalized scattering matrix to have these properties also hold in the normalized case). □

An additional, and more significant, invariant of the scattering matrix normalization is the existence of the matrix. The proof of this will be delayed till Chapter 4, when additional algebraic tools are available.

2-6 General Representations

In this section the most general network matrices, capable of characterizing essentially all linear networks, are formulated. Since solvability is not assumed, it is not possible to write the output as an explicit function of the input (since

86   TIME DOMAIN REPRESENTATION OF LINEAR NETWORKS

the output due to a given input might be multiple-valued or undefined), as in the case of the scattering and immittance matrices. Rather, the admissible pairs (a,b) in N are written as $2n$ vectors in D_+^{2n},

$$\begin{bmatrix} a \\ b \end{bmatrix} \text{ in } D_+^{2n} \tag{2-149}$$

with the set of all admissible pairs characterized as the null space (π matrix) or image (Q matrix) of an integral operator, the existence of these operators being assured by the kernels theorem.

Definition 2-8

Let N be the set of admissible pairs for an n-port. Then a π *matrix* for the n-port is an $m \times 2n$ (m arbitrary) matrix of distributions in two variables, $\pi(t,q)$, such that the null space of the integral operator

$$\int_{-\infty}^{\infty} \pi(t,q)(\cdot)dq$$

is exactly the set of $2n$ vectors, $\begin{bmatrix} a \\ b \end{bmatrix}$, where (a,b) is in N. □

$\pi(t,q)$ is clearly nonunique (if it exists). In fact, it can be premultiplied by any $k \times m$ matrix (of rank m) without changing the null space it generates.

Since the immittance variables are linear combinations of the scattering variables (and conversely), a π matrix which characterizes the scattering variables of a network induces a π matrix on the immittance variables. The relationship between these two π matrices is most easily discerned by partitioning π as

$$\pi(t,q) = [\pi_a(t,q) \mid \pi_b(t,q)] \tag{2-150}$$

where π_a and π_b are $m \times n$ matrices corresponding to the a and b portions of π. Upon invoking the definition of the scattering variables,

$$\begin{bmatrix} a \\ b \end{bmatrix} = \begin{bmatrix} 1 & 1 \\ 1 & -1 \end{bmatrix} \begin{bmatrix} v \\ i \end{bmatrix} \tag{2-151}$$

and substituting Equation (2-151) into the integral operator defining π, one obtains

$$\begin{aligned}\int_{-\infty}^{\infty} \pi(t,q) \begin{bmatrix} a(q) \\ b(q) \end{bmatrix} dq &= \int_{-\infty}^{\infty} [\pi_a(t,q) \quad \pi_b(t,q)] \begin{bmatrix} 1 & 1 \\ 1 & -1 \end{bmatrix} \begin{bmatrix} v(q) \\ i(q) \end{bmatrix} dq \\ &= \int_{-\infty}^{\infty} [\pi_a(t,q) + \pi_b(t,q) \mid \pi_a(t,q) - \pi_b(t,q)] \begin{bmatrix} v(q) \\ i(q) \end{bmatrix} dq \\ &\equiv \int_{-\infty}^{\infty} [\pi_v(t,q) \mid \pi_i(t,q)] \begin{bmatrix} v(q) \\ i(q) \end{bmatrix} dq \end{aligned} \tag{2-152}$$

Thus, the immittance formulation for the π matrix defines an integral operator

$$\int_{-\infty}^{\infty} [\pi_v(t,q) \mid \pi_i(t,q)](\cdot)dq \qquad (2\text{-}153)$$

where

$$\pi_v(t,q) = \pi_a(t,q) + \pi_b(t,q) \qquad (2\text{-}154)$$

$$\pi_i(t,q) = \pi_a(t,q) - \pi_b(t,q) \qquad (2\text{-}155)$$

for which the null space is the admissible pairs of voltage and current variables. The immittance formulation of the π matrix is commonly referred to as the A–B matrix [31], [110]. The π notation is used here so as to avoid confusion with the topological matrices, A and B [155].

A constant resistor is characterized by the equation

$$b(t) = sa(t) \qquad (2\text{-}156)$$

or, equivalently, by

$$[s \mid -1]\begin{bmatrix}a(t)\\b(t)\end{bmatrix} = 0 \qquad (2\text{-}157)$$

Observing that $\delta(t-q)$ is the identity under the integral operator,

$$0 = \int_{-\infty}^{\infty} [s\delta(t-q) \mid -\delta(t-q)]\begin{bmatrix}a(q)\\b(q)\end{bmatrix}dq \qquad (2\text{-}158)$$

if and only if (a,b) is an admissible pair for the resistor with scattering parameter s. The resistor is thus characterized by the π matrix,

$$\pi(t,q) = [s\delta(t-q) \mid -\delta(t-q)] \qquad (2\text{-}159)$$

Since $\dot\delta(-q)$ is representative of the derivative under the integral operator, a capacitor having admissible pairs (a,b) in N where

$$(1 + cD)b(t) = (1 - cD)a(t) \qquad (2\text{-}160)$$

(where D is the derivative operator, $D = d/dt$) may be characterized by the matrix

$$\pi(t,q) = [c\dot\delta(t-q) - \delta(t-q) \quad c\dot\delta(t-q) + \delta(t-q)] \qquad (2\text{-}161)$$

The fact that the π matrix for the capacitor is not unique, may be confirmed by operating on $\pi(t,q)$ with any invertible kernel. For instance,

$$U(t-q)o\pi(t,q) = [c\delta(t-q) - U(t-q) \quad c\delta(t-q) + U(t-q)] \qquad (2\text{-}162)$$

is another π matrix for the capacitor.

If one observes that inversion of the immittance variables corresponds to negation of the scattering variables, then a π matrix for the inductor may be obtained from that of the capacitor simply by negating $\pi_b(t,q)$, since this has the effect of negating the reflected wave associated with a given incident wave. One π matrix for the inductor is thus

$$\pi(t,q) = [l\dot\delta(t-q) - U(t-q) \quad -l\dot\delta(t-q) - U(t-q)] \qquad (2\text{-}163)$$

Though the resistor and capacitor have π representations, they also may be characterized by the more classical immittance and scattering formulations. To the contrary, the nullator and norator have π matrices

$$\pi(t,q) = \begin{bmatrix} \delta(t-q) & 0 \\ 0 & \delta(t-q) \end{bmatrix} \qquad (2\text{-}164)$$

and

$$\pi(t,q) = [0 \mid 0] \qquad (2\text{-}165)$$

respectively, but do not possess admittance, impedance, or even scattering matrices.

The existence of a π matrix for such pathological n-ports as the nullator and the norator is the rule, not the exception. In fact, except for a closure requirement, every linear n-port possesses a π matrix representation.

Theorem 2-8

A necessary and sufficient condition for an n-port to possess a $\pi(t,q)$ matrix is that it be linear and closed. □

The proof of Theorem 2-8 relies rather heavily on function analytic concepts and, therefore, only a sketch of the proof will be given. The reader who is not familiar with the terminology of functional analysis [22], [140] is advised to skip the remainder of this paragraph. The basic approach to the proof is to define an operator π as the projection of D_+^{2n} onto the orthogonal compliment of the linear space of admissible pairs, N, in D_+^{2n}. Unfortunately, since D_+^{2n} is not complete, $N^\perp$ may not exist [140] and some subtrafuge is required. We therefore embed $N \cap L_2$ into L_2 and let $\overline{N \cap L_2}$ be its closure in the Hilbert space L_2. Here, the superscript indicating that the spaces are composed of $2n$ vectors has been dropped to simplify the notation. Now one lets

$$P : L_2 \to L_2 \qquad (2\text{-}166)$$

be the projection of L_2 onto $\overline{N \cap L_2}^\perp$ [140]. Because of the closure requirement in the hypothesis, it may be verified that

$$\text{Ker}(P) \cap D_+ = N \cap L_2 \qquad (2\text{-}167)$$

since any D_+ vectors in the closure of this intersection must be in $N \cap L_2$ itself. Since N is not complete, there may, however, be vectors from $L_2 \backslash D_+$ in Ker (P) and not in N, but these cause no difficulty since we will eventually restrict P to D_+. The functions in L_2 are locally integrable; hence, we may embed L_2 into D' via the mapping j of Proposition B-1. The operator

$$\bar{P} : D_+ \cap L_2 \to D' \qquad (2\text{-}168)$$

defined by

$$\bar{P} = jP\,|_{D_+ \cap L_2} \qquad (2\text{-}169)$$

is thus well defined and, from Equation (2-167) it follows that

$$\text{Ker}(\bar{P}) = N \cap L_2 \qquad (2\text{-}170)$$

Now, P is a projection on L_2 and j is continuous (in the sense of weak convergence); therefore, $\bar{P}$ is also continuous (in the sense of weak convergence). Moreover, in the sense of weak convergence, $D_+ \cap L_2$ is dense in D_+; hence, $\bar{P}$ may be extended uniquely to a (weakly) continuous linear operator π, taking all of D_+ into D' [63], this operator having the representation $\pi(t,q)$, via the corollary to the kernels theorem ([157], Corollary B-1). Finally, Equation (2-170), together with the closure condition, assures that the kernel of π is N, as required. Hence, linearity, together with the closure condition, assures the existence of $\pi(t,q)$. The converse of the theorem follows from the fact that the kernel of a linear operator is a linear space; hence, N must be linear, and such an operator is (weakly) continuous, which in turn implies that its kernel, being the inverse image of a closed set $\{0\}$, is also closed.

The Q matrix is essentially the dual of the π matrix, representing the admissible pairs of a linear n-port as the image rather than null space of an integral operator.

Definition 2-9

Let N be the set of admissible pairs for an n-port. A Q *matrix* for the n-port is a $2n \times m$ (m arbitrary) matrix of distributions in two variables, $Q(t,q)$, such that the image of the integral operator

$$\int_{-\infty}^{\infty} Q(t,q)(\cdot) dq$$

is exactly the set of $2n$ vectors $\begin{bmatrix} a \\ b \end{bmatrix}$ where (a,b) is in N. □

As with the π matrix, the Q matrix is not unique and it can be converted to an immittance representation by an appropriate transformation of variables. In partitioned form, it may be written as

$$Q(t,q) = \begin{bmatrix} Q_a(t,q) \\ \hline Q_b(t,q) \end{bmatrix} \quad \text{(2-171)}$$

and in the immittance formulation,

$$Q_{\mathrm{IM}}(t,q) = \begin{bmatrix} Q_v(t,q) \\ \hline Q_i(t,q) \end{bmatrix} \quad \text{(2-172)}$$

The Q matrix formulation for the resistor is found by observing that

$$b(t) = s\, a(t) \quad \text{(2-173)}$$

implies

$$\begin{bmatrix} a(t) \\ b(t) \end{bmatrix} = \begin{bmatrix} 1 \\ \hline s \end{bmatrix} [a(t)] \quad \text{(2-174)}$$

where $a(t)$ spans $D_+{}^n$. Thus the Q matrix representation for the resistor is

$$Q(t,q) = \begin{bmatrix} \delta(t-q) \\ \text{---------} \\ s\delta(t-q) \end{bmatrix} \tag{2-175}$$

The duality of π and Q is possibly best illustrated by the nullator and norator, the Q matrices of which are given by

$$Q(t,q) = \begin{bmatrix} 0 \\ --- \\ 0 \end{bmatrix} \tag{2-176}$$

and

$$Q(t,q) = \begin{bmatrix} \delta(t-q) & 0 \\ \text{---------} & \text{---------} \\ 0 & \delta(t-q) \end{bmatrix} \tag{2-177}$$

respectively. Thus the Q matrix for the nullator is the transpose of the π matrix for the norator, and vice-versa.

Since $\delta(t-q)$ represents the identity operator, and $\delta(t+T-q)$ the predictor (delay), the Q matrix

$$\begin{bmatrix} a \\ b \end{bmatrix} = \int_{-\infty}^{\infty} \begin{bmatrix} \delta(t-q) \\ \text{---------} \\ \delta(t+T-q) \end{bmatrix} x(q)]dq \tag{2-178}$$

defines a 1-port for which

$$a(t) = x(t) \tag{2-179}$$

while

$$b(t) = x(t+T) \tag{2-180}$$

b is thus identical to a, except for an advance of T units; hence, the 1-port of Equation (2-178) is a predictor.

An argument similar to that used for Theorem 2-8 yields the following theorem.

Theorem 2-9

A necessary and sufficient condition for an n-port to possess a $Q(t,q)$ matrix is that it be linear and closed. □

Although the formulation of π and Q matrices for general n-ports often proves to be a rather complex problem, in the special case of an n-port which is memoryless and linear, the input-output pairs for the network may be characterized by a matrix relation of the form

$$M_1(t)b = M_2(t)a \tag{2-181}$$

which can be converted to the π matrix characterization,

$$\pi(t,q) = [M_2(t)\delta(t-q) \mid -M_1(t)\delta(t-q)] \tag{2-182}$$

since $\delta(t - q)$ is just the identity operator. In essence, the $\delta(t - q)$ in Equation (2-182) cancels the integral; hence, placing $\pi(t,q)$ under the integral sign leaves Equation (2-181) as the characterization of the network. Although this is certainly a "roundabout" means of writing Equation (2-181), it allows one to establish a unified representation theory for all n-ports. Even though the distributional weighting function could be dispensed with, in the memoryless special case it is required for time-dependent n-ports, such as the delay and differentiator.

In the preceding development, the number of rows in π and columns in Q has been completely arbitrary. While there are a number of cases where this parameter is of no significance, it is useful to consider the subclass of these matrices where the number of rows in π (or columns in Q) is minimized.

*Definition 2-10

A π matrix, the number of rows of which is minimal (over all possible π matrices for the n-port), is a $\boldsymbol{\pi}$ matrix. □

*Definition 2-11

A Q matrix, the number of columns of which is minimal (over all possible Q matrices for the n-port), is a $\boldsymbol{Q}$ matrix. □

Since the minimization of Definitions 2-10 and 2-11 is carried out over the positive integers, minimal π and Q matrices exists if any such matrices exist at all. Hence, every linear closed n-port possesses both $\boldsymbol{\pi}$ and $\boldsymbol{Q}$ matrices.

Not surprisingly, the number of rows in $\boldsymbol{\pi}$ and columns in $\boldsymbol{Q}$ are related.

*Proposition 2-14

Let a time-invariant n-port be characterized by $\boldsymbol{\pi}$ and $\boldsymbol{Q}$ matrices. Then the sum of the number of rows in $\boldsymbol{\pi}$ and the number of columns in $\boldsymbol{Q}$ is equal to $2n$.

Note that in applying Proposition 2-14, the $\boldsymbol{Q}$ matrix for the nullator and the $\boldsymbol{\pi}$ matrix for the norator have zero columns and rows, respectively (since they generate zero-dimensional vector spaces). For notational convenience, however, these are usually written with a single row or column of zeros.

A nullor is a 2-port that looks like a nullator at one port and like a norator at the other (and is often used as a model for an operational amplifier). Its π matrix representation is thus

$$\pi(t,q) = \begin{bmatrix} \delta(t-q) & 0 & 0 & 0 \\ 0 & 0 & \delta(t-q) & 0 \end{bmatrix} \quad \text{(2-183)}$$

Since the first and third columns of this π matrix combine to form a 2×2 identity operator, no fewer than 2 rows will suffice to characterize the nullor. The matrix of Equation (2-183) is, therefore, a $\boldsymbol{\pi}$ matrix.

Seven different network matrices have now been defined. The relationship between them is given by the following proposition.

*Proposition 2-15

Whenever the appropriate inverses exist, the relationship between the various network matrices is as shown in Table 2-2.

2-7 Normality

The resistor, capacitor, and most "real" n-ports have π matrices with n rows. In fact, the only devices thus far considered which do not satisfy this condition are the nullator and norator. A further study of this property, termed normality [49], [201], [203], sheds some light on these pathological devices, and, in fact, serves to distinguish the devices which are either "physical" or idealizations of physical devices from those which are purely mathematical. Since the theory of normality is more completely developed in the time-invariant case we restrict ourselves to such n-ports in this section.

Definition 2-12

A linear, time-invariant, closed n-port is *normal* if its π matrices have n rows. A non-normal n-port is *plus-normal* if π has less than n rows and *minus-normal* if π has more than n rows. □

Proposition 2-14 implies that the Q matrix is plus-normal if it has more than n columns, and minus-normal if it has less than n columns, which is the reason for the apparent reversal of plus and minus normality in the definition [47]. Normality could equally have been defined using the immittance variables, since the transformation between the immittance and scattering variables is nonsingular (therefore not changing the rank of π when it is converted from the scattering to immittance formulation).

The basic theorem on normality follows.

Theorem 2-10

The limit (in norm) of a convergent sequence of time-invariant normal n-ports is normal.

Proof

If the limit of such a sequence were minus-normal, the limit of a sequence of matrices with n rows would converge to a π matrix with more than n rows, which is, of course, impossible. If the limit were plus-normal, then the number

Table 2.2 RELATIONSHIP BETWEEN NETWORK MATRICES

	π	Q	S	Y	Z	Y^a	Z_a
S	$-\pi_b^{-L}\pi_a$	$Q_b Q_a^{-R}$	S	$(1+Y)^{-1}(1-Y)$	$(Z+1)^{-1}(Z-1)$	$1-2Y^a$	$2Z_a - 1$
Y	$-(\pi_a - \pi_b)^{-L}(\pi_a + \pi_b)$	$(Q_a - Q_b)(Q_a + Q_b)^{-R}$	$(1+S)^{-1}(1-S)$	Y	Z^{-1}	$(Y^{a-1} - 1)^{-1}$	$Z_a^{-1} - 1$
Z	$-(\pi_a + \pi_b)^{-L}(\pi_a - \pi_b)$	$(Q_a + Q_b)(Q_a - Q_b)^{-R}$	$(1-S)^{-1}(1+S)$	Y^{-1}	Z	$Y^{a-1} - 1$	$(Z_a^{-1} - 1)^{-1}$
Y^a	$\frac{1}{2}(1 + \pi_b^{-L}\pi_a)$	$\frac{1}{2}(1 - Q_b Q_a^{-R})$	$\frac{1}{2}(1-S)$	$Y(Y+1)^{-1}$	$(Z+1)^{-1}$	Y^a	$1 - Z_a$
Z_a	$\frac{1}{2}(1 - \pi_b^{-L}\pi_a)$	$\frac{1}{2}(1 + Q_b Q_a^{-R})$	$\frac{1}{2}(S+1)$	$(Y+1)^{-1}$	$Z(1+Z)^{-1}$	$1 - Y^a$	Z_a

of columns in the limiting Q matrix would exceed n, also an impossibility. Since it is impossible for the limit of a sequence of normal networks to be either plus-normal or minus-normal, it must be normal. □

This theorem, due to Carlin [49], has a number of interesting ramifications. Since all "physical" devices are normal, the theorem implies that the non-normal n-ports are neither "physical" nor idealizations of "physical" devices. It is for this reason that some have deigned to accept the non-normal n-ports [175].

The two examples thus far given of non-normal n-ports, the nullator and norator, are actually the only non-normal 1-ports.

*Proposition 2-16

The nullator and norator are the only time-invariant non-normal 1-ports. □

The fact that the nullator and norator are non-normal does not mean that they cannot be "constructed" from real devices. The reader is, in fact, invited to confirm that the 1-ports illustrated in Figure 2-8 are realizations for the

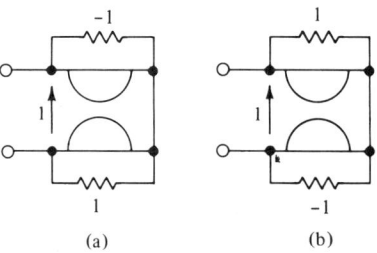

Figure 2-8 Realizations of the (a) nullator and (b) norator.

nullator and norator. The theorem, however, implies that if the component values of either realization are perturbed by an arbitrarily small amount, then the n-ports of Figure 2-8 represent neither the original non-normal components nor even approximations thereof. Due to the uncertainty inherent in any real device, this implies that the non-normal n-ports are not "physically" realizable.

Surprisingly, even though it is not possible to realize the nullator and norator alone, it is possible to realize (within an approximation) the nullor which looks like a nullator at one port and a norator at the other. This is due to the fact that the nullor is normal, the plus normality of its norator cancelling the minus-normality of the nullator.

Finally, we observe that if an n-port has a scattering matrix $S(t - q)$ it is normal since

$$[S(t - q) \overset{!}{\vdots} -\delta(t - q)] \quad \text{(2-184)}$$

is a π matrix for it, which has n rows. Since a similar argument holds for the immittance matrices we have the following.

Definition 2-13

If a time-invariant n-port posesses a scattering matrix $S(t-q)$, an impedance matrix $Z(t-q)$, or an admittance matrix $Y(t,q)$, it is normal. □

2-8 Discussion

In this chapter we have attempted to achieve a compromise between the generality of Chapter 1 and the operational convenience needed in the sequel. This is achieved through the integral operator representation for a network,

$$f(t) = \int_{-\infty}^{\infty} k(t,q)g(q)dq \qquad \text{(2-185)}$$

These operators have appeared in several forms. The π and Q matrices represent any linear closed n-port as the null space or range, respectively, of the integral operator they define, while the scattering matrix yields an explicit description of the reflected wave, due to a given incident wave. Furthermore, various immittance matrices were defined, but the usefulness of these is restricted by the limited class of networks to which they are applicable.

One advantage of the integral operator characterization of a network is the ease with which the various network axioms may be related to the integral operator kernel. In particular, the causality, reciprocity, and time-invariance of a network may be determined by inspection of the scattering matrix.

CHAPTER
3
Frequency Domain Representation of Linear Networks

3-1 Introduction

Network theory is often identified entirely with the frequency domain representation. In fact, this representation is often (naively) taken as the definition for a network [143]. This is due, in part, to the convenient description of the filtering and phase shifting characteristics of the network inherent in its system function, and also to the fact that the system function of an RLC network is isomorphic to its differential operator [207]. This latter property permits many of the algebraic characteristics of the differential operator to be transformed to the (seemingly) simpler system function. In the case of generalized networks, this is not the case; in fact "interesting" networks can only rarely be represented as the input–output pairs of an ordinary differential operator. The generalized frequency domain, however, still retains its advantages as a description of the filtering characteristics of the network, and, in addition, serves as a stepping stone between the classical RLC theory and the generalized theory. The non-RLC frequency domain, in fact, proves to be a powerful tool for the design of distributed devices, possibly with lumped discontinuities [48].

The system function is possibly the most "ill defined" concept of network theory, a characteristic which renders it an ideal subject for examination questions. In the special case of RLC networks, it may be defined as the complex function obtained by replacing D (d/dt) with the complex variable p. A sounder approach (which does not assume the existence of an ordinary differential operator representation for the network) is to let

$$H(p) = \frac{\text{response to } e^{pt}}{e^{pt}} \tag{3-1}$$

This, though not dependent on the characteristics of the network, does assume that the response to e^{pt} is defined, which may not be the case. A similar approach, to be employed here, is to define the system function as the distributional Laplace transform of the impulse response matrix (with respect to the age variable). Since a number of existence theorems have already been obtained for the various impulse response matrices (in Chapter 2), and a theory for the Laplace transformation of distributions is formulated in Appendix B, the system function may be "well defined" via this approach.

In dealing with the system function formulation of a network, one is interested, as with the time domain representation, in the relationship between the various operator characteristics and the frequency domain representation. In particular, if one is given the frequency domain representation for operators K and L [$K(p)$ and $L(p)$] then how does one calculate the response

$$Kf \tag{3-2}$$

to an input f, the sum

$$K + L \tag{3-3}$$

and the product

$$KL \tag{3-4}$$

of the operators within the frequency domain formalism? As one might expect from RLC experience, the frequency domain analogs of Equations (3-2) through (3-4) are

$$K(p)F(p) \tag{3-5}$$

$$K(p) + L(p) \tag{3-6}$$

and

$$K(p)L(p) \tag{3-7}$$

respectively. Here we adopt the notation $K(p)$ for the Laplace transform of the distribution $k(v)$, rather than $L(k)(p)$ as used in the development of Appendix B. These results are considered in Appendix B, and are rederived in their more general time-variable form in Section 3-2 of this chapter.

In addition to the above formulas, one would like to obtain a characterization for the frequency representation of the adjoint operator

$$K^* \tag{3-8}$$

given that of K. This is most readily obtained by translating the inner product itself into the frequency domain. In that case the classical result of Parseval (Proposition B-3) yields

$$\langle f, g \rangle \equiv \int_{-\infty}^{\infty} f(u)'g(u)du = \text{Re} \int_{-\infty}^{\infty} F(j\omega)^{\dagger} G(j\omega) d\omega \tag{3-9}$$

where $F(p)$ and $G(p)$ are the transforms of $f(t)$ and $g(t)$, respectively, and † denotes the complex conjugate transpose. It is significant here to note that one can calculate the inner product wholly from a knowledge of the characteristics of $F(p)$ and $G(p)$ on the $j\omega$ axis, the remainder of the complex plane playing no role whatsoever. If one now represents the operator K in the frequency domain via $K(p)$, and the inputs $f(t)$ and $g(t)$ by $F(p)$ and $G(p)$, then

$$\langle f, Kg \rangle = \operatorname{Re} \int_{-\infty}^{\infty} F(j\omega)^{\dagger}(K(j\omega)G(j\omega))d\omega$$
$$= \operatorname{Re} \int_{-\infty}^{\infty} (K(j\omega)^{\dagger}F(j\omega))^{\dagger}G(j\omega)d\omega = \langle K^*f, g \rangle \tag{3-10}$$

The adjoint K^* is thus represented in the frequency domain by

$$K(p)^{\dagger} \tag{3-11}$$

Consistent with this result, we will henceforth adopt the usual adjoint notation, *, for the complex conjugate transpose, in which case the adjoint is denoted by

$$K(p)^* \equiv K(p)^{\dagger} \tag{3-12}$$

in the frequency domain.

After formalizing the system function concept and considering its relationship to the various classes of network matrices, it is employed to prove a new existence theorem for the scattering matrices of time-invariant networks. The passivity and losslessness conditions of Chapter 2 are readily translated into the frequency domain, yielding the classical positive reality and bounded reality theorems [31], [44], [92], [110], [183]. Since these theorems result entirely from the axioms of Chapter 1, no recourse to the circuit elements from which the network might be constructed is made, nor is rationality of the system function assumed. Passivity and losslessness conditions are thus obtained simultaneously for distributed, lumped, 1-port, multiport, reciprocal, and nonreciprocal networks.

The final section of the chapter is devoted to the proofs of a number of theorems from classical network theory. This material, though relevant to many of the applications of the general theory, is not needed in the remainder of the text; hence, the results are only briefly outlined, with most of the proofs being left to the reader as exercises.

A number of the results, especially those in the first section, are applicable to any of the network matrices formulated in Chapter 2. In obtaining these results, we thus adopt the practice of denoting a general impulse response by

$$H(t,q) \tag{3-13}$$

and its corresponding system function by

$$H(p) \tag{3-14}$$

or, in the more general time-variable case,

$$H(p,t) \tag{3-15}$$

Of course, as is required by the application, one may replace H by any of the specific network matrices of Chapter 2. Results derived under this notation thus hold for scattering and immittance matrices, as well as for the more general π and Q matrices.

3-2 The System Function

In the case of RLC networks, the system function is the Laplace transform of the impulse response with respect to the age variable, that is,

$$H(p) = \int_{-\infty}^{\infty} h(v)e^{-pv}dv = \int_{-\infty}^{\infty} h(t-q)e^{-p(t-q)}dq \tag{3-16}$$

For the purposes of this work, a similar definition suffices, after taking account of the second variable in the "impulse response" matrix for a time-variable network.

Definition 3-1

Let $H(t,q)$ be the (distributional) weighting function of an integral operator. Then its associated *system* function (if it exists) is

$$H(p,t) = \int_{-\infty}^{\infty} H(t,q)e^{-p(t-q)}dq$$

where p is a complex variable. □

Upon making the substitution

$$q = t - v \tag{3-17}$$

the system function may be written as

$$H(p,t) = \int_{-\infty}^{\infty} H(t, t-v)e^{-pv}dv$$
$$= \int_{-\infty}^{\infty} \mathbf{H}(t,v)e^{-pv}dv \tag{3-18}$$

which is exactly the (distributional) Laplace transform of the modified kernel

$$\mathbf{H}(t,v) \equiv H(t, t-v) \tag{3-19}$$

with respect to the age variable [40], [210] if the conditions of Theorem B-2 are satisfied.

Though distributional inputs are not allowed, the above development shows that the distributional Laplace transform defined in the Appendix B still plays a significant role, since the integral operator kernels are, in general, distributions. Of course, the Laplace transform of a distribution is not assured to exist; hence, it is necessary to demonstrate the existence of the integral in Definition

3-1 before the system function can be employed. The $H(p,t)$ function in its full generality was first defined by Zadeh [208], and is thus sometimes called "Zadeh's function."

Applying Definition 3-1 to the impulse response

$$S(t,q) = \delta(t + T - q) \tag{3-20}$$

of the predictor (delay), one obtains

$$S(p,t) = \int_{-\infty}^{\infty} \delta(t + T - q)e^{-p(t-q)}dq = e^{pT} \tag{3-21}$$

for the required system function. More generally, if one considers a predictor with time-variable gain characterized by

$$S(t,q) = s(t)\delta(t + T - q) \tag{3-22}$$

one obtains the system function

$$\begin{aligned} S(p,t) &= \int_{-\infty}^{\infty} s(t)\delta(t + T - q)e^{-p(t-q)}dq \\ &= s(t)\int_{-\infty}^{\infty} \delta(t + T - q)e^{-p(t-q)}dq = s(t)e^{pT} \end{aligned} \tag{3-23}$$

which is a function of both t and p, rather than just the p variable. The variable gain of the impulse response of Equation (3-22) is a function of t, the time at which the response is measured. Alternatively, one could define a predictor, the gain of which is a function of the time the input was applied. In this case the time domain scattering matrix is

$$S(t,q) = \delta(t + T - q)s(q) \tag{3-24}$$

and the system function is

$$S(p,t) = s(t - T)e^{pT} \tag{3-25}$$

Finally, one might be interested in a predictor, the gain of which is a function of the time elapsed between the application of the input and the measurement of the output. Such a network might represent the effect of transmitting a wave (of almost anything) through a lossy medium. Under such circumstances, the gain (or loss) is a function of the length of time during which the wave is in the medium, rather than the actual time at which it is applied or measured. We thus have time and frequency domain scattering matrices of

$$S(t,q) = s(t - q)\delta(t + T - q) \tag{3-26}$$

and

$$S(p,t) = s(-T)e^{pT} \tag{3-27}$$

Here, $S(p,t)$ is once again independent of t, as might be expected from the time-invariant character of the scattering matrix $S(t,q)$.

Another class of scattering matrices which appear quite often in practice are the memoryless n-ports which are characterized by matrices of the form

$$H(t,q) = H(t)\delta(t-q) \tag{3-28}$$

Here

$$H(p,t) = \int_{-\infty}^{\infty} H(t)\delta(t-q)e^{-p(t-q)}dq = H(t) \tag{3-29}$$

showing, as may be expected, that the memoryless n-ports have system functions which are independent of p and, in fact, coincide with the matrix algebraic characterizations for these devices. That is, if the matrix $H(t)$ characterizes an n-port via

$$x(t) = H(t)y(t) \tag{3-30}$$

then the corresponding system function is just $H(t)$.

Of course, in network studies two of the most important impulse responses are

$$H(t,q) = \dot{\delta}(t-q) \tag{3-31}$$

and

$$H(t,q) = U(t-q) \tag{3-32}$$

which represent the differential and integral operators, respectively. Substituting into the definition of the system function one obtains

$$H(p,t) = \int_{-\infty}^{\infty} \dot{\delta}(t-q)e^{-p(t-q)}dq = p \tag{3-33}$$

and

$$H(p,t) = \int_{-\infty}^{\infty} U(t-q)e^{-p(t-q)}dq = \frac{1}{p} \tag{3-34}$$

which are the usual rational system functions representing these operators.

From the above examples it is seen that the system function associated with the kernel $\delta(t-q)$ is p times the system function associated with $U(t-q)$, and that associated with $\dot{\delta}(t-q)$ is p times that of $\delta(t-q)$. Since $\dot{\delta}(t-q)$ is the distributional derivative of $\delta(t-q)$, which in turn is the distributional derivative of $U(t-q)$ (both with respect to the age variable), this is consistent with the result of classical transform theory that the transform of the derivative of a function is p times the transform of the function itself [131]. Indeed this is the case for all of our system functions.

Proposition 3-1

Let $H(t,q) = \boldsymbol{H}(t,v)$ have system function $H(p,t)$. Then the system function of $H(t,q)^{(n)}$, that is, the nth distributional derivative of $H(t,q) = \boldsymbol{H}(t,v)$ with respect to v, exists and is given by $p^n H(p,t)$.

Proof

Using the definition of a distributional derivative of Appendix B, we have

$$\int_{-\infty}^{\infty} H(t,q)^{(n)} e^{-p(t-q)} dq = \int_{-\infty}^{\infty} \boldsymbol{H}(t,v)^{(n)} e^{-pv} dv$$

$$= (-1)^n \int_{-\infty}^{\infty} \boldsymbol{H}(t,v) \left(\frac{d^n e^{-pv}}{dv^n} \right) dv = (-1)^n \int_{-\infty}^{\infty} \boldsymbol{H}(t,v)(-p)^n e^{-pv} dv \quad \text{(3-35)}$$

$$= p^n \int_{-\infty}^{\infty} \boldsymbol{H}(t,v) e^{-pv} dv = p^n H(p,t)$$

as required. □

The existence theorem for $H(p,t)$ follows.

Theorem 3-1

Let $H(t,q)$ be the kernel of a causal integral operator with finite norm. Then, for all p in the right half plane (RHP) (that is, complex p such that Re $p > 0$), $H(p,t)$ exists and is analytic.

Proof

The proof of this theorem will be based on a number of results from distribution theory, none of which we will prove here. Since $H(t,q)$ has finite norm it maps inputs from $D \subset L_2$ into L_2 and, hence, by a theorem of Schwartz [40], [156],

$$\boldsymbol{H}(t,v) = H(t, t - v) \quad \text{(3-36)}$$

is in S' for each fixed t (viewed as a function of v). (See Section B-4 for a definition of S'. Actually, Schwartz proves that $\boldsymbol{H}(t,v)$ is in D'_{L_2} [40], but this is contained in S', which suffices for our purposes.) Since $H(t,q)$ is causal, $\boldsymbol{H}(t,v)$ is zero for $v < 0$, while for $\sigma > 0$, $e^{-\sigma v}$ decays exponentially; hence

$$\boldsymbol{H}(t,v) e^{-\sigma v} \quad \text{(3-37)}$$

is also in S' for each $\sigma > 0$, whence Theorem B-2 [40] implies that for all t, $\boldsymbol{H}(t,v)$ has a distributional Laplace transform which is defined and analytic for all $p = \sigma + j\omega$ (that is, p in the RHP), given by

$$L[\boldsymbol{H}(t,v)] = \int_{-\infty}^{\infty} \boldsymbol{H}(t,v) e^{-pv} dv = H(p,t) \quad \text{(3-38)}$$

This is, however, just the system function which we are required to exhibit. □

Combining Proposition 3-1 with Theorem 3-1 yields a stronger form of the theorem, as follows.

Corollary 3-1

Let $H(t,q)$ be the kernel of a causal integral operator which has at least one derivative or primitive (that is, inverse distributional derivative) with finite norm. Then for all p in the RHP, $H(p,t)$ exists and is analytic. □

The corollary implies that a kernel such as $\dot{\delta}(t - q)$, which does not have finite norm, but which has a primitive $\delta(t - q)$ which does have finite norm, possesses a system function. One special, but important, case where the above applies is a kernel which is a finite linear combination of (possibly shifted) δ functions and their derivatives or primatives.

Corollary 3-2

If
$$H(t,q) = \sum_{i=1}^{n} a_i(t)\delta^{(n_i)}(t - q + T_i)$$
then $H(p,t)$ exists and is analytic for all p in the RHP. □

In fact, the system function for the kernel of the corollary is
$$H(p,t) = \sum_{i=1}^{n} a_i(t) p^{n_i} e^{T_i p} \tag{3-39}$$

Although $H(p,t)$ may exist and even be analytic for some p such that
$$\text{Re } p < 0 \tag{3-40}$$
[as in the system function of Equation (3-39)], we are actually only interested in RHP p since this region is entirely sufficient to describe the network. This may appear to be contradictory to experience in RLC network theory, where the poles and zeros which completely describe the network are never in the RHP. The reason for considering only the RHP may, however, be discerned by considering the inverse Laplace transform integral, wherein the integration is taken about the right half plane only. Thus the output from the network is determined only by the properties of the system function in the RHP.

$H(t,q)$ in Theorem 3-1 may represent any of the network matrices defined in Chapter 2, and $H(p,t)$ exists whenever the hypotheses are satisfied. Two special cases where this is assured to be the case are the following.

Corollary 3-3

Every linear, closed, causal n-port has a $\pi(p,t)$ and a $Q(p,t)$ representation which exists and is analytic for all p in the RHP. □

Corollary 3-4

Every linear, solvable, nonenergetic (hence passive) n-port has a scattering matrix $S(p,t)$ which exists and is analytic for all p in the RHP. □

For the usual RLC networks the complex system function does not change with time, rather it is a function of p only. This is, in fact, the case for all time-invariant networks.

Theorem 3-2

For a linear n-port, $H(p,t)$ is not a function of time [that is, $H(p,t) = H(p)$] if and only if $H(t,q) = H(t - q)$.

Proof

If $H(t,p) = H(t-q)$,

$$H(p,t) = \int_{-\infty}^{\infty} H(t-q)e^{-p(t-q)}dq$$
$$= \int_{-\infty}^{\infty} H(v)e^{-pv}dv \tag{3-41}$$

Now, v is the variable of integration and t does not appear independently; hence, $H(p,t) = H(p)$ is not a function of t. Conversely, if $H(p,t) = H(p)$,

$$H(p) = \int_{-\infty}^{\infty} H(t,t-v)e^{pv}dv = \int_{-\infty}^{\infty} H(0,0-v)e^{-pv}dv \tag{3-42}$$

where the last equality holds since the left side of the equation is not a function of t, thereby allowing us to choose its value arbitrarily. Now, since Equation (3-42) holds for all t,

$$H(t,q) = H(t,t-v) = H(0,0-v) \tag{3-43}$$

which yields the required result, since the right side of Equation (3-43) is a function only of the age variable. □

Note that the theorem does not state that the n-port is time-invariant, since it is possible for a π or Q matrix, of a time-invariant n-port, not to have the form of the theorem. For instance,

$$\pi(t,q) = [f(t)\delta(t-q) - f(t)\delta(t-q)] \tag{3-44}$$

is a matrix for an open circuit, a time-invariant device, which does not have the $t-q$ form. On the other hand, Theorem 2-7 and Proposition 2-10 assure the following.

Corollary 3-5

For a linear n-port, $S(p,t)$ $[Z(p,t), Y(p,t), Z_a(p,t), Y^a(p,t)]$ is not a function of time if and only if the n-port is time invariant. □

The capacitor has the time domain admittance matrix

$$Y(t,q) = c(q)\dot{\delta}(t-q) \tag{3-45}$$

and, hence, the system function

$$Y(p,t) = \int_{-\infty}^{\infty} c(q)\dot{\delta}(t-q)e^{-p(t-q)}dq$$
$$= \int_{-\infty}^{\infty} (t-q)(c(q)e^{-p(t-q)})dq$$
$$= \int_{-\infty}^{\infty} \delta(t-q)(-c(q)e^{-p(t-q)})dq \tag{3-46}$$
$$= -\int_{-\infty}^{\infty} \delta(t-q)c(\dot{q})e^{-p(t-q)}dq + \int_{-\infty}^{\infty} \delta(t-q)c(q)pe^{-p(t-q)}dq$$
$$= c(\dot{t}) + c(t)p$$

Now, clearly, this is a function only of p if and only if the capacitance is constant and, in that case, the system function for the capacitor is

$$Y(p,t) = cp \tag{3-47}$$

which is the classical result.

It is worthy of note here that the system function for a time-variable capacitor cannot be obtained from the differential equation of the capacitor simply by substitution of the complex variable p for the differential operator, as is usually done for the RLC networks (see Proposition 3-2, [71], [181], [208]).

The power of the system function is that it permits the Laplace transform of the output to be calculated from the transform of the input upon multiplication by the system function. This is indeed the case, in general.

Theorem 3-3

Let an n-port be represented by the system function $H(p,t)$ and have input $f(t)$ with transform $F(p)$. Then the inverse Laplace transform of $H(p,t)F(p)$ is the output $g(t)$.

Proof

Since $f(t)$ is in D_+ (all inputs are in D_+) and its Laplace transform is $F(p)$, $f(t)$ is given by the inverse transform integral [71]

$$f(t) = \frac{1}{2\pi j} \int_B F(p)e^{pt}dp \tag{3-48}$$

where B represents the Bromwich contour [71] about the RHP. Now, the output of the network with input $f(t)$ is given by

$$g(t) = \int_{-\infty}^{\infty} H(t,q)f(q)dq \tag{3-49}$$

Substituting Equation (3-48) into Equation (3-49) now yields

$$\begin{aligned}
g(t) &= \frac{1}{2\pi j}\int_{-\infty}^{\infty} H(t,q)\left(\int_B F(p)e^{pq}dp\right)dq \\
&= \frac{1}{2\pi j}\int_B \left(\int_{-\infty}^{\infty} H(t,q)e^{pq}dq\right)F(p)dp \\
&= \frac{1}{2\pi j}\int_B \left(\int_{-\infty}^{\infty} H(t,q)e^{-p(t-q)}dq\right)F(p)e^{pt}dp \\
&= \frac{1}{2\pi j}\int_B H(p,t)F(p)e^{-pt}dp
\end{aligned} \tag{3-50}$$

Thus,

$$H(p,t)F(p) \tag{3-51}$$

is the Laplace transform (parameterized by t) of the output. □

Theorem 3-3 readily permits the operations of composition, o, and addition,

+, for the integral operator kernels to be converted to the frequency domain. For the sake of simplicity, we consider only the time-invariant case (where the system function concept proves to be most useful).

Corollary 3-6

Let $H(t-q)$ and $K(t-q)$ be integral operator kernels with system functions $H(p)$ and $K(p)$, respectively. Then the system function of

$$H(t-q) + K(t-q)$$

is

$$H(p) + K(p)$$

and the system function of

$$H(t-q) \circ K(t-q)$$

is

$$H(p)K(p)$$

Proof

The addition formula follows from the linearity of the integral, while the multiplication formula may be obtained by applying Theorem 3-3 twice, the first time with input e^{pt}. □

Although Corollary 3-6 can be extended to the time-variable case [208], the system function interpretation of "o" is sufficiently complex in this case so as to render it of little practical value. For this reason, in the sequel the system function is employed only in the causal time-invariant case, the integral operator approach of Chapter 2 being employed in the general case. The argument of Corollary 3-6 does not hold in the case of a time variable $K(p,t)$, since this would induce an input $K(p,t)F(p)$ on the second application of Theorem 3-3, which is not admissible (since it is a function of t). In fact, Theorem 3-3 cannot be readily extended to this class of inputs, since the interchange of integrals employed in the proof of the theorem is not valid in that case.

In Chapter 2, and in the present section, three distinct types of network characterization have been defined; the impulse response, $H(t,q)$; the impulse response modified by a change of variable, $\boldsymbol{H}(t,v)$; and $H(p,t)$, the Laplace transform of $\boldsymbol{H}(t,v)$ with respect to the age variable. It would be possible to go on almost indefinitely defining new representations via these techniques. Gersho and Liskov [68], [97] have established twelve such representations, all derived from $H(t,q)$ by a change of variable and/or Laplace transformation (with respect to one or both variables). For the purposes of this work, however, $H(t,q)$ and $H(p,t)$ suffice (with $\boldsymbol{H}(t,v)$ being used as an intermediary).

In the special case of networks, the input-output relations of which are described by time-invariant ordinary differential operators, $H(p)$ is essentially identical to the differential operator. It has even been claimed by some (including your author) that the usefulness of the system function in classical network theory is entirely due to this characteristic.

*Proposition 3-2

Let a time-invariant n-port be represented by the ordinary differential operator

$$\left(\sum_{i=0}^{n} a_i D^i\right)^{-1}\left(\sum_{i=0}^{m} b_i D^i\right)$$

Then the system function for the n-port is given by

$$H(p) = \frac{\sum_{i=0}^{m} b_i p^i}{\sum_{i=0}^{n} a_i p^i}$$ □

In the case of a time-variable differential operator, the system function retains the rational character of the operator, but does not have the same coefficients (except for the leading coefficients, which are always the same in $H(p,t)$, as in the corresponding operator [145]).

Although a nonstandard definition for the system function has been employed, it does yield the expected results for the common circuit elements.

*Proposition 3-3

The frequency domain scattering matrices for the constant resistor, capacitor, and inductor are the following.
Resistor:
$$S(p) = \frac{r-1}{r+1}$$
Inductor:
$$S(p) = \frac{lp-1}{lp+1}$$
Capacitor:
$$S(p) = \frac{1-cp}{1+cp}$$

*Proposition 3-4

The frequency domain immittance matrices for the constant resistor, capacitor, and inductor are the following.
Resistor:
$$Z(p) = r, \qquad Y(p) = \frac{1}{r}$$

Capacitor:
$$Z(p) = \frac{1}{cp}, \qquad Y(p) = cp$$
Inductor:
$$Z(p) = lp, \qquad Y(p) = \frac{1}{lp} \qquad \square$$

The system function for the various memoryless devices (gyrator, circulator, ideal transformer, and so forth) can be obtained from their integral operator kernels by dropping the $\delta(t - q)$. Thus the scattering matrices of the open and short circuits are 1 and -1, respectively, and similarly for the other memoryless devices.

As an illustration of the difficulties encountered in using the system function in the time-variable case we have the following proposition.

*Proposition 3-5

The frequency domain immittance matrices for the time-variable resistor, capacitor, and inductor are the following.

Resistor:
$$Z(p,t) = r(t), \qquad Y(p,t) = \frac{1}{r(t)}$$
Inductor:
$$Z(p,t) = l(t)p + \dot{l}(t), \qquad Y(p,t) = \frac{1}{l(t)p}$$
Capacitor:
$$Z(p,t) = \frac{1}{c(t)p}, \qquad Y(p,t) = c(t)p + \dot{c}(t) \qquad \square$$

It is noteworthy that, in general,
$$Z(p,t) \neq \frac{1}{Y(p,t)} \tag{3-52}$$

which is one of the properties of the time-variable system function which inhibits its application. This pathological behavior is due to the fact that "o" does not correspond to multiplication of the system functions in the time-variable case.

3-3 Existence of the Scattering Matrix

This section is devoted to the final, but most powerful, existence theorem for the scattering matrix: that every linear, nonenergetic (hence, passive), time-invariant, normal n-port possesses a frequency domain scattering matrix $S(p)$. This theorem differs from the one of Chapter 2 in that solvability is not assumed, rather it is implied by the other hypotheses.

Though it may at first appear that the theorem is overly burdened with hypotheses, its power lies in the extraction techniques to be employed in the sequel.

This is a technique wherein the effect of various components on a network is studied by viewing them as loads on artificial ports defined at their terminals. This is illustrated in Figure 3-1, where the four distinguished components of the

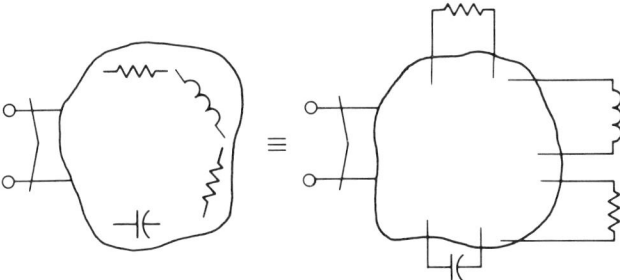

Figure 3-1 Component extraction.

n-port are viewed as loads on an $n + 4$-port coupling network. Of course, the network has not been changed (as seen from the original n-ports) by the process, but its properties are often more easily discerned by a study of the coupling network than of the original network.

The difficulty often encountered with this approach is that the coupling network may not have a given matrix representation (even if the original network had such a representation), thereby rendering the study of the coupling network difficult. This is usually the case when a large number of components are extracted, leaving a coupling network which may be composed entirely of wires. Fortunately, the extracted components usually include all of the activity, non-normality, and time-variability in the network, in which case the existence theorem assures that the resultant coupling network does have a scattering matrix. This is the case even when the coupling network is composed entirely of wires.

The existence theorem which follows is due to Carlin [47].

Theorem 3-4

Every linear, nonenergetic, normal, and time-invariant n-port possesses a scattering matrix $S(p)$.

Proof

Corollary 3-3 together with the definition of normality (Definition 2-12), assures that every linear, time-invariant, normal n-port is characterized by a $Q(p)$ matrix. Now, for normal networks, $Q(p)$ is $2n \times n$ and

$$\begin{bmatrix} a \\ b \end{bmatrix} = \begin{bmatrix} Q_a(p) \\ \hline Q_b(p) \end{bmatrix} [x] \tag{3-53}$$

where x spans D_+^n. Now, $S(p)$ exists if and only if $Q_a(p)^{-1}$ exists and, in that case, is given by

$$S(p) = Q_b(p)Q_a(p)^{-1} \qquad (3\text{-}54)$$

for each p in the RHP (for, otherwise, some input a is not admissible, in which case the n-port is not solvable and, therefore, cannot have a scattering matrix).

Assume that $Q_a(p)^{-1}$ does not exist for some p. Then, after appropriate column operations are carried out, an equivalent Q matrix

$$\begin{bmatrix} a \\ \hline b \end{bmatrix} = \begin{bmatrix} Q_a^1(p) & \vdots & 0 \\ \hline Q_b^1(p) & \vdots & Q_b^2(p) \end{bmatrix} \begin{bmatrix} x^1 \\ x^2 \end{bmatrix} \qquad (3\text{-}55)$$

is obtained. Here, $Q_a^2(p)$ can always be made to be zero, since the failure of $Q_a(p)^{-1}$ to exist means that the rank of $Q_a(p)$ is less than n. Now, if

$$Q_b^2(p) = 0 \qquad (3\text{-}56)$$

the rank of Q must be strictly less than n, which implies that the given network was non-normal, contradicting the hypothesis. On the other hand, if

$$Q_b^2(p) \neq 0 \qquad (3\text{-}57)$$

we may let

$$x^1 = 0 \qquad (3\text{-}58)$$

and choose x^2 such that

$$Q_b^2(p)x^2 \neq 0 \qquad (3\text{-}59)$$

which yields the admissible pair for the network of

$$\begin{bmatrix} 0 \\ \hline Q_b^2(p)x^2 \end{bmatrix} = \begin{bmatrix} Q_a^1(p) & \vdots & 0 \\ \hline Q_b^1(p) & \vdots & Q_b^2(p) \end{bmatrix} \begin{bmatrix} 0 \\ x^2 \end{bmatrix} \qquad (3\text{-}60)$$

For this pair we have

$$M\langle a,a \rangle - \langle b,b \rangle = -\langle Q_b^2(p)x^2, Q_b^2(p)x^2 \rangle < 0 \qquad (3\text{-}61)$$

$\square$

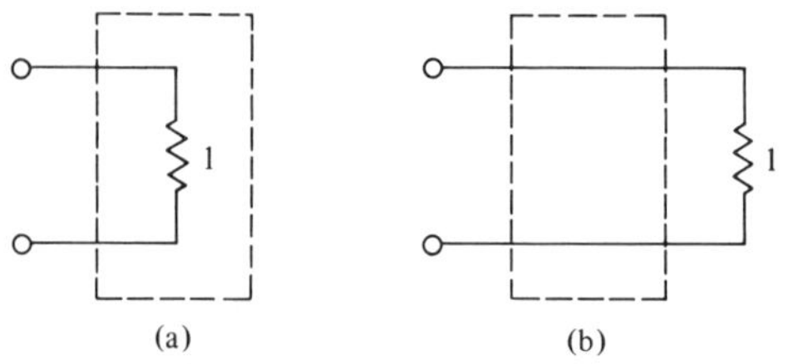

Figure 3-2 (a) 1-ohm resistor. (b) Its equivalent.

As an illustration of the theorem consider the network of Figure 3-2(a). This 1-port is a 1-ohm resistor, and thus has scattering and immittance matrices,

$$S(p) = 0 \tag{3-62}$$
$$Z(p) = 1 \tag{3-63}$$
and
$$Y(p) = 1 \tag{3-64}$$

If, however, one extracts the resistor at "the back port" of a coupling network, this, being composed only of the two wires, has neither admittance nor impedance matrix. Fortunately, the wires are linear, passive, normal, and time invariant; hence Carlin's theorem assures the existence of a scattering matrix

$$S(p) = \begin{bmatrix} 0 & 1 \\ 1 & 0 \end{bmatrix} \tag{3-65}$$

Like all powerful results, Carlin's theorem has a number of useful corollaries.

*Corollary 3-7

Every linear, nonenergetic, time-invariant, normal n-port possesses a (time domain) scattering matrix $S(t - q)$. ☐

*Corollary 3-8

Every linear, nonenergetic, time-invariant, normal n-port possesses augmented admittance and impedance matrices. ☐

An argument similar to that used in the proof of Theorem 3-4 can be employed for the following corollary.

*Corollary 3-9

Every linear, time-invariant, plus-normal n-port is energetic. ☐

The requirement of time invariance in the theorem is necessary because the finite dimensionality of $Q(p)$ (for a fixed p) was used in the proof. If time invariance is not assumed, networks which are linear, passive, and normal, but do not have a scattering matrix, can be exhibited. Anderson and Newcomb [9] have, in fact, exhibited linear passive networks which have none, one, two, or three of the basic network matrices. These are shown in Figure 3-3, and listed in Table 3-1. In the table, a dash is used to indicate that the network does not have the specified matrix.

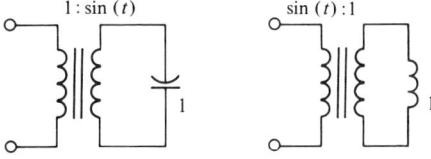

Figure 3-3 Degenerate networks.

Table 3-1 EXISTENCE OF NETWORK MATRICES

	$S(t,q)$	$Z(t,q)$	$Y(t,q)$
1 OHM RESISTOR	0	$\delta(t-q)$	$\delta(t-q)$
OPEN CIRCUIT	$\delta(t-q)$	—	0
SHORT CIRCUIT	$-\delta(t-q)$	0	—
IDEAL TRANSFORMER	exists	—	—
CAPACITOR-LOADED TRANSFORMER	—	$\sin(t)\sin(q)$ $\delta^{(1)}(t-q)$	—
INDUCTOR-LOADED TRANSFORMER	—	—	$\sin(t)\sin(q)$ $\delta^{(1)}(t-q)$
NULLATOR	—	—	—

Although Carlin's theorem does not, in general, hold for the time-variable case, as we have illustrated, it does hold for memoryless time-variable networks.

*Corollary 3-10

Every linear, nonenergetic, normal, memoryless n-port possesses time and frequency domain scattering matrices

$$S(t)\delta(t-q)$$

and

$$S(t)$$

respectively (and their corresponding augmented immittance matrices). □

The 1-ports of Figure 3-3 are examples of passive time-variable networks which have immittance but not scattering matrices. It follows directly from Carlin's theorem that these pathological n-ports cannot occur for the time-invariant case. In that case, a linear passive network with either an admittance or impedance matrix also has a scattering matrix.

*Corollary 3-11

Let a linear, nonenergetic, time-invariant n-port have either an admittance or an impedance matrix. Then it also has a scattering matrix, which is given by

$$S(p) = [1 + Y(p)]^{-1}[1 - Y(p)] = [Z(p) + 1]^{-1}[Z(p) - 1]$$ □

3-4 Passive and Lossless Scattering Matrices

In this section consideration is restricted to time-invariant networks for which necessary and sufficient conditions for passivity in terms of the scattering and immittance matrices are obtained. The results coincide with the traditional positive real (PR) bounded real (BR) conditions [44], [110], [183] for an RLC network, with the exception of rationality, this weakening of the restriction resulting from the fact that we consider distributed as well as lumped networks.

The property of a system function that its adjoint is given by

$$H(p)* \equiv H(\bar{p})' \equiv H(p)^\dagger \tag{3-66}$$

(derived in the Section 3-1) will be employed, as well as Parseval's relation (Proposition B-9) equating the time and frequency domain inner products via

$$\langle f,g \rangle = \int_{-\infty}^{\infty} f(t)'g(t)dt = \int_{-\infty}^{\infty} F(j\omega)*G(j\omega)d\omega \tag{3-67}$$

where $F(j\omega)$ and $G(j\omega)$ are the transforms of $f(t)$ and $g(t)$, respectively. This equality, together with an argument analogous to that used in the proof of Corollary 2-2, yields the following.

Proposition 3-6

Let $S(p)$ be the scattering matrix of a passive network. Then

$$1 - S(p)*S(p) \geq 0$$

for all p in the RHP.

Proof

Since the network is passive,

$$0 \leq \langle a,a \rangle - \langle b,b \rangle = \int_{-\infty}^{\infty} a(j\omega)*a(j\omega)d\omega - \int_{-\infty}^{\infty} b(j\omega)*b(j\omega)d\omega$$
$$= \int_{-\infty}^{\infty} a(j\omega)*(1 - S(j\omega)*S(j\omega))a(j\omega)d\omega \tag{3-68}$$

which holds for all ω if and only if

$$(1 - S(j\omega)*S(j\omega)) \geq 0 \tag{3-69}$$

Since the $j\omega$ axis is the boundary of the RHP and $S(p)$ is analytic in the RHP, application of the maximum modulus theorem [144] yields

$$1 - S(p)*S(p) \geq 0 \tag{3-70}$$

for all p in the RHP (that is, since Equation (3-69) implies that $\|S(p)\| \leq 1$ on the boundary of the RHP, the $j\omega$ axis, the maximum modulus theorem implies that the bound also holds in the interior of the RHP). □

As we have defined a network, real inputs (the only ones allowed) yield real outputs. Now, in the time domain representation this is "obviously" the case; but in the frequency domain an additional constraint is needed on the complex system function to insure that the inverse transform of the output, $H(p)F(p)$, is real. This constraint, called "reality," (not surprisingly) is defined as follows.

Definition 3-2

A complex matrix (and the n-port it represents) is *real* if

$$H(\bar{p}) = \overline{H(p)}$$

for all p in the RHP. Here, the overbar denotes complex conjugation. □

That system functions are real is due to the following proposition.

Proposition 3-7

Let $H(p)$ be the system function corresponding to a kernel $H(t-q)$. Then $H(p)$ is real.

Proof

Let

$$H(p) = \int_{-\infty}^{\infty} H(t-q)e^{-p(t-q)}dq \tag{3-71}$$

Now,

$$\begin{aligned} H(\bar{p}) &= \int_{-\infty}^{\infty} H(t-q)e^{-\bar{p}(t-q)}dq \\ &= \int_{-\infty}^{\infty} H(t-q)\overline{e^{-p(t-q)}}dq \\ &= \int_{-\infty}^{\infty} \overline{H(t-q)e^{-p(t-q)}}dq \\ &= \overline{H(p)} \end{aligned} \tag{3-72}$$

This last equality holds since $H(t-q)$ is real valued, and hence its own complex conjugate. □

A similar argument yields the following corollary.

Corollary 3-12

Let $F(p)$ be the Laplace transform of a function $f(t)$. Then $F(p)$ is real if and only if $f(t)$ is a real-valued function. □

Combining Proposition 3-7 and Corollary 3-12 now yields the desired result that the output due to a real function is real.

Corollary 3-13

Let $H(p)$ be the system function of an n-port and $F(p)$ be the transform of a real-valued function $f(t)$. Then the output time function, given by the inverse Laplace transform of $H(p)F(p)$, is a real-valued time function. □

Corollary 3-13 is actually unnecessary, since in an n-port the output is, by definition, real valued. It does, however, serve as a check on the frequency domain theory. More importantly, the corollary shows that the class of system functions does not include all functions which are analytic in the RHP; rather, it is restricted to the real functions satisfying

$$H(\bar{p}) = \overline{H(p)} \tag{3-73}$$

Reality is the source of the complex conjugate arrangement of the poles and zeros for the system functions of RLC networks, and, more generally, for the complex conjugate character of all singularities in the general system function.

To see this, consider the zeros of the complex function $H(p)$. Now, if

$$H(p_0) = 0 \tag{3-74}$$

and $H(p)$ is real, then

$$H(\bar{p}_0) = \overline{H(p_0)} = \bar{0} = 0 \tag{3-75}$$

Hence, if p_0 is a zero of $H(p)$, then so is its complex conjugate, $\bar{p}_0$. Similarly, if

$$\lim_{p \to p_0} H(p) = \infty \tag{3-76}$$

then

$$\lim_{p \to \bar{p}_0} H(p) = \infty \tag{3-77}$$

Thus both the poles and the zeros of $H(p)$ appear in complex conjugate pairs and, moreover, one can show by similar arguments that all singularities of $H(p)$ appear in complex conjugate pairs. In fact, $H(p)$ is symmetric about the real axis of the complex plane.

Consider the system function

$$H(p) = ap^n \tag{3-78}$$

Now,

$$H(\bar{p}) = a\bar{p}^n \tag{3-79}$$

whereas

$$\overline{H(p)} = \bar{a}\bar{p}^n \tag{3-80}$$

Hence, $H(p)$ is real if and only if

$$a = \bar{a} \tag{3-81}$$

is real. The reader may readily verify that the sum of real functions is real, as are the product and quotient of such; hence, it follows from the above argument that a polynomial or rational function is real if and only if all of its coefficients are real. Of course, this is the case with the rational functions of RLC network theory. Moreover, any system function $H(p)$ which has a Taylor or Laurant series expansion (in the RHP)

$$H(p) = \sum_{i=0}^{\infty} a_i p^i \tag{3-82}$$

or

$$H(p) = \sum_{i=0}^{\infty} \frac{a_i}{p^i} \tag{3-83}$$

with real coefficients is itself real. Since one can expand the system function of the delay (T negative) as

$$e^{pT} = \sum_{i=0}^{\infty} \frac{(-pT)^i}{i} \qquad (3\text{-}84)$$

it is indeed real, as is predicted by Proposition 3-7.

With the concept of reality added to our repertory, it is now possible to consider the character of the system functions of passive time-invariant networks. In the case of the scattering matrix this is summed up via the concept of a bounded real matrix.

Definition 3-3

An $n \times n$ matrix $H(p)$ is *bounded real* (BR) if the following conditions are satisfied (including either 3 or 3'):
  1. $H(p)$ is analytic for all p in the RHP
  2. $H(\bar{p}) = \overline{H(p)}$ for all p in the RHP
  3. $1 - H(p)^*H(p) \geq 0$ (that is, a positive matrix) for all p in the RHP
  3'. $\|H(p)\| \leq 1$ for all p in the RHP. □

The fact that conditions 3 and 3' are equivalent follows from the corresponding time domain condition (together with Parseval's theorem and the maximum modulus theorem) and will not be proven again. The significance of BR matrices is due to the following theorem.

Theorem 3-5

A necessary and sufficient condition for an n-port to be linear, passive, normal, and time invariant (hence causal) is that it possess a bounded real scattering matrix $S(p)$.

Proof

The necessity of the theorem follows upon combination of the necessary conditions of Propositions 3-6 and 3-7, together with Theorems 3-1 and 3-4. Conversely, if $S(p)$ is given, its input-output pairs define a linear, passive (by Parseval's theorem), normal, and time-invariant network. □

Consider the scattering matrix

$$S(p) = p \qquad (3\text{-}85)$$

This is clearly real and analytic in the RHP (in fact, it is analytic everywhere); but for

$$p = 2 \qquad (3\text{-}86)$$

in the RHP,

$$\|S(p)\| = |S(2)| = 2 \qquad (3\text{-}87)$$

which fails to satisfy condition 3' of the theorem. The function is thus not passive.

For condition 3' of the theorem to be satisfied, the norm of $H(p)$ must be bounded by 1 for all p in the RHP. Thus, in particular, $H(p)$ must be bounded by 1 at infinity. That is,

$$\lim_{p \to \infty} |H(p)| \leq 1 \qquad (3\text{-}88)$$

Although Equation (3-88) is only a necessary condition for passivity, it is readily tested and thus yields a convenient means of exhibiting the nonpassivity of various system functions.

Consider now the scattering matrix

$$S(p) = \frac{p}{p+1} \qquad (3\text{-}89)$$

which is clearly real and analytic in the RHP. Moreover, for any p in the RHP,

$$\|S(p)\| = \frac{(\text{Re } p)^2 + (\text{Im } p)^2}{(\text{Re } p + 1)^2 + (\text{Im } p)^2} < 1 \qquad (3\text{-}90)$$

since in the RHP,

$$(\text{Re } p)^2 < (\text{Re } p + 1)^2 \qquad (3\text{-}91)$$

Similarly, the delay is analytic everywhere and, as we have seen, real. Moreover, for T negative and p in the RHP,

$$\|S(p)\| = |e^{Tp}| = |e^{\text{Re } Tp}| \leq 1 \qquad (3\text{-}92)$$

Hence, the scattering matrix for the delay is bounded real and passive. It is worthy of note here that this device, when operating in its predictor mode (that is, T positive), is not passive, for in that case, the last inequality of Equation (3-92) is not valid and, in fact, the norm of $S(p)$ will be unbounded in this mode of operation. Of course, the reality and analyticity conditions on this function are independent of the value chosen for T.

Finally, it should be observed that for memoryless n-ports, the scattering matrix will simply be a matrix of real numbers which is, of course, real and analytic in the RHP (since it is not a function of p it is trivially analytic). The passivity conditions for this special case, which, in fact, are also valid in the more general time-variable case, thereby reduce to the simple requirement that

$$\|S(t)\| \leq 1 \qquad (3\text{-}93)$$

A frequency domain criterion for an n-port to be linear, nonenergetic, time-invariant, and normal, analogous to the above BR criterion, can be obtained by similar techniques [150], where condition 3 is weakened to a requirement that the norm be finite. The details of such a derivation are left to the reader as an exercise.

Losslessness is a stronger condition than passivity; hence, an additional constraint is added to the matrices of such networks.

Definition 3-4

An $n \times n$ matrix $H(p)$ is *bounded real-paraunitary* (LBR) if it is bounded real and
$$1 - H(j\omega)^*H(j\omega) = 0$$
for all imaginary axis $j\omega$ where $H(j\omega)$ is defined. □

Theorem 3-6

A necessary and sufficient condition for an n-port to be linear, lossless, normal, and time invariant is that it possess a bounded real-paraunitary scattering matrix $S(p)$.

Proof

The requirement that $S(p)$ be BR follows from Theorem 3-5. To obtain the paraunitary condition, Parseval's equality is applied, yielding

$$\begin{aligned} 0 = \langle a,a \rangle - \langle b,b \rangle &= \int_{-\infty}^{\infty} (a'a - b'b)dt \\ &= \int_{-\infty}^{\infty} (a(j\omega)^*a(j\omega) - b(j\omega)^*b(j\omega))d\omega \\ &= \int_{-\infty}^{\infty} a(j\omega)^*(1 - S(j\omega)^*S(j\omega))a(j\omega)d\omega \end{aligned} \quad (3\text{-}94)$$

Now, Equation (3-94) holds if and only if

$$(1 - S(j\omega)^*S(j\omega)) = 0 \quad (3\text{-}95)$$

for all ω where $S(j\omega)$ is defined. Conversely, given an LBR $S(p)$, an n-port satisfying the conditions of the theorem can be constructed from its input-output pairs. □

As an illustration of the theorem, consider the (1-farad) capacitor with scattering matrix

$$S(p) = \frac{1-p}{1+p} \quad (3\text{-}96)$$

which is analytic except at $p = -1$ (which is not in the RHP).

$$S(\bar{p}) = \frac{1-\bar{p}}{1+\bar{p}} = \overline{\left(\frac{1-p}{1+p}\right)} = \overline{S(p)} \quad (3\text{-}97)$$

Hence, the capacitor has a real scattering matrix. Finally,

$$\begin{aligned} 1 - S(p)^*S(p) &= 1 - \left(\frac{1-p}{1+p}\right)^*\left(\frac{1-p}{1+p}\right) \\ &= 1 - \frac{1 + \bar{p}p - 2\,\text{Re}\,p}{1 + \bar{p}p + 2\,\text{Re}\,p} \end{aligned} \quad (3\text{-}98)$$

Now, this is a real number (since $\bar{p}p$ is real), and for p in the RHP,

$$\operatorname{Re} p > 0 \qquad (3\text{-}99)$$

Hence, the quotient in Equation (3-98) is less than or equal to 1 and the difference is greater than or equal to 0. $S(p)$ is thus BR and the capacitor is passive. Now, for

$$p = j\omega \qquad (3\text{-}100)$$

Equation (3-98) becomes

$$1 - S(j\omega)^*S(j\omega) = 1 - \frac{1+\omega^2}{1+\omega^2} = 1 - 1 = 0 \qquad (3\text{-}101)$$

Hence, the capacitor is also lossless.

A more interesting, if less physical, example of an LBR function is

$$S(p) = e^{-1/p} \qquad (3\text{-}102)$$

This function has a singularity on the $j\omega$ axis (at 0) but is analytic for

$$\operatorname{Re} p > 0 \qquad (3\text{-}103)$$

where $e^{-1/p}$ has the Laurent series expansion

$$e^{-1/p} = \sum_{i=0}^{\infty} \left(\frac{l}{\lfloor i}\right)\left(\frac{-1}{p}\right)^i \qquad (3\text{-}104)$$

Equation (3-104) also demonstrates the reality of the function (since all polynomials, rational functions, and series with real coefficients are real). The norm condition is obtained by considering

$$\|e^{-1/p}\| = \|e^{-\bar{p}/|p|^2}\| \leq \|e^{-\operatorname{Re} p/|p|^2}\| \, \|e^{j\operatorname{Im} p/|p|^2}\| \qquad (3\text{-}105)$$

For p in the RHP,

$$\|e^{-\operatorname{Re} p/|p|^2}\| = |e^{-\operatorname{Re} p/|p|^2}| < 1 \qquad (3\text{-}106)$$

while

$$\|e^{j\operatorname{Im} p/|p|^2}\| = |e^{j\operatorname{Im} p/|p|^2}| = 1 \qquad (3\text{-}107)$$

Substitution of Equations (3-106) and (3-107) into Equation (3-105) yields

$$\|e^{-1/p}\| \leq 1 \qquad (3\text{-}108)$$

for p in the RHP, which shows that the function is BR. It is actually LBR, for if

$$\operatorname{Re} p = 0 \qquad p \neq 0 \qquad (3\text{-}109)$$

then

$$\|e^{-\operatorname{Re} p/|p|^2}\| = 1 \qquad (3\text{-}110)$$

in which case

$$\|e^{-1/p}\| = |e^{j/\omega}| = 1 \qquad (3\text{-}111)$$

We have thus exhibited one classical device, the capacitor, which is LBR, and another device, the significance of which is not immediately evident, which is also LBR. Both, therefore, correspond to linear, lossless, normal, and time-invariant networks. Significantly, the rationality of the scattering matrix of the capacitor did not simplify the calculations; hence, the study of generalized networks may proceed simultaneously with the RLC special case.

A major difficulty encountered in dealing with BR and LBR system functions results from the fact that the adjoint of a system function which is analytic in the RHP is not itself analytic in the RHP. This anomaly results from the, possibly surprising, fact that the operation of complex conjugation is not analytic. Consider, for example, the function

$$H(p) \equiv p \tag{3-112}$$

Here

$$H(p)^* = \overline{H(p)'} = \bar{p} \tag{3-113}$$

Now

$$\frac{d \operatorname{Re} \bar{p}}{d \operatorname{Re} p} = 1 \tag{3-114}$$

whereas

$$\frac{d \operatorname{Im} \bar{p}}{d \operatorname{Im} p} = -1 \tag{3-115}$$

Hence, the Cauchy–Reimann condition that requires

$$\frac{d \operatorname{Re} H(p)^*}{d \operatorname{Re} p} = \frac{d \operatorname{Im} H(p)^*}{d \operatorname{Im} p} \tag{3-116}$$

is not satisfied.

A partial solution to the difficulties indicated above is to work with the Hurwitz conjugate rather than the adjoint. This is obtained by replacing the complex conjugation operation of the adjoint with the analytic operation of negation (that is, p is replaced by $-p$ rather than $\bar{p}$). Such an operation transforms a function which is analytic in the RHP to one analytic in the LHP, which is easier to deal with than the nonanalytic function induced by the adjoint. Moreover, $\bar{p}$ and $-p$ coincide on the $j\omega$ axis, which is in fact, the most important part of the p plane for our applications (re: Parseval's theorem).

Definition 3-5

Let $H(p)$ be a system function. Then

$$H(p)_* = H(-p)'$$

is the *Hurwitz conjugate* of $H(p)$. □

Clearly,

$$H(j\omega)_* = H(j\omega)^* \tag{3-117}$$

so that a matrix has the paraunitary property if and only if

$$1 - H(j\omega)_* H(j\omega) = 0 \tag{3-118}$$

The advantage of the Hurwitz conjugate is that, unlike complex conjugation (employed in the adjoint), negation of p is an analytic operation. It is, therefore, possible to deal with the equality

$$1 - H(p)_* H(p) = 0 \qquad (3\text{-}119)$$

in the RHP, rather than the paraunitary condition of Equation (3-118). In fact, we will show that in the RLC case (Chapter 6), the equalities (3-118) and (3-119) are equivalent. Upon invoking the preceding relationship, the LBR character of $e^{-1/p}$ is obtained from

$$1 - S(p)_* S(p) = 1 - e^{1/p} e^{-1/p} = 1 - 1 = 0 \qquad (3\text{-}120)$$

As with any complex number or matrix, the real and imaginary parts of $H(p)$ are given by

$$\operatorname{Re} H(p) = \tfrac{1}{2}(H(p) + H(p)^*) \qquad (3\text{-}121)$$

and

$$j \operatorname{Im} H(p) = \tfrac{1}{2}(H(p) - H(p)^*) \qquad (3\text{-}122)$$

If the adjoint in Equations (3-121) and (3-122) is replaced by the Hurwitz conjugate, the even and odd parts of the system function are obtained.

*Definition 3-6

Let $H(p)$ be a system function. Then its *even part* is

$$\operatorname{Ev} H(p) = \tfrac{1}{2}(H(p) + H(p)_*)$$

and its *odd part* is

$$\operatorname{Od} H(p) = \tfrac{1}{2}(H(p) - H(p)_*) \qquad \square$$

Although the passivity and losslessness conditions in the frequency domain are completely different than those in the time domain, most of the other properties formulated in Chapter 2 are readily translated from the time to the frequency domain. These include time invariance, reciprocity, and the various algebraic relationships between the different network matrices.

The positive real concept of RLC network theory readily generalizes to the case of distributed networks, but, as is usually the case with immittance matrices, the question of existence is not answered.

*Definition 3-7

An $n \times n$ matrix $H(p)$ is *positive real* (PR) if the following conditions are satisfied:

1. $H(p)$ is analytic for all p in the RHP
2. $H(\bar{p}) = \overline{H(p)}$ for all p in the RHP
3. $H(p) + H(p)^* \geq 0$ for all p in the RHP. $\qquad \square$

*Proposition 3-8

Let a linear, time-invariant network be represented by an immittance matrix. Then it is passive if and only if the immittance matrix is PR. □

A time-invariant capacitor is characterized by the admittance

$$Y(p) = cp \qquad (3\text{-}123)$$

Now, this is real for c real and analytic in the RHP. Moreover, if

$$\operatorname{Re} p > 0 \qquad (3\text{-}124)$$

then

$$\operatorname{Re} Y(p) = c \operatorname{Re} p > 0 \qquad (3\text{-}125)$$

if and only if c is non-negative. Hence, we have verified our previous result concerning the passivity of a capacitor.

The positive reality condition of Proposition 3-8 is not, as is the RLC PR condition [183], restricted to rational functions; thus, a fractional power impedance

$$Z(p) = p^{1/2} \qquad (3\text{-}126)$$

where the "square root" is taken as the unique RHP square root of p, is positive real. By definition, $Z(p)$ is in the RHP for all p, and hence, certainly for those p in the RHP. To confirm that $Z(p)$ is analytic in the RHP and real, one may expand it in an infinite series with real coefficients. It is left as an exercise to the reader to determine if such a series exists and that its coefficients are indeed real. This being the case, it then follows from Proposition 3-8 that the fractional impedance

$$Z(p) = p^{1/2} \qquad (3\text{-}127)$$

is PR.

The strengthening of positive reality to cover the lossless case results in an LPR matrix.

*Definition 3-8

An $n \times n$ matrix $H(p)$ is *lossless-positive real* (LPR) if it is PR and

$$H(j\omega) + H(j\omega)^* = 0$$

for all imaginary axis $j\omega$ where $H(j\omega)$ is defined. □

*Proposition 3-9

Let a linear, time-invariant network be represented by an immittance matrix. Then it is lossless if and only if its immittance matrix is LPR. □

As with the scattering matrix, the $j\omega$ axis condition for the losslessness of a PR matrix is conveniently tested by replacing the adjoint with the Hurwitz conjugate. This yields the equality

$$H(p) + H(p)_* = 0 \tag{3-128}$$

for all RHP p.

Applying the Hurwitz conjugate test to the time-invariant capacitor characterized by

$$Y(p) = cp \tag{3-129}$$

one obtains

$$Y(p) + Y(p)_* = cp + (-cp) = 0 \tag{3-130}$$

verifying the fact that such a capacitor is lossless. To the contrary, the fractional power impedance

$$Z(p) = p^{1/2} \tag{3-131}$$

is not lossless since

$$Z(p) + Z(p)_* = p^{1/2} + jp^{1/2} = (1+j)p^{1/2} \neq 0 \tag{3-132}$$

Two useful results on positive reality are the following.

*Proposition 3-10

If $H(p)$ is PR, then $H(p)'$ and $H(p)^{-1}$ (if they exist) are PR. □

*Proposition 3-11

Let a network have a scattering matrix $S(p)$ and augmented immittance matrices $Z_a(p)$ and $Y^a(p)$. Then, if $S(p)$ is BR, $Z_a(p)$ and $Y^a(p)$ are PR. □

3-5 Network Theorems

In this section a number of "well known" results of RLC network theory are extended to the general case. Since they are primarily of historical interest (at least as far as this work is concerned), the results are only outlined, the reader being expected to fill in the details.

In Chapter 1, Lorentz reciprocity was defined by the requirement that

$$\int_{-\infty}^{t} a(t-u)'b(u)\,du = \int_{-\infty}^{t} \boldsymbol{a}(t-u)'\boldsymbol{b}(u)\,du \tag{3-133}$$

for any two admissible pairs, (a,b) and $(\boldsymbol{a},\boldsymbol{b})$. Equation (3-133) is translated to the frequency domain via the following proposition.

Proposition 3-12

Let an n-port have admissible pairs $(a(t),b(t))$ and $(\boldsymbol{a}(t),\boldsymbol{b}(t))$ with corresponding transform representations $(a(p),b(p))$ and $(\boldsymbol{a}(p),\boldsymbol{b}(p))$. Then the n-port is Lorentz reciprocal if and only if

$$a(p)'\boldsymbol{b}(p) = \boldsymbol{a}(p)'b(p)$$

for all (a,b) and $(\boldsymbol{a},\boldsymbol{b})$ in N.

Proof

The result follows from the fact that convolution in the time domain corresponds to multiplication in the frequency domain [71]. □

The classical reciprocity condition [176] now follows readily from Proposition 3-12.

Corollary 3-14

A time-invariant network characterized by a scattering matrix $S(p)$ is Lorentz reciprocal if and only if it is reciprocal (that is, symmetric).

Proof

For any two admissible pairs

$$(a(p),b(p)) \qquad (3\text{-}134)$$

and

$$(\boldsymbol{a}(p),\boldsymbol{b}(p)) \qquad (3\text{-}135)$$

we have

$$a(p)'\boldsymbol{b}(p) = a(p)'S(p)\boldsymbol{a}(p) \qquad (3\text{-}136)$$

and

$$\boldsymbol{a}(p)'b(p) = \boldsymbol{a}(p)'S(p)a(p) \equiv a(p)'S(p)'\boldsymbol{a}(p) \qquad (3\text{-}137)$$

where Equation (3-137) is valid because the 1×1 matrix

$$\boldsymbol{a}(p)'S(p)a(p) \qquad (3\text{-}138)$$

is symmetric (as are all 1×1 matrices). Now, Equations (3-136) and (3-137) represent the same quantity if and only if

$$S(p) = S(p)' \qquad (3\text{-}139)$$

which is exactly the condition for reciprocity. Hence, the network is Lorentz reciprocal if and only if it is reciprocal. □

Combining the above result with that obtained on Lorentz reciprocity in Chapter 2, yields the following proposition.

*Proposition 3-13

Let an n-port be characterized by a scattering matrix

$$S(t,q)$$

Then a necessary and sufficient condition for the network to be Lorentz reciprocal is that it be time-invariant and reciprocal (that is, have a symmetric scattering matrix). □

Foster's reactance theorem, the scattering formulation of which is given in Proposition 3-14, is one of the oldest theorems of network theory [46], [67].

*Proposition 3-14

Let $S(p)$ be LPR. Then for all imaginary axis $j\omega$ where $S(j\omega)$ is defined,

$$(jS(j\omega)^*)\frac{dS(j\omega)}{d\omega} = (jS(j\omega)_*)\frac{dS(j\omega)}{d\omega}$$

is a positive matrix. □

*Corollary 3-15

Let ϕ be the phase shift between the incident and reflected waves of a 1-port LBR scattering matrix. Then

$$\frac{d\phi(\omega)}{d\omega}$$

is greater than or equal to zero. □

For the predictor-delay characterized by the scattering matrix

$$S(p) = e^{pT} \quad \text{(3-140)}$$

$$(jS(j\omega)_*)\frac{dS(j\omega)}{d\omega} = (je^{-j\omega T})\frac{de^{j\omega T}}{d\omega} = je^{-j\omega T}jTe^{j\omega T}$$

$$= j(jT) = -T \quad \text{(3-141)}$$

Now, this is positive when T is negative, or equivalently, when the device is operating in its lossless delay mode, but not when it is operating as an active predictor. This is consistent with Foster's theorem, which requires that Equation (3-41) be positive only in the lossless case. The more commonly encountered immittance formulation for Foster's reactance theorem is the following.

*Proposition 3-15

Let $H(p)$ be LPR. Then for all imaginary axis $j\omega$ where it is defined,

$$-j\frac{d}{d\omega}H(j\omega)$$

is a positive matrix. □

3-6 Discussion

In this chapter a frequency domain representation for the generalized networks which naturally extends the RLC rational functions has been formulated. The approach, though circumferential, by-passes the usual difficulties associated with the system function, this being made possible only by the previous development of the time domain network representation.

It should be noted that a direct derivation of the frequency domain scattering matrix for linear passive networks is possible. In this case, the Bochner theorems [140], [192] are used to assure the existence of $S(p)$ directly, rather than the indirect application of the kernels theorem, together with a Laplace transformation. Of course, it is desired to have representations in both the time and frequency domains, in which case the approach employed is the more efficient.

CHAPTER
4
Topological Analysis

4-1 Introduction

In the preceding chapters we have considered the characteristics of individual n-ports without reference to the network of which they are a part. Here we return to the problem, considered heuristically in Chapter 1, of determining the characteristics of a network and/or the n-port it realizes from the properties of its component n-ports and the connection. Of course, the intuitive viewpoint taken in Chapter 1 must give way to a more orderly process if general network analysis techniques are to be successfully formulated. For this reason, a topological formulation of the problem is adopted. Here the connection constraints induced by the network graph are characterized by various graph matrices (see Appendix C) which are combined with the matrix characterizations for the component n-ports (as per Chapters 2 and 3) to obtain a matrix characterization for the network variables and/or the n-port being realized.

Since many different n-ports may be realized by the same network, depending on the location of the ports, it does not prove to be convenient to obtain a characterization of the realized n-port directly by topological means. Rather, one would like to apply the topological formulation to calculate the characteristics of some "large" n-port canonically associated with the network, the characteristics of this n-port then being used to obtain the properties of the n-port actually realized. In this way, the topological techniques are rendered functions of the network only and are independent of the particular n-port actually being realized by the network.

There are two alternative approaches to defining the canonical n-port associated with a network. First [155], if one observes that the graph of the network

$$(G, N_1, N_2, \cdots, N_m) \tag{4-1}$$

has Σn_i edges, each of which is characterized by admissible incident and reflected waves for the network variables, one may define a (Σn_i)-port associated with the network simply by letting its ports have admissible pairs which coincide with the admissible network variables. In essence, then, a canonical n-port, so defined, would be a single n-port which simultaneously characterizes the constraints imposed on the network by its components and the connection. Such an n-port, though it has no explicit connection constraints, includes the connection information of the network as part of its admissible pairs structure (and thereby differs from the composite n-port of Chapter 1 which combines the characteristics of the components but carries no connection information). Remembering that the admissible variables for the network are determined by the intersection of the set of admissible pairs allowed by the components, N, and the set of admissible pairs allowed by the graph, $\hat{G}$ (as per Section 1-4), the canonical n-port as defined above is thus given by

$$(G_{(\Sigma n_i)}, N \cap \hat{G}) \tag{4-2}$$

Clearly, it suffices if our analysis procedures can determine the properties of this canonical n-port (say, one of its matrices) since the characteristics of any n-port realized by the network may readily be determined from those of the canonical n-port.

A second approach to the problem of associating a canonical n-port with a network is to insert ports in the network in some standardized manner and allow the canonical n-port to be that realized at these ports. Of course, the number of ports must be sufficiently large so as to assure that the canonical n-port carries all relevant information about the network. Since the cotree currents and tree voltages associated with any tree of the network graph completely determine the network variables [83], it is convenient to place our ports in series with cotree edges and in parallel with tree edges (for some arbitrary tree of the network graph). This procedure for defining a canonical n-port is, in some sense, more complicated than the previous one in that one must take cognizance of the perturbations of the network induced by the ports (that is, cutting an edge to insert a series port); but the resultant canonical n-port is usually more closely related to the n-port actually being realized by the network. In fact, if one chooses the appropriate tree, the realized n-port can usually be made to coincide with a subset of the ports of the canonical n-port.

Mathematically, both approaches to the definition of the canonical n-port are equivalent, since both include all relevant information regarding the network, and both allow one to characterize the realized n-port. The choice of the particular approach to be employed is essentially a matter of convenience and, in fact, both are employed to advantage in the analysis procedures of this chapter. In essence, one could say that the first procedure characterizes the internal variables of the network, whereas the second procedure is external in nature.

Unlike the simple configurations considered in Chapter 1, here we want to consider essentially arbitrary interconnections of arbitrary component n-ports. To achieve such a goal, one must invoke the powerful algebraic mechanism of the graph matrices considered in Appendix C. These permit the constraints imposed on the "internal" variables of the network to be described by two matrix equations. Moreover, when one allows "external port variables," the graph matrices serve as an intermediary between the port variables and the components. In either case, the canonical n-port associated with a network may be readily obtained upon combining the equations induced by the graph, through the graph matrices, and those induced by the component n-ports, through a composite network matrix. This latter matrix is usually a scattering, π, or immittance matrix for the composite n-port of the network (which is obtained by taking the direct product of the corresponding component matrices).

The origins of topological network analysis lie in the classical RLC theory (as does most of network theory) with which it is often associated. In this case, an edge of a graph is readily associated with each component, from which point the analysis begins [83], [155]. These techniques can be extended by a number of procedures [57], [59], [79], [150] to include multiport and multi-terminal devices; the one employed here, of identifying an edge with each port of each component, generalizes the RLC theory with a minimum of modification. Although we identify an edge with each port, there is no restriction on the connections. Port-to-port connections and cross connections, as well as open and short circuited ports, are allowed.

The validity of this approach is due to the requirement on an n-port that the current entering one terminal of a port must equal that leaving the other terminal of the port. This being the case, the current entering and leaving the terminal pair of a port may, without loss of generality, be viewed as flowing through an edge connected between the two terminals.

Though the connection constraints on a network are described by a graph, this is not readily amenable to the required analysis procedures. Rather, as a first step in the procedure, the graph is replaced by a graph matrix. Since the graph is completely determined by its matrix, and conversely [83], no information is lost in the process; the graph matrix, however, reduces a hitherto topological problem to a (simpler ?) algebraic problem. The graph matrices may take various forms. For the purposes of this work, however, we consider only the fundamental circuit, cocircuit, and connection matrices defined in Appendix C. This latter square matrix is characterized by the fact that it is its own inverse, and proves to be useful in analyzing networks characterized by scattering matrices.

The first section of the chapter is devoted to a review of the graph matrices, together with an exposition of the constraints they impose on the network variables, and the effect of external sources thereon. Two alternative approaches to the topological analysis problem are taken. The first, which is applicable to all linear networks (in either the time or frequency domain), permits the calculation of the π matrix of the canonical n-port associated with a network from that of its arbitrarily connected components. This procedure is, as with most π matrix techniques, easily carried out, but of little value in describing the properties of specific classes of networks. The second approach permits the calculation of the scattering matrix of the canonical n-port associated with a network from those of its (arbitrarily connected) parts. This technique is possibly our most powerful analysis tool, since it combines the desirable existence properties and explicit input-output character of the scattering matrix. The tools developed for the π and scattering analysis are such as to readily allow the formulation of an n-port extension of the classical RLC immittance analyses [83], [155]. However, the derivation of these is, for the most part, left to the reader as an exercise.

The scattering analysis, though initially derived in the unnormalized case, can be modified to account for the possibility that the component parts and/or the overall network have scattering matrices with arbitrary (real or complex) normalization. Once formulated, the resultant analysis allows a number of results on the existence of the scattering matrix with arbitrary normalization and a change of normalization formula to be derived.

The final two sections are devoted to a study of the cascade loading concept [18] and its application to the study of certain elementary n-ports. This approach is applicable only to the very special configuration, illustrated in Figure 4-1,

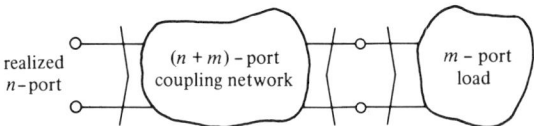

Figure 4-1 Cascade loading configuration.

where an n-port is realized at the first n ports of an $(n + m)$-port coupling network, the last m ports of which are loaded in some specified m-port. The power of the cascade loading configuration is due, in part, to the extraction techniques considered in Section 3-3, where the m extracted components may be viewed as a load on the resultant $(n + m)$-port coupling network. Moreover, most of the synthesis procedures in the sequel may be viewed as sequential processes, wherein an n-port is realized by an $(n + m)$-port (which is, in some sense, simpler than the given n-port) loaded in m "elementary" devices. Upon repeated application of such techniques, one eventually arrives at an $(n + \Sigma m_i)$-port coupling network which is itself "elementary" and, hence, a complete synthesis employing "elementary" devices. Of course, in each step of such a procedure one is dealing with the cascade loading configuration; hence its importance in our catalog of analysis formulas.

The techniques of this chapter are essentially algebraic in nature and, hence, are universally applicable to both the time and frequency domain characterizations. We, therefore, adopt the operator notation which was briefly employed in Chapter 2. That is, a network matrix is denoted simply by a capital italic letter (without explicitly showing its dependence on either the p or t,q variables), operator composition is denoted by multiplication, addition by a plus sign, and the identity by 1. The techniques developed may then be applied to (time-invariant) frequency domain operators simply by replacing the general operator H by $H(p)$. On the other hand, to apply the results to (possibly time-variable) time domain operators, H is replaced by $H(t,q)$ multiplication by "o," and 1 by $\delta(t-q)$. This notational approach permits a unified development in both domains and, at the same time, allows the notation of both to be simplified. Although the general theory is formulated in the operator notation, various examples are given wherein either time or frequency domain techniques are employed.

4-2 Graph Matrices and Sources

In this section the properties of the three fundamental graph matrices B_f, S_f, and F (defined in Appendix C) are reviewed and the equations they determine among the network and source variables are considered.

The three matrices are

$$B_f = [C \ \vdots \ 1] \tag{4-3}$$

$$S_f = [1 \ \vdots \ -C'] \tag{4-4}$$

and

$$F = \left[\begin{array}{c|c} -1 & C' \\ \hline 0 & 1 \end{array}\right] \tag{4-5}$$

where the columns (and rows of F) have a standard order, with columns to the left corresponding to tree edges, and those to the right corresponding to cotree edges (for some predetermined tree and cotree). Basic to all three matrices is the submatrix C which carries all necessary information about the graph (and, in fact, can be used to recover the graph from any of its matrices [83]). The four equations which characterize the connection constraint upon the network variables are

$$B_f v = 0 \tag{4-6}$$

$$S_f i = 0 \tag{4-7}$$

$$i = B'_f i^c \tag{4-8}$$

and

$$v = S'_f v^t \tag{4-9}$$

where i and v are vectors of all network currents and voltages, while i^c and v^t are vectors of cotree currents and tree voltages, respectively. These equations are derived in Appendix C, where it is shown that the last two equations are derivable from the first two, this being achieved via an exploitation of the orthogonality of S_f and B_f [83].

If it is desired to allow for the possibility of external ports in the canonical n-port, they may be placed in parallel with tree edges and in series with cotree

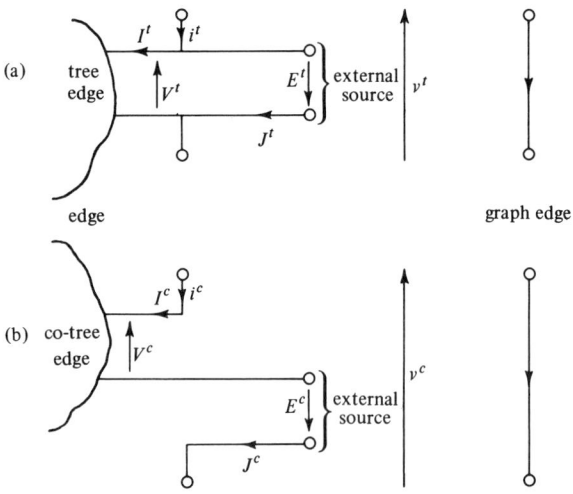

Figure 4-2 Scheme for connecting "external" sources to (**a**) tree edges and (**b**) cotree edges.

edges (of the same tree and cotree which was used to define the graph matrices). This is shown schematically in Figure 4-2, where the variables in the tree and cotree edges are related by the equations

$$V^t = v^t = -E^t \tag{4-10}$$

$$I^t = i^t - J^t \tag{4-11}$$

$$V^c = v^c + E^c \tag{4-12}$$

and
$$I^c = i^c = J^c \tag{4-13}$$

Here
$$V = \begin{bmatrix} V^t \\ V^c \end{bmatrix} \tag{4-14}$$

and
$$I = \begin{bmatrix} I^t \\ I^c \end{bmatrix} \tag{4-15}$$

are vectors of voltages and currents measured at the component ports,

$$E = \begin{bmatrix} E^t \\ E^c \end{bmatrix} \quad (4\text{-}16)$$

and

$$J = \begin{bmatrix} J^t \\ J^c \end{bmatrix} \quad (4\text{-}17)$$

are the external port vectors for voltage and current, respectively, and

$$v = \begin{bmatrix} v^t \\ v^c \end{bmatrix} \quad (4\text{-}18)$$

and

$$i = \begin{bmatrix} i^t \\ i^c \end{bmatrix} \quad (4\text{-}19)$$

are the effective branch variables which are "seen" by the graph (and to which Equations (4-6) through (4-9) apply).

The perturbations of Equations (4-6) and (4-7) caused by the presence of external ports are described by the following proposition.

Proposition 4-1

Let a network have voltage sources, characterized by the vector E^c, in series with its cotree edges and current sources, characterized by the vector J^t, in parallel with the tree edges (and polarized as shown in Figure 4-2). Then

$$E^c = B_f V$$

and

$$J^t = -S_f I$$

Proof

From Equation (4-6),

$$B_f v = 0 \quad (4\text{-}20)$$

which, after partitioning and substitution of Equations (4-10) and (4-12), becomes

$$0 = [C \vdots 1] \begin{bmatrix} v^t \\ v^c \end{bmatrix} = [C \vdots 1] \begin{bmatrix} V^t \\ V^c - E^c \end{bmatrix}$$

$$= [C \vdots 1] \begin{bmatrix} V^t \\ V^c \end{bmatrix} - E^c = B_f V - E^c \quad (4\text{-}21)$$

Upon rearranging Equation (4-21), the desired result that

$$E^c = B_f V \qquad (4\text{-}22)$$

is obtained.

Starting with Equation (4-7), and substituting Equations (4-11) and (4-13), yields the dual result that

$$J^t = -S_f I \qquad (4\text{-}23)$$

□

The essence of Proposition 4-1 is the elimination of the branch variables v and i, these having served as an intermediary in obtaining a relationship between the I and V variables at the component ports and the E and J variables measured at the external ports of the canonical n-port associated with the network.

The results of Proposition 4-1, though useful, are restricted in generality because they only consider the current measured in the cotree edges and the voltage at the tree edges. If one desires to obtain a complete description of the canonical n-port, we must determine the voltage and current (or incident and reflected waves) at all of the external ports which determine this n-port. The $n \times n$ nature of the fundamental connection matrix, F, allows one to achieve this end. Furthermore, its nonsingularity permits one to also obtain an inverse relation.

Proposition 4-2

Let a network have source vectors E and J (connected in series with co-tree edges and in parallel with tree edges, as per Figure 4-2). Then

$$E = F'V$$

and

$$J = FI$$

Proof

Beginning with the partitioned form of E and applying Proposition 4-1 results in

$$E = \begin{bmatrix} E^t \\ E^c \end{bmatrix} = \begin{bmatrix} -V^t \\ B_f V \end{bmatrix} = \begin{bmatrix} -V^t \\ CV^t + V^c \end{bmatrix} = \begin{bmatrix} -1 & 0 \\ C & 1 \end{bmatrix} \begin{bmatrix} V^t \\ V^c \end{bmatrix} \qquad (4\text{-}24)$$

$$= F'V$$

which is the desired equality. Of course, a similar argument beginning with J will result in the equality

$$J = FI \qquad (4\text{-}25)$$

completing the proof of the proposition. □

Due to the self-inverse character of F and F', the inverse relations to those in the proposition are given by the following corollary.

*Corollary 4-1

Under the hypotheses of Proposition 4-2,
$$V = F'E$$
and
$$I = FJ \qquad \square$$

4-3 Topological Analysis Via the π Matrix

In this section our most general analysis technique is formulated. Here the π matrices for the components of a linear network are combined with the graph matrices to obtain a single π matrix describing the canonical n-port realized by the network. In essence, we desire to find a canonical n-port associated with the network (and characterized by a π matrix) whose port behavior is identical to that of the given network (characterized by several π matrices and a graph). This is illustrated schematically in Figure 4-3, where the network, together with its associated canonical n-port, are shown.

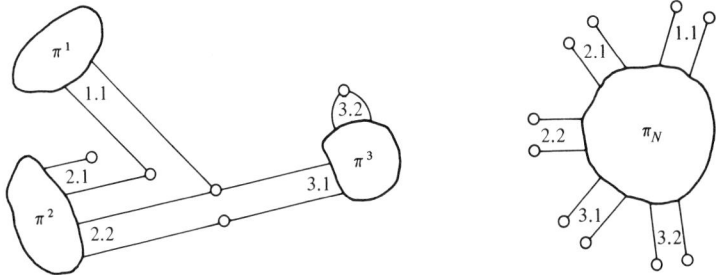

Figure 4-3 A network and its associated canonical n-port.

It should be noted that the canonical n-port has no explicit connection constraints; rather, the effect of connection in the given network is built into the π matrix characterizing the canonical n-port. For the purposes of this section, we will assume that the n-port's characteristics coincide with the edge variables of the original network, whereas in the following section the canonical n-port will be taken as the one whose variables coincide with those "seen" at the external ports (in series with cotree edges and in parallel with tree edges). The two approaches are essentially equivalent, since the edge variables can be obtained from the "external port" variables, and conversely; thus they may be interchanged as is convenient. In Figure 4-3, port 3.2 is shorted; hence, the π matrix of the canonical n-port must allow only zero voltage across port 3.2 (even though it is not "externally" shorted). This is because the π matrix of the realized n-port carries all connection as well as component information from the original network.

136 TOPOLOGICAL ANALYSIS

The first step in the analysis procedure is to combine the π matrices of the various components into a single composite π matrix which simultaneously characterizes all of the components, but is independent of the connection. That is, the composite π matrix is a π matrix for the composite n-port associated with the network (as per definition 1-6).

Definition 4-1

Let $(G, N_1, N_2, \cdots, N_m)$ be a network composed of linear n_i-ports,

$$(G_{n_i}, N_i)$$

each characterized by π matrices

$$\pi^i = [\pi_a{}^i \mid \pi_b{}^i]$$

Then the *composite π matrix* for the network is

$$\hat{\pi} = [\hat{\pi}_a \mid \hat{\pi}_b]$$

where

$$\hat{\pi}_a = \Sigma_i \oplus \pi_a{}^i$$

and

$$\hat{\pi}_b = \Sigma_i \oplus \pi_b{}^i$$

are the direct products of $\pi_a{}^i$ and $\pi_b{}^i$; that is,

$$\hat{\pi} = [\hat{\pi}_a \quad \hat{\pi}_b] = \begin{bmatrix} \pi_a{}^1 & & & \mid & \pi_b{}^1 & & \\ & \pi_a{}^2 & \underline{0} & \mid & & \pi_b{}^2 & \underline{0} \\ & & \cdot & \mid & & & \cdot \\ & & \cdot & \mid & & & \cdot \\ & \underline{0} & \pi_a{}^m & \mid & & \underline{0} & \pi_b{}^m \end{bmatrix} \quad \square$$

$\hat{\pi}$ carries exactly the same information as the collection of π matrices for the components, but is more easily manipulated.

Lemma 4-1

Let $(G, N_1, N_2, \cdots, N_m)$ be a network composed of linear n_i-ports (G_{n_i}, N_i) with admissible pairs

$$(a^i, b^i) \text{ in } N_i \quad i = 1, 2, \cdots, m$$

Then $\hat{\pi}$ is the matrix characterizing the (Σn_i)-port with admissible pairs

$$(a, b) = (\text{col } (a^i), \text{col } (b^i))$$

where

$$(a^i, b^i) \text{ is in } N_i \quad i = 1, 2, \cdots, m$$

Here,

$$a = \text{col } (a^i) = \begin{bmatrix} a^1 \\ a^2 \\ \cdot \\ \cdot \\ \cdot \\ a^m \end{bmatrix}$$

and similarly for col (b^i).

Proof

$$\hat{\pi}\begin{bmatrix} a \\ b \end{bmatrix} = \hat{\pi}_a a + \hat{\pi}_b b$$

$$= \begin{bmatrix} \pi_a^1 & & & \boxed{0} \\ & \pi_a^2 & & \\ & & \cdot & \\ \boxed{0} & & & \cdot \\ & & & \pi_a^m \end{bmatrix} \begin{bmatrix} a^1 \\ a^2 \\ \cdot \\ \cdot \\ \cdot \\ a^m \end{bmatrix} + \begin{bmatrix} \pi_b^1 & & & \boxed{0} \\ & \pi_b^2 & & \\ & & \cdot & \\ \boxed{0} & & & \cdot \\ & & & \pi_b^m \end{bmatrix} \begin{bmatrix} b^1 \\ b^2 \\ \cdot \\ \cdot \\ \cdot \\ b^m \end{bmatrix} \quad \textbf{(4-26)}$$

Now, the m rows of the partitioned matrix of Equation (4-26) are, after deletion of the zero terms,

$$\pi_a^i a^i + \pi_b^i b^i = \pi^i \begin{bmatrix} a^i \\ b^i \end{bmatrix} \qquad i = 1, 2, \cdots, m \quad \textbf{(4-27)}$$

These are exactly the π matrices for the individual components; hence,

$$\hat{\pi}\begin{bmatrix} a \\ b \end{bmatrix} = 0 \quad \textbf{(4-28)}$$

if and only if

$$\pi^i \begin{bmatrix} a^i \\ b^i \end{bmatrix} = 0 \qquad i = 1, 2, \cdots, m \quad \textbf{(4-29)}$$

□

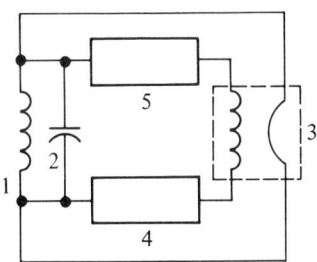

Figure 4-4 A network composed of five components.

The network of Figure 4-4 has five components characterized by frequency domain π matrices,

$$\pi^1(p) = [p - 1 \:\vdots\: p + 1] \tag{4-30}$$

$$\pi^2(p) = [1 - p \:\vdots\: 1 + p] \tag{4-31}$$

$$\pi^3(p) = \begin{bmatrix} 0 & e^{-p} & \vdots & 1 & 0 \\ e^{-p} & 0 & \vdots & 0 & 1 \end{bmatrix} \tag{4-32}$$

$$\pi^4(p) = [e^{-(1/p)} \:\vdots\: 1] \tag{4-33}$$

and

$$\pi^5(p) = \begin{bmatrix} 0 & \vdots & 1 \\ -1 & \vdots & 0 \end{bmatrix} \tag{4-34}$$

whence the π matrix for the composite 6-port associated with the network is

$$\hat{\pi}(p) = [\hat{\pi}_a(p) \quad \hat{\pi}_b(p)]$$

where

$$\hat{\pi}_a(p) = \begin{bmatrix} p - 1 & & & & & \\ & 1 - p & & & & 0 \\ & & 0 & e^{-p} & & \\ & & e^{-p} & 0 & & \\ & 0 & & & e^{-(1/p)} & \\ & & & & & 0 \\ & & & & & -1 \end{bmatrix} \tag{4-35}$$

and

$$\hat{\pi}_b(p) = \begin{bmatrix} p + 1 & & & & \\ & 1 + p & & & 0 \\ & & 1 & 0 & \\ & & 0 & 1 & \\ & 0 & & & 1 \\ & & & & 1 \\ & & & & 0 \end{bmatrix} \tag{4-36}$$

Lemma 4-1 allows the characteristics of all of the network's components to be combined into one $\hat{\pi}$ matrix. Upon combining the composite π matrix thus defined with the graph matrices, a new π matrix for the n-port realized by the network is obtained. The technique employed to achieve this goal is a modification of Reed's primary analysis [139].

Theorem 4-1

Let $(G,N_1,N_2,\cdots,N_m)$ be a network composed of linear n_i-ports,

$$(G_{n_i},N_i)$$

characterized by a composite π matrix, $\hat{\pi}$, and graph matrices S_f and B_f. Then a π matrix for the canonical n-port characterized by

$$(G_{(\Sigma n_i)}, N \cap \hat{G})$$

is

$$\pi = \begin{bmatrix} B_f & | & B_f \\ \hline S_f & | & -S_f \\ \hline \hat{\pi}_a & | & \hat{\pi}_b \end{bmatrix}$$

Proof

It must be shown that the null space of π_N is exactly the vectors

$$\begin{bmatrix} a \\ b \end{bmatrix} \tag{4-37}$$

which simultaneously satisfy the component constraints for each of the n-ports (G_{n_i},N_i) and the connection constraints imposed by G. If

$$0 = \pi \begin{bmatrix} a \\ b \end{bmatrix} = \begin{bmatrix} B_f & | & B_f \\ \hline S_f & | & -S_f \\ \hline \hat{\pi}_a & | & \hat{\pi}_b \end{bmatrix} \begin{bmatrix} a \\ b \end{bmatrix} \tag{4-38}$$

the three rows of Equation (4-38) yield

$$0 = \hat{\pi}_a a + \hat{\pi}_b b = \hat{\pi} \begin{bmatrix} a \\ b \end{bmatrix} \tag{4-39}$$

and

$$0 = B_f a + B_f b = B_f(a+b) = 2B_f v \tag{4-40}$$

$$0 = S_f a - S_f b = S_f(a-b) = 2S_f i \tag{4-41}$$

Now, Equations (4-39) through (4-41) are exactly the three required equations for (a,b) to be admissible to the network. Conversely, if (a,b) is in N, the three equations are satisfied; hence, so is Equation (4-31). □

The analysis of Theorem 4-1 has the distinction of being the most general and also the simplest of the linear analyses [139]. Its weakness lies in the inherent difficulty of interpreting the characteristics of a network from its π matrix.

The immittance form of the π matrix analysis may be obtained via a similar argument to that employed above, the details of which are left to the reader as an exercise.

*Proposition 4-3

Let $(G, N_1, N_2, \cdots, N_m)$ be a network composed of linear n_i-ports,

$$(G_{n_i}, N_i)$$

characterized (in immittance variables) by a composite π matrix, $\hat{\pi}$, and graph matrices S_f and B_f. Then a π matrix for the canonical n-port realized by the network is

$$\pi = \begin{bmatrix} B_f & 0 \\ \hline 0 & S_f \\ \hline \hat{\pi}_v & \hat{\pi}_i \end{bmatrix} \qquad \square$$

The generality of the primary analysis can be illustrated by the nullator-norator network of Figure 4-5. Since these components are non-normal, only

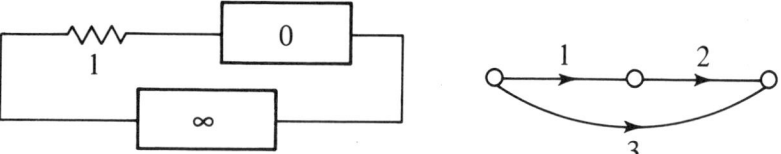

Figure 4-5 Nullator-norator network and graph.

analysis techniques using π or Q matrices are applicable. In this case,

$$S_f = \begin{bmatrix} 1 & 0 & 1 \\ 0 & 1 & 1 \end{bmatrix} \qquad (4\text{-}42)$$

$$B_f = [-1 \quad -1 \;\vdots\; 1] \qquad (4\text{-}43)$$

and

$$\hat{\pi} = \begin{bmatrix} 0 & 0 & 0 & 1 & 0 & 0 \\ 0 & 1 & 0 & 0 & 0 & 0 \\ 0 & 0 & 0 & 0 & 1 & 0 \\ 0 & 0 & 0 & 0 & 0 & 0 \end{bmatrix} \qquad (4\text{-}44)$$

Hence, π is given by

$$\pi = \left[\begin{array}{cccccc} -1 & -1 & 1 & -1 & -1 & 1 \\ \hline 1 & 0 & 1 & -1 & 0 & -1 \\ 0 & 1 & 1 & 0 & -1 & -1 \\ \hline 0 & 0 & 0 & 1 & 0 & 0 \\ 0 & 1 & 0 & 0 & 0 & 0 \\ 0 & 0 & 0 & 0 & 1 & 0 \\ 0 & 0 & 0 & 0 & 0 & 0 \end{array}\right] \quad (4\text{-}45)$$

It is noteworthy that in the primary analysis, if the components are characterized by π matrices, the canonical characterization may not be a π matrix. In fact, this is the case with the above example, where the trivial last row can be dropped without changing the constraints imposed by the matrix.

Although the above example is memoryless, the analysis may be carried out just as readily for networks of arbitrary complexity, this because the analysis is carried out without employing any algebraic operations on the matrices; rather, one forms π simply by substitution of the appropriate component and graph matrices into a partitioned form of π. The reader may verify this for himself by forming the π matrix for the network of Figure 4-4, which contains an inductor, a capacitor, a nullator, an LC transmission line (component 3), and a lossless 1-port (see Chapter 3) characterized by the π matrix

$$\pi^4(p) = [e^{-(1/p)} \quad 1] \quad (4\text{-}46)$$

Of these components, only the nullator is memoryless, and that is nonnormal. Moreover, the network is not composed wholly of components from a single class (RLC, transmission line, and so forth).

4-4 Topological Analysis Via the Scattering Matrix

In this section the techniques of topological analysis are applied to the problem of calculating the scattering matrix of the canonical n-port realized by a network at its external ports from those of its components and its graph matrices. Unlike the π matrix analysis, wherein the port variables of the realized n-port coincide with the edge variables of the network, the canonical n-port realized via the scattering analysis is that observed at the "external" ports. These are arranged as shown in Figure 4-2, with the externally accessible ports in series with the cotree edges and in parallel with the tree edges of some specified tree.

As in the preceding section, the formulation is simplified if the scattering matrices of the network components are combined into a single composite scattering matrix characterizing the composite n-port associated with the network.

Definition 4-2

Let $(G, N_1, N_2, \cdots, N_m)$ be a network composed of linear n_i-ports,

$$(G_{n_i}, N_i)$$

each characterized by a scattering matrix

$$S^i$$

Then the *composite scattering matrix* for the network $\hat{S}$, is

$$\hat{S} = \sum_i \oplus S^i$$

that is,

$$\hat{S} = \begin{bmatrix} S^1 & & & 0 \\ & S^2 & & \\ & & \ddots & \\ 0 & & & S^m \end{bmatrix}$$

□

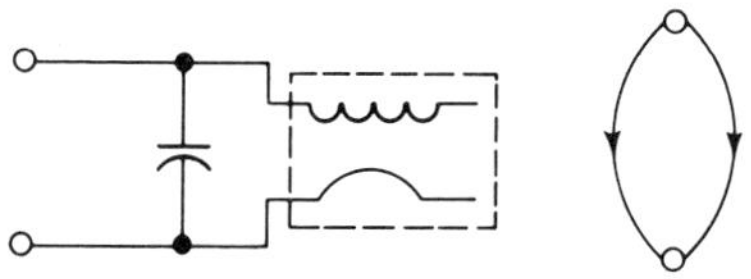

Figure 4-6 (a) Network. (b) Its graph.

The network of Figure 4-6(a) contains two components, a capacitor and a delay characterized by scattering matrices

$$S^1(p) = \frac{1-p}{1+p} \qquad (4\text{-}47)$$

and

$$S^2(p) = e^{-p} \qquad (4\text{-}48)$$

for which the corresponding composite scattering matrix is

$$\hat{S}(p) = \begin{bmatrix} \frac{1-p}{1+p} & 0 \\ 0 & e^{-p} \end{bmatrix} \qquad (4\text{-}49)$$

As in the previous section (and via a similar argument, which is left for the reader), the composite scattering matrix is just the scattering matrix of the composite n-port associated with the network.

*Lemma 4-2

Let $(G, N_1, N_2, \cdots, N_m)$ be a network composed of linear n_i-ports,

$$(G_{n_i}, N_i)$$

Then S is the scattering matrix of the (Σn_i)-port with admissible pairs

$$(a, b) = (\text{col } (a^i), \text{col } (b^i))$$

where

$$(a^i, b^i) \text{ is in } N_i \qquad i = 1, 2, \cdots, m \qquad \square$$

Equivalently,

$$b = \hat{S}a$$

In the following development let

$$A = E + J \tag{4-50}$$

and

$$B = E - J \tag{4-51}$$

be the incident and reflected waves seen at the external ports, while

$$a = V + I \tag{4-52}$$

and

$$b = V - I \tag{4-53}$$

are the incident and reflected waves of the components. From Lemma 4-2 we have

$$b = \hat{S}a \tag{4-54}$$

The goal of the analysis is to find a new scattering matrix S such that

$$B = SA \tag{4-55}$$

This is most easily accomplished when the fundamental connection matrix F is used to characterize the graph (rather than the circuit and cocircuit matrices). The following lemmas are necessary tools.

Lemma 4-3

Let F be the fundamental connection matrix of a graph. Then

$$(F' + F)$$

is invertible.

Proof

The fact that the inverse is

$$(F' + F)^{-1} = \begin{bmatrix} -2(4 + C'C)^{-1} & (4 + C'C)^{-1}C' \\ \hdashline (4 + CC')^{-1}C & 2(4 + CC')^{-1} \end{bmatrix} \quad (4\text{-}56)$$

can be verified by multiplication. Here

$$(4 + CC') \quad (4\text{-}57)$$

and

$$(4 + C'C) \quad (4\text{-}58)$$

are positive definite (since they are the sums of a positive and a positive definite operator); hence their inverses exist [132]. □

Lemma 4-4

Let a matrix A have norm less than or equal to one. Then

$$M = ((F' + F) + (F' - F)A)$$

is invertible.

Proof

Since $(F' + F)$ differs from $(F' - F)$ only by the addition of plus and minus two entries on the diagonal and some sign changes,

$$\|(F' + F)^{-1}(F' - F)\| < 1 \quad (4\text{-}59)$$

Now, for A such that

$$\|A\| \leq 1 \quad (4\text{-}60)$$

$$\|(F' + F)^{-1}(F' - F)A\| \leq \|(F' + F)^{-1}(F' - F)\| \, \|A\| < 1 \quad (4\text{-}61)$$

Writing M as

$$\begin{aligned} M &= [(F' + F) + (F' - F)A] \\ &= (F' + F)[1 + (F' + F)^{-1}(F' - F)A] \end{aligned} \quad (4\text{-}62)$$

its inverse is given by

$$M^{-1} = \sum_{i=0}^{\infty} [-(F' + F)^{-1}(F' - F)A]^i (F' + F)^{-1} \quad (4\text{-}63)$$

where the infinite sum converges [132], since

$$\|(F' + F)^{-1}(F' - F)A\| < 1 \quad (4\text{-}64)$$

and $(F' + F)^{-1}$ is assured to exist by Lemma 4-3. □

Consistent with the preceding lemmas, the scattering analysis is obtained via Theorem 4-2, the proof of which is due to DeClaris and Saeks [59].

Theorem 4-2

Let a network $(G, N_1, N_2, \cdots, N_m)$ be composed of linear n_i-ports,

$$(G_{n_i}, N_i)$$

each characterized by a scattering matrix

$$S^i$$

Then the scattering matrix of the canonical n-port associated with the network (at its "external" ports in series with co-tree edges and in parallel with tree edges of G) is

$$S = [(F' - F) + (F' + F)\hat{S}] [(F' + F) + (F' - F)\hat{S}]^{-1}$$

where $\hat{S}$ is the composite scattering matrix of the network and F is its fundamental connection matrix. Furthermore, if each component n_i-port is passive, the inverse is assured to exist.

Proof

For the component n_i-ports, Lemma 4-2 yields

$$b = \hat{S} a \tag{4-65}$$

where

$$a = V + I \tag{4-66}$$

and

$$b = V - I \tag{4-67}$$

It is desired to find S such that

$$B = S A \tag{4-68}$$

where

$$A = E + J \tag{4-69}$$

and

$$B = E - J \tag{4-70}$$

Substituting the equations

$$E = F'V \tag{4-71}$$

and

$$J = FI \tag{4-72}$$

derived in Proposition 4-2 into Equation (4-69) now yields

$$A = E + J = F'V + FI = [F' \mid F] \begin{bmatrix} V \\ I \end{bmatrix} \tag{4-73}$$

From the definition of V and I we have

$$\begin{bmatrix} V \\ I \end{bmatrix} = \frac{1}{2}\begin{bmatrix} 1 & \vdots & 1 \\ \hdashline 1 & \vdots & -1 \end{bmatrix}\begin{bmatrix} a \\ b \end{bmatrix} = \frac{1}{2}\begin{bmatrix} 1 & \vdots & 1 \\ \hdashline 1 & \vdots & -1 \end{bmatrix}\begin{bmatrix} a \\ \hat{S}a \end{bmatrix}$$

$$= \frac{1}{2}\begin{bmatrix} 1 & \vdots & 1 \\ \hdashline 1 & \vdots & -1 \end{bmatrix}\begin{bmatrix} 1 \\ \hat{S} \end{bmatrix}[a] = \frac{1}{2}\begin{bmatrix} (1 + \hat{S}) \\ \hdashline (1 - \hat{S}) \end{bmatrix}[a]$$

(4-74)

Hence, substitution of Equation (4-74) into Equation (4-73) yields

$$A = [F' \vdots F]\begin{bmatrix} V \\ I \end{bmatrix} = \frac{1}{2}[F' \vdots F]\begin{bmatrix} (1+\hat{S}) \\ \hdashline (1-\hat{S}) \end{bmatrix}[a]$$

$$= \frac{1}{2}[F'(1+\hat{S}) + F(1-\hat{S})]a] = \frac{1}{2}[(F'+F) + (F'-F)\hat{S}]a]$$ (4-75)

$$\equiv \frac{1}{2}Ma$$

Applying a similar argument to

$$B = E - J = F'V - FI = [F' \vdots F]\begin{bmatrix} V \\ I \end{bmatrix}$$ (4-76)

yields

$$B = \frac{1}{2}[(F'-F) + (F'+F)\hat{S}]a \equiv \frac{1}{2}La$$ (4-77)

Combining Equations (4-75) and (4-77) now results is

$$B = (LM^{-1})A$$ (4-78)

or, equivalently,

$$S = LM^{-1}$$ (4-79)

if M^{-1} exists.

If each component n_i-port for the network is passive,

$$\|\hat{S}\| \leq 1$$ (4-80)

in which case Lemma 4-4 assures that the inverse exists. $\square$

Figure 4-7 Ideal transformer network and graph.

As an example of the scattering analysis, consider the ideal transformer loaded (?) by an open circuit, as shown in Figure 4-7. Choosing edges 1.1 and 1.2 as the tree,

$$B_f = [\,0 \quad 1 \;\vdots\; 1\,]$$
$$= [C \;\vdots\; 1]\qquad(4\text{-}81)$$

$$C = [0 \;\vdots\; 1]\qquad(4\text{-}82)$$

$$F = \begin{bmatrix} -1 & 0 & \vdots & 0 \\ 0 & -1 & \vdots & 1 \\ \hline 0 & 0 & \vdots & 1 \end{bmatrix}\qquad(4\text{-}83)$$

and

$$F' = \begin{bmatrix} -1 & 0 & \vdots & 0 \\ 0 & -1 & \vdots & 0 \\ \hline 0 & 1 & \vdots & 1 \end{bmatrix}\qquad(4\text{-}84)$$

The scattering matrix of the open circuit is

$$S^2 = 1\qquad(4\text{-}85)$$

whereas that of the ideal transformer is

$$S^1 = \begin{bmatrix} -\left(\frac{4}{5}\right) & \left(\frac{3}{5}\right) \\ \left(\frac{3}{5}\right) & \left(\frac{4}{5}\right) \end{bmatrix}\qquad(4\text{-}86)$$

Hence,

$$\hat{S} = \begin{bmatrix} -\left(\frac{4}{5}\right) & \left(\frac{3}{5}\right) & \vdots & 0 \\ \left(\frac{3}{5}\right) & \left(\frac{4}{5}\right) & \vdots & 0 \\ \hline 0 & 0 & \vdots & 1 \end{bmatrix}\qquad(4\text{-}87)$$

Carrying out the operations indicated in the theorem now yields

$$M = [(F' + F) + (F' - F)\hat{S}] = \begin{bmatrix} -2 & 0 & 0 \\ 0 & -2 & 0 \\ \frac{3}{5} & \frac{9}{5} & 2 \end{bmatrix}\qquad(4\text{-}88)$$

$$M^{-1} = \left(\frac{1}{20}\right)\begin{bmatrix} -10 & 0 & 0 \\ 0 & -10 & 0 \\ 3 & 9 & 10 \end{bmatrix}\qquad(4\text{-}89)$$

and

$$L = [(F' - F) + (F' + F)\hat{S}] = \begin{bmatrix} \frac{8}{5} & -\frac{6}{5} & 0 \\ -\frac{6}{5} & -\frac{8}{5} & 0 \\ -\frac{3}{5} & \frac{9}{5} & 2 \end{bmatrix}\qquad(4\text{-}90)$$

The desired scattering matrix is thus

$$S = LM^{-1} = \begin{bmatrix} -\frac{4}{5} & \frac{3}{5} & 0 \\ \frac{3}{5} & \frac{4}{5} & 0 \\ 0 & 0 & 1 \end{bmatrix}\qquad(4\text{-}91)$$

Recognizing that the external ports illustrated in Figure 4-8 for the tree edges (the edges of the transformer, 1.1 and 1.2 in this case) are in parallel with the transformer ports and that the open circuit does not effect the behavior of the transformer, it is to be expected (and, in fact, is the case) that the scattering

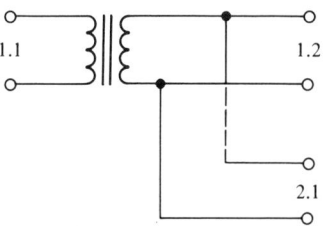

Figure 4-8 External port configuration for the network of Figure 4-7.

matrix of the transformer appears in the 1.2 and 1.2 entries of S. Since the external port for the cotree edge (the open circuit) is in series with the edge, only an open circuit appears at this external port, and hence the "1" in the "3 − 3" position of S. The scattering matrix obtained via the theorem is thus consistent with that obtained via an intuitive argument.

It is noteworthy that the analysis formula of Theorem 4-2 always gives a scattering matrix seen from all external source ports. In many cases, some of these ports are of no interest and may be eliminated by short or open circuiting them (if they are in series with cotree edges or in parallel with tree edges, respectively). In the preceding example, if one is interested only in the characteristics of the 1-port realized at the transformer input (external port 1), the two remaining ports may be killed by opening and shorting. Now, in the analysis, port 2 is in parallel with a tree edge and thus should be opened, whereas port 3 is in series with a cotree edge and hence should be shorted. Mathematically, the process of opening and shorting ports is equivalent to the constraints

$$a_o = b_o \tag{4-92}$$

and

$$a_s = -b_s \tag{4-93}$$

for the open and short circuit cases, respectively. Applying these formulas to ports 2 and 3, respectively, of the 3-port scattering matrix obtained in the previous example [Equation (4-87)] results in the matrix equality

$$\begin{bmatrix} b_1 \\ a_2 \\ -a_3 \end{bmatrix} = \begin{bmatrix} -(\tfrac{4}{5}) & (\tfrac{3}{5}) & 0 \\ (\tfrac{3}{5}) & (\tfrac{4}{5}) & 0 \\ 0 & 0 & 1 \end{bmatrix} \begin{bmatrix} a_1 \\ a_2 \\ a_3 \end{bmatrix} \tag{4-94}$$

which, when solved for a relation between b_1 and a_1, yields

$$b_1 = a_1 \tag{4-95}$$

Hence, the scattering matrix realized at port 1 by the network of Figure 4-7 is
$$S = 1 \qquad (4\text{-}96)$$
that is, an open circuit.

As with the preceding analysis, the technique is applicable in either the time or frequency domain to arbitrary networks, so long as each component n-port is characterized by a scattering matrix. Unlike the π matrix analysis, however, the complex algebraic manipulations required by Theorem 4-2 may yield rather involved scattering matrices even for simple components. Consider, for example, the network of Figure 4-6. Here the two 1-ports are characterized by the composite scattering matrix of Equation (4-49), whereas the fundamental connection matrix is

$$F = \begin{bmatrix} -1 & \vdots & -1 \\ \hdashline 0 & \vdots & 1 \end{bmatrix} \qquad (4\text{-}97)$$

with edge 1, the capacitor, taken as the tree, and edge 2, the open circuited transmission line, taken as the cotree. Upon carrying out the operations indicated by the theorem, one obtains the scattering matrix

$$S(p) = \frac{1}{2e^{-p} - 4p - 6}\begin{bmatrix} 2e^{-p} + 4p - 6 & 2(p+1)(1 - e^{-2p}) \\ \dfrac{8p}{p+1} & -4e^{-p}p - 6e^{-p} + 2 \end{bmatrix} \qquad (4\text{-}98)$$

for the canonical 2-port associated with the network. Of course, as in the preceding example, one may "kill" an undesired port by open circuiting or short circuiting it. This can be achieved by the "informal" technique used in the preceding example, or alternatively, the cascade loading formula obtained in Section 4-7 may be invoked. This latter approach is considered in more detail later.

4-5 Analysis with Normalized Scattering Matrices

The analysis of the preceding section can be extended to include the case where the scattering matrices of both the components and the canonical n-port have arbitrary normalization. This serves both to generalize the analysis procedure and, additionally, to allow the derivation of a number of useful results on normalized scattering matrices. The derivation follows essentially the same line of argument as in the unnormalized case, once the fundamental connection matrix has been appropriately modified. For this reason most of the proofs are left to the reader as exercises.

It is desired to consider a network, each of whose components is characterized by scattering matrix S^i, normalized by R_i.

$$a_i = R_i^{-1/2} V_i + R_i^{1/2} I_i = V_i + I_i \qquad (4\text{-}99)$$
and
$$b_i = R_i^{-1/2} V_i - R_i^{1/2} I_i = V_i - I_i \qquad (4\text{-}100)$$

We must calculate a scattering matrix S for the (Σn_i)-port realized at the external ports and normalized by R, for which

$$A = R^{-1/2}E + R^{1/2}J = \mathbf{E} + \mathbf{J} \qquad (4\text{-}101)$$

and

$$B = R^{-1/2}E - R^{1/2}J = \mathbf{E} - \mathbf{J} \qquad (4\text{-}102)$$

Since the development of the preceding section was entirely algebraic once the three equations

$$b = \hat{S}a \qquad (4\text{-}103)$$
$$E = F'V \qquad (4\text{-}104)$$

and

$$J = FI \qquad (4\text{-}105)$$

were established, a similar analysis can be obtained in the normalized case if Equations (4-103) through (4-105) are replaced by new equations relating the normalized variables.

The scattering matrices for the components are given as normalized by R_i; hence, $\hat{S}$ defined as per Definition 4-2 satisfies Equation (4-103) if a and b are normalized by

$$\hat{R} = \Sigma \oplus R_i = \begin{bmatrix} R_1 & & & & \\ & R_2 & & \lfloor 0 & \\ & & \cdot & & \\ & \overline{0\rfloor} & & \cdot & \\ & & & & R^m \end{bmatrix} \qquad (4\text{-}106)$$

The analog of Equations (4-104) and (4-105) may be achieved via the "normalized graph matrices."

Proposition 4-4

Let a graph be characterized by a fundamental connection matrix F and normalized immittance variables

$$\mathbf{V} = \hat{R}^{-1/2}V$$
$$\mathbf{I} = \hat{R}^{1/2}I$$
$$\mathbf{E} = R^{-1/2}E$$
$$\mathbf{J} = R^{1/2}J$$

Then

$$\mathbf{E} = \mathbf{F}'\mathbf{V}$$

and

$$\mathbf{J} = \mathbf{F}\,\mathbf{I}$$

where

$$\mathbf{F}' = R^{-1/2}F'\hat{R}^{1/2}$$

and

$$\mathbf{F} = R^{1/2}F\hat{R}^{-1/2}$$

Proof

By Proposition 4-2,
$$E = F'V \tag{4-107}$$

Multiplication of Equation (4-107) by $R^{-1/2}$ now yields
$$\mathbf{E} = R^{-1/2}E = R^{-1/2}F'V \tag{4-108}$$

But
$$\mathbf{V} = R^{-1/2}V \tag{4-109}$$

Hence,
$$E = R^{1/2}F'\hat{R}^{1/2}\mathbf{V} = F'V \tag{4-110}$$

A similar argument holds for
$$J = FI \tag{4-111}$$

□

Note that, in general, F' is not the transpose of F, the prime notation being used only to indicate the parallel between the normalized and unnormalized cases.

In the network of Figure 4-9, if the first resistor is normalized to 2 and the

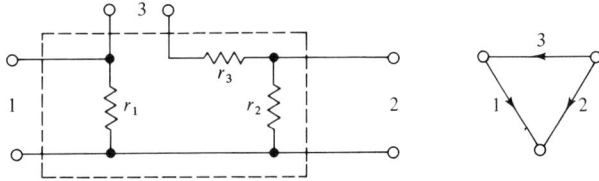

Figure 4-9 Network of resistors with normalized scattering matrices and graph.

second and third to 3, while one desires that the canonical n-port realized by the network have all of its ports normalized to 4, we have

$$R_1 = 2 \tag{4-112}$$

and
$$R_2 = R_3 = 3 \tag{4-113}$$

Hence,
$$\hat{R} = \begin{bmatrix} 2 & & \underline{0} \\ & 3 & \\ \overline{0} & & 3 \end{bmatrix} \tag{4-114}$$

while
$$R = \begin{bmatrix} 4 & & \underline{0} \\ & 4 & \\ \overline{0} & & 4 \end{bmatrix} \tag{4-115}$$

Now, from the graph of Figure 4-9(b), one may obtain the fundamental connection matrix

$$F = \begin{bmatrix} -1 & 0 & -1 \\ 0 & -1 & 1 \\ 0 & 0 & 1 \end{bmatrix} \qquad (4\text{-}116)$$

Invoking the formulas of Proposition 4-4 now results in the normalized connection matrices

$$F' = R^{-1/2} F' \hat{R} = \begin{bmatrix} \dfrac{-1}{\sqrt{2}} & 0 & 0 \\ 0 & \dfrac{-\sqrt{3}}{\sqrt{2}} & 0 \\ \dfrac{-1}{\sqrt{2}} & \dfrac{3}{\sqrt{2}} & \dfrac{3}{\sqrt{2}} \end{bmatrix} \qquad (4\text{-}117)$$

and

$$F = R \, F \hat{R}^{-1/2} = \begin{bmatrix} \dfrac{-1}{\sqrt{2}} & 0 & \dfrac{-2}{\sqrt{3}} \\ 0 & \dfrac{-2}{\sqrt{3}} & \dfrac{2}{\sqrt{3}} \\ 0 & 0 & \dfrac{2}{\sqrt{3}} \end{bmatrix} \qquad (4\text{-}118)$$

Clearly,

$$F' \neq (F)' \qquad (4\text{-}119)$$

That is, F' is not the transpose of F.

Upon replacing $\hat{S}$ by the normalized composite scattering matrix and F and F' by F and F', respectively, the analysis formulas for the normalized case may be formulated via the same argument that was employed in the unnormalized case.

*Theorem 4-3

Let a network $(G, N_1, N_2, \cdots, N_m)$ be composed of linear n_i-ports,

$$(G_{n_i}, N_i)$$

each characterized by a scattering matrix

$$S^i$$

normalized to R_i. Then the scattering matrix S_R of the canonical n-port realized by the network (at its "external" ports in series with cotree edges and in parallel with tree edges of G) is, when normalized to R,

$$S_R = [(F' - F) + (F' + F)\hat{S}_{\hat{R}}][(F' + F) + (F' - F)\hat{S}_{\hat{R}}]^{-1}$$

where $\hat{S}$ is the composite scattering matrix of the network normalized to R and F' and F are the "normalized" connection matrices of Proposition 4-4. Furthermore, if each component n_i-port is passive, the inverse is assured to exist. □

An immediate corollary to the analysis technique is a formula for changing the normalization of a scattering matrix. This is obtained by analyzing a network with only one component and letting $\hat{R}$ be the old normalization and R be the new normalization. In this case,

$$\hat{S}_{\hat{R}} = S_{\hat{R}} \tag{4-120}$$

and

$$F = -1 \tag{4-121}$$

(since the entire graph is a tree). Upon substituting Equations (4-120) and (4-121) into the analysis formula of Theorem 4-3 one obtains the following Corollary.

*Corollary 4-2

Let an n-port have scattering matrix $S_{\hat{R}}$ when normalized by $\hat{R}$ and scattering matrix S_R when normalized by R. Then

$$S_R = ((R^{-1/2}\hat{R}^{1/2} - R^{1/2}\hat{R}^{-1/2}) + (R^{-1/2}\hat{R}^{1/2} + R^{1/2}\hat{R}^{-1/2})S_{\hat{R}})$$
$$\cdot ((R^{-1/2}\hat{R}^{1/2} + R^{1/2}\hat{R}^{-1/2}) + (R^{-1/2}\hat{R}^{1/2} - R^{1/2}\hat{R}^{-1/2})S_{\hat{R}})^{-1} \quad □$$

Since the inverse in Corollary 4-2 is assured to exist for passive n-ports, we have the following.

*Corollary 4-3

If a passive n-port has a scattering matrix for one strictly positive real normalization, then it has a scattering matrix for all strictly positive real normalizations (that is, the existence of a scattering matrix is an invariant of the choice of strictly positive real normalization). □

Corollary 4-3 justifies our emphasis on the unnormalized scattering matrice, since it assures that if a scattering matrix exists for any positive real normalization, then an unnormalized matrix also exists. Normalization, however, often yields a certain convenience which is not achieved in the unnormalized case. This is illustrated by the transmission line, for which the characteristic impedance is most easily defined as the normalization of its scattering matrix.

In a number of the previous examples we have employed the transmission line as a network element. This element, which will be encountered in the succeeding chapters with increasing frequency, is characterized by two parameters, the (electrical) length τ, and the characteristic impedance R_0. Fortunately, however, the latter can be built into the normalization of the scattering variables with which the device is described, thereby removing it from explicit consideration. In fact, historically, scattering matrix normalization developed as a tool of transmission line theory [44].

Definition 4-3

A *transmission line* of length τ and characteristic impedance R_0 is a 2-port, the admissible pairs of which are characterized by the scattering variables

$$\left(\begin{pmatrix} a_1(t) \\ a_2(t) \end{pmatrix}, \begin{matrix} b_1(t) = a_2(t - \tau) \\ b_2(t) = a_1(t - \tau) \end{matrix} \right)$$

when normalized to R_0. □

Of course, a time-variable transmission line (with either variable length or characteristic impedance) could readily be defined. We, however, do not find need for such and leave the definition to the reader as an exercise. Clearly, the time domain scattering matrix of the transmission line when normalized to R_0 is

$$S_{R_0}(t,q) = \begin{bmatrix} 0 & \delta(t - \tau - q) \\ \delta(t - \tau - q) & 0 \end{bmatrix} \quad \text{(4-122)}$$

In the frequency domain the reader may verify the following proposition.

*Proposition 4-5

A transmission line of length τ and characteristic impedance R_0 is a 2-port with scattering matrix

$$S_{R_0} = \begin{bmatrix} 0 & e^{-\tau p} \\ e^{-\tau p} & 0 \end{bmatrix}$$

normalized to

$$R_0 = \begin{bmatrix} R_0 & 0 \\ 0 & R_0 \end{bmatrix}$$

□

The scattering matrix of the transmission line is extremely simple if normalized to its characteristic impedance. In fact, in that case, R_0 does not even appear explicitly in $S(p)$. Of course, Corollary 4-3 assures that a scattering matrix will exist for the transmission line in the unnormalized case, but it will not, in general, have the simple off-diagonal form of the definition. Taking R_0 to be 4 and representing such a line by an unnormalized scattering matrix, we obtain

$$S_1 = \frac{1}{(25 - 9e^{-2\tau p})} \begin{bmatrix} 15(1 - e^{-2\tau p}) & 16e^{-\tau p} \\ 16e^{-\tau p} & 15(1 - e^{-2\tau p}) \end{bmatrix} \quad \text{(4-123)}$$

This result is obtained via Corollary 4-2 with

$$S_{\hat{R}} = \begin{bmatrix} 0 & e^{-\tau p} \\ e^{-\tau p} & 0 \end{bmatrix} \tag{4-124}$$

$$\hat{R} = \begin{bmatrix} 4 & 0 \\ 0 & 4 \end{bmatrix} \tag{4-125}$$

and

$$R = \begin{bmatrix} 1 & 0 \\ 0 & 1 \end{bmatrix} \tag{4-126}$$

4-6 Topological Analysis Via the Immittance Matrices

The tools which have been thus far developed to implement the π and S matrix analysis may also be applied to the problem of calculating the immittance matrices of a canonical n-port realized by a network from those of the network components. Of course, as with most immittance matrix theorems, the usefulness of the technique is restricted by the failure of many networks to possess immittance matrices. The complexity of the analysis lies between that of the straight forward π matrix analysis and the rather complex scattering formula.

As in the preceding sections, the first step is to form a composite immittance matrix.

*Definition 4-4

Let $(G, N_1, N_2, \cdots, N_m)$ be a network composed of linear n_i-ports,

$$(G_{n_i}, N_i)$$

each characterized by an admittance (impedance) matrix

$$Y^i$$

Then the composite admittance (impedance) matrix for the network is

$$\hat{Y} = \sum_i \oplus Y^i$$

That is,

$$\hat{Y} = \begin{bmatrix} Y^1 & & & & \\ & Y^2 & & \underline{|0} & \\ & & \cdot & & \\ & & & \cdot & \\ & \overline{0|} & & & \cdot \\ & & & & Y^m \end{bmatrix}$$

□

156 TOPOLOGICAL ANALYSIS

*Lemma 4-5

Let $(G, N_1, N_2, \cdots, N_m)$ be a network composed of linear n_i ports,

$$(G_{n_i}, N_i)$$

with admissible pairs

$$(v^i, i^i) \text{ in } N_i \qquad i = 1, 2, \cdots, m$$

Then $\hat{Y}$ is the admittance (impedance) matrix of the (Σn_i)-port with admissible pairs

$$(v, i) = (\text{col } (v^i), \text{col } (i^i))$$

where

$$(v^i, i^i) \text{ is in } N_i \qquad i = 1, 2, \cdots, m$$

that is,

$$i = \hat{Y} v \qquad \square$$

Combining $\hat{Y}$ with the fundamental connection matrix for G now yields an admittance matrix for the canonical n-port realized at the "external" ports of the network.

*Theorem 4-4

Let $(G, N_1, N_2, \cdots, N_m)$ be a network composed of linear n_i-ports,

$$(G_{n_i}, N)$$

each characterized by an admittance (impedance) matrix

$$Y^i$$

Then the admittance (impedance) matrix of the n-port realized (at the "external" ports in series with the cotree edges and in parallel with the tree edges) by the network is

$$Y = F\hat{Y}F'$$

(and in the impedance case, $Z = F'\hat{Z}F$). $\square$

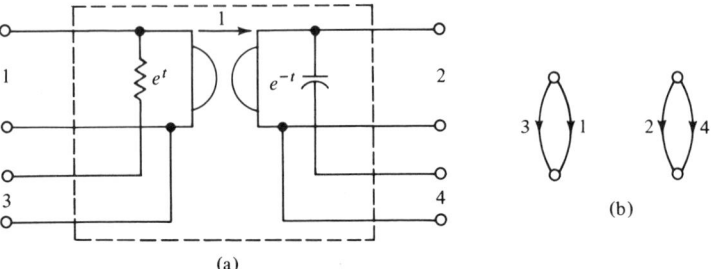

Figure 4-10 (a) Network of admittances. (b) Its graph.

In the network of Figure 4-10, the gyrator is characterized by the admittance

$$Y^1(t,q) = \begin{bmatrix} 0 & -\delta(t-q) \\ \delta(t-q) & 0 \end{bmatrix} \quad (4\text{-}127)$$

while the resistor and capacitor are characterized by

$$Y^2(t,q) = e^t \delta(t-q) \quad (4\text{-}128)$$

and

$$Y^3(t,q) = e^{-q} \dot\delta(t-q) \quad (4\text{-}129)$$

respectively. Upon combining these, one obtains the composite admittance matrix

$$\hat Y(t,q) = \left[\begin{array}{cc|c|c} 0 & -\delta(t-q) & 0 & 0 \\ \delta(t-q) & 0 & 0 & 0 \\ \hline 0 & 0 & e^t\delta(t-q) & 0 \\ \hline 0 & 0 & 0 & e^{-q}\dot\delta(t-q) \end{array}\right] \quad (4\text{-}130)$$

Remembering that in the time domain the identity operator is represented by $\delta(t-q)$, the fundamental connection matrix

$$F = \left[\begin{array}{cc|cc} -1 & 0 & -1 & 0 \\ 0 & -1 & 0 & -1 \\ \hline 0 & 0 & 1 & 0 \\ 0 & 0 & 0 & 1 \end{array}\right] \quad (4\text{-}131)$$

for the graph of Figure 4-10(b) may be represented by

$$F(t,q) = F\delta(t-q) = \left[\begin{array}{cc|cc} -\delta(t-q) & 0 & -\delta(t-q) & 0 \\ 0 & -\delta(t-q) & 0 & -\delta(t-q) \\ \hline 0 & 0 & \delta(t-q) & 0 \\ 0 & 0 & 0 & \delta(t-q) \end{array}\right] \quad (4\text{-}132)$$

in the time domain.

Applying the admittance analysis formula of the theorem, one obtains the canonical 4-port

$$Y(t,q) = F(t,q) \circ \hat Y(t,q) \circ F'(t,q)$$

$$= \begin{bmatrix} -e^t\delta(t-q) & -\delta(t-q) & e^t\delta(t-q) & 0 \\ -\delta(t-q) & -e^{-q}\dot\delta(t-q) & 0 & e^{-q}\dot\delta(t-q) \\ -e^t\delta(t-q) & 0 & e^t\delta(t-q) & 0 \\ 0 & -e^{-q}\dot\delta(t-q) & 0 & e^{-q}\dot\delta(t-q) \end{bmatrix} \quad (4\text{-}133)$$

Of course, as with the previous analysis, one can "kill" any unwanted ports by opening or shorting them. In the immittance analysis, this implies the algebraic constraints

$$i_o = 0 \qquad (4\text{-}134)$$

and

$$v_s = 0 \qquad (4\text{-}135)$$

rather than the scattering variable equalities used previously.

This approach to the immittance analysis [59] allows arbitrary inputs to the network (as long as they are in series with cotree edges and in parallel with tree edges). A more commonly encountered analysis [57], [83], [155], restricts the inputs to current sources in parallel with tree edges or (not and) voltage sources in series with cotree edges. Its advantage (?) is that it permits the analysis to be carried out using only the fundamental circuit and cocircuit matrices. In this case, the analysis formulas are

$$J^t = S_f Y S'_f E^t \qquad (4\text{-}136)$$

and

$$E^c = B_f \hat{Z} B'_f J^c \qquad (4\text{-}137)$$

which are obtained by combining the equations characterizing the composite immittance matrices with those characterizing the Kirchoff laws under external excitation [Proposition 4-1 and Equations (4-8) and (4-9)].

If the analysis formula of Equation (4-136) is applied to the network of Figure 4-10, a canonical 2-port observed in parallel with the tree edges (1 and 2) is realized as

$$Y_n = S_f(t,q) \circ \hat{Y}(t,q) \circ S'_f(t,q)$$
$$= \begin{bmatrix} -e^t \delta(t-q) & -\delta(t-q) \\ -\delta(t-q) & -e^{-q}\hat{\delta}(t-q) \end{bmatrix} \qquad (4\text{-}138)$$

4-7 Cascade Loading

In the preceding sections, the analysis procedures were such as to allow arbitrarily connected components. Here we consider a more restrictive configuration, cascade loading, as illustrated in Figure 4-11, but one which arises naturally

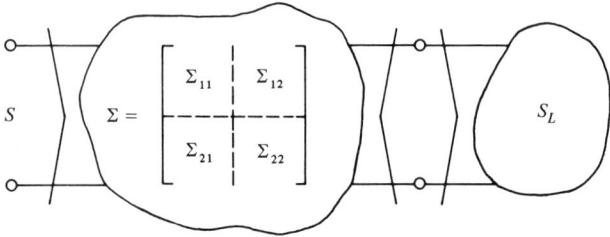

Figure 4-11 Cascade loading configuration.

in many network theoretic contexts [18]. In fact, the cascade loading configuration is predominant among the topologies found in the following chapters. In dealing with this configuration, the $(n + m)$-port, Σ, is usually termed the coupling network, whereas the m-port, S_L, is the load. Of course, one is interested primarily in the n-port realized at the "first" n ports of the coupling network.

Our main result on cascade loading is the following proposition.

Proposition 4-6

Let a scattering matrix Σ be partitioned as

$$\Sigma = \begin{bmatrix} \Sigma_{11} & \Sigma_{12} \\ \Sigma_{21} & \Sigma_{22} \end{bmatrix} \tag{4-139}$$

where the "2" ports are loaded with an m-port whose scattering matrix is S_L. Then the scattering matrix realized at the "1" ports of Σ is

$$S = \Sigma_{11} + \Sigma_{12}(1 - S_L\Sigma_{22})^{-1}S_L\Sigma_{21}$$

Proof

The coupling network Σ is characterized by the equation

$$\begin{bmatrix} b_1 \\ b_2 \end{bmatrix} = \begin{bmatrix} \Sigma_{11} & \Sigma_{12} \\ \Sigma_{21} & \Sigma_{22} \end{bmatrix} \begin{bmatrix} a_1 \\ a_2 \end{bmatrix} \tag{4-140}$$

while the load satisfies

$$b_L = S_L a_L \tag{4-141}$$

Now, in addition, the port-to-port connection between the "2" ports of Σ and the load yields the connection constraints

$$a_2 = b_L \tag{4-142}$$

and

$$a_L = b_2 \tag{4-143}$$

Combining these four equations now results in

$$a_2 = S_L b_2 \tag{4-144}$$

and the equations

$$b_1 = \Sigma_{11}a_1 + \Sigma_{12}a_2 \tag{4-145}$$

and

$$\begin{aligned} a_2 &= S_L b_2 \\ &= S_L\Sigma_{21}a_1 + S_L\Sigma_{22}a_2 \end{aligned} \tag{4-146}$$

Solving Equation (4-146) for a_2 and substituting into Equation (4-145) yields

$$a_2 = (1 - S_L\Sigma_{22})^{-1}S_L\Sigma_{21}a_1 \quad (4\text{-}147)$$

and

$$b_1 = [\Sigma_{11} + \Sigma_{12}(1 - S_L\Sigma_{22})^{-1}S_L\Sigma_{21}]a_1$$
$$= Sa_1 \quad (4\text{-}148)$$

which is the desired result. □

From the proof, it should be apparent that the proposition is, in fact, valid for scattering variables with arbitrary normalization, though in this case, it is necessary to assure that the "2" ports of Σ and the load be identically normalized. The algebraic manipulations are, however, normalization independent; hence, the proof is valid under this restriction.

The cascade loading formula of Proposition 4-6 can alternatively be written as

$$S = \Sigma_{11} + \Sigma_{12}S_L(1 - \Sigma_{22}S_L)^{-1}\Sigma_{21} \quad (4\text{-}149)$$

In verifying this, one requires the operator identity of the following lemma [164].

Lemma 4-6

For any operators X and Y,

$$(1 + XY)^{-1} = 1 - X(1 + YX)^{-1}Y$$

where both inverses exist if one does.

Proof

The equality will be verified by showing that $(1 + XY)$ is the inverse of $1 - X(1 + YX)^{-1}Y$, this by showing that their product is one. Now,

$$[1 - X(1 + YX)^{-1}Y][(1 + XY)]$$
$$= 1 - X(1 + YX)^{-1}Y + XY - X(1 + YX)^{-1}YXY$$
$$= 1 + X[1 - (1 + YX)^{-1} - (1 + YX)^{-1}YX]Y \quad (4\text{-}150)$$
$$= 1 + X(1 + YX)^{-1}[(1 + YX) - 1 - YX]Y$$
$$= 1 + X[0]Y = 1 + 0 = 1 \quad □$$

Proposition 4-7

Under the hypotheses of Proposition 4-6,

$$S = \Sigma_{11} + \Sigma_{12}S_L(1 - \Sigma_{22}S_L)^{-1}\Sigma_{21}$$

Proof

Clearly, it suffices to show that

$$S_L(1 - \Sigma_{22}S_L)^{-1} = (1 - S_L\Sigma_{22})^{-1}S_L \quad (4\text{-}151)$$

CASCADE LOADING

Applying the lemma with

$$X = -\Sigma_{22} \tag{4-152}$$

and

$$Y = S_L \tag{4-153}$$

we have

$$\begin{aligned}
S_L(1 - \Sigma_{22}S_L)^{-1} &= S_L + S_L\Sigma_{22}(1 - S_L\Sigma_{22})^{-1}S_L \\
&= (1 + S_L\Sigma_{22}(1 - S_L\Sigma_{22})^{-1})S_L \\
&= ((1 - S_L\Sigma_{22})S_L\Sigma_{22})(1 - S_L\Sigma_{22})^{-1}S_L \\
&= (1 - S_1\Sigma_{22})^{-1}S_L
\end{aligned} \tag{4-154}$$

which was to be shown. □

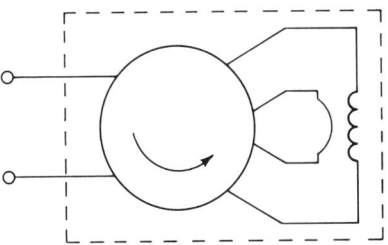

Figure 4-12 Circulator loaded by a transmission line.

Consider the circulator loaded by a transmission line (of unit length and characteristic impedance) illustrated in Figure 4-12. Applying the cascade loading formula of Proposition 4-6, we have

$$S_L = \begin{bmatrix} 0 & e^{-p} \\ e^{-p} & 0 \end{bmatrix} \tag{4-155}$$

and

$$\Sigma = \begin{bmatrix} \Sigma_{11} & \Sigma_{12} \\ \hline \Sigma_{21} & \Sigma_{22} \end{bmatrix} = \begin{bmatrix} 0 & 0 & 1 \\ \hline 1 & 0 & 0 \\ 0 & 1 & 0 \end{bmatrix} \tag{4-156}$$

and

$$\begin{aligned}
S &= \Sigma_{11} + \Sigma_{12}(1 - S_L\Sigma_{22})^{-1}\Sigma_{21} \\
&= 0 + [0\ 1]\left(\begin{bmatrix} 1 & 0 \\ 0 & 1 \end{bmatrix} - \begin{bmatrix} 0 & e^{-p} \\ e^{-p} & 0 \end{bmatrix}\begin{bmatrix} 0 & 0 \\ 1 & 0 \end{bmatrix}\right)^{-1}\begin{bmatrix} 0 & e^{-p} \\ e^{-p} & 0 \end{bmatrix}\begin{bmatrix} 1 \\ 0 \end{bmatrix} \\
&= [0\ 1]\begin{bmatrix} (1 - e^{-p})^{-1} & 0 \\ 0 & 1 \end{bmatrix}\begin{bmatrix} 0 & e^{-p} \\ e^{-p} & 0 \end{bmatrix}\begin{bmatrix} 1 \\ 0 \end{bmatrix} \\
&= [0\ 1]\begin{bmatrix} 0 \\ e^{-p} \end{bmatrix} = e^{-p}
\end{aligned} \tag{4-157}$$

for the scattering parameter seen at the "input" port of the coupling network.

One advantage of the cascade loading analysis formulation is that the scattering matrix of the n-port realized by the network may be calculated directly without the intermediary of the canonical n-port, which usually has more ports than are actually needed. In fact, one application of the cascade loading configuration is to obtain a formula for "killing" unwanted ports of the canonical n-port by opening or shorting them. Consider the scattering matrix Σ for the canonical n-port

$$\Sigma = \begin{bmatrix} \Sigma_{11} & \Sigma_{12} & \Sigma_{13} \\ \Sigma_{21} & \Sigma_{22} & \Sigma_{23} \\ \Sigma_{31} & \Sigma_{32} & \Sigma_{33} \end{bmatrix} \tag{4-158}$$

where the "1" ports are the desired ports for the n-port being realized, the "2" ports are the undesired ports in series with the cotree edges, and the "3" ports are the undesired ports in parallel with tree edges. To kill the "2" ports, they must be shorted or equivalently loaded with short circuits. Similarly, the "3" ports must be loaded with open circuits. Applying the cascade loading formula with the partitioned load

$$S_L = \begin{bmatrix} -1 & 0 \\ 0 & 1 \end{bmatrix} \tag{4-159}$$

corresponding to the "2" and "3" undesired ports, and with Σ as in Equation (4-158) (partitioned between the "1" and "2" rows and columns), the n-port realized at the "1" ports is found to be

$$S = \Sigma_{11} + [\Sigma_{12} \ \Sigma_{13}] \begin{bmatrix} 1 + \Sigma_{22} & \Sigma_{23} \\ -\Sigma_{32} & 1 - \Sigma_{33} \end{bmatrix}^{-1} \begin{bmatrix} -\Sigma_{21} \\ \Sigma_{31} \end{bmatrix} \tag{4-160}$$

In classical electric network theory, the rational character of the RLC system function plays a significant role [183]. In fact, this rationality is not restricted to the RLC case; rather, rationality (in some sense) is a characteristic

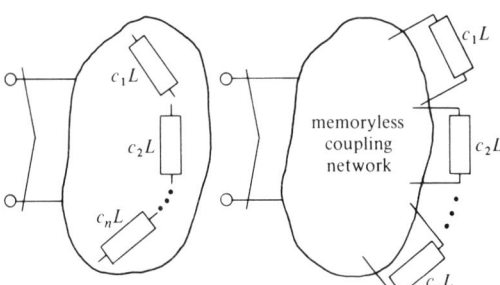

Figure 4-13 Memoryless network with extracted components.

associated with finiteness, but independent of the character of any particular class of networks. Consider, for example, a class of networks composed of time-invariant memoryless components and a finite number of 1-ports with scattering matrices of the form

$$S_i = c_i L \tag{4-161}$$

where L is some specified (but arbitrary) operator and the c_i are real constants. Now, if one extracts the 1-ports, S_i, from such a network as a load on a time-invariant coupling network (as illustrated in Figure 4-13), then if the coupling network has a scattering matrix (this may be assured by the existence theorems of Chapter 3 if certain normality and passivity conditions are required for the memoryless network components), the cascade loading formula may be invoked to exhibit the scattering matrix for the network, this being rational in L.

Of course, if one applies this result to an RLC network,

$$L = \frac{p-1}{p+1} \tag{4-162}$$

corresponds to an inductor and

$$-L = \frac{1-p}{1+p} \tag{4-163}$$

to a capacitor. Now, the system function is rational in L, which is itself rational in p; hence, the system function for an RLC network is rational in the complex variable p. Note that one need not consider reactors (that is, capacitors and inductors) of other than unit value, since a general reactor may be obtained from the unit reactor through the artifice of an ideal transformer which may be taken as part of the memoryless network.

A generalization of the above development which allows several types of elementary 1-port to be included within the network is as follows.

*Proposition 4-8

Let a network be composed of time-invariant memoryless components and 1-ports with scattering matrices of the form

$$S_i = c_i L_i \qquad i = 1, 2, \cdots, m$$

where the c_i are arbitrary real constants, and the L_i are arbitrary pairwise commutative operators. That is,

$$L_i L_j = L_j L_i$$

for any i and j. Then the scattering matrix of the network, if it exists, is rational in the operators

$$L_i \qquad i = 1, 2, \cdots, m \qquad \square$$

The proof for the proposition follows readily from the cascade loading formula after an appropriate extraction of components. The details are, however, left to the reader as an exercise. We should note that in the above proposition it is quite possible (and, in fact, to be expected) for several of the L_i to coincide, that is,

$$L_i = L_j \qquad i \neq j \qquad (4\text{-}164)$$

This causes no difficulty since every operator commutes with itself and, in fact, if one takes all of the L_i to be identical, the proposition reduces to the special case considered previously.

The immittance form of the cascade loading formula is readily obtained as per Proposition 4-9. Of course, its usefulness is dependent on the existence of an immittance representation for the coupling network, which is not, in general, assured.

*Proposition 4-9

Let an $(n+m)$-port coupling network with impedance matrix

$$X = \begin{bmatrix} X_{11} & X_{12} \\ \hline X_{21} & X_{22} \end{bmatrix}$$

be loaded at its "2" ports with an m-port characterized by the admittance matrix

$$Y_L$$

Then the n-port realized at the "1" ports of X is characterized by the impedance matrix

$$Z = X_{11} - X_{12}Y_L(1 + X_{22}Y_L)^{-1}X_{21} \qquad \square$$

Of course, as with the scattering formula, the alternate form of Z is

$$Z = X_{11} - X_{12}(1 + Y_L X_{22})^{-1} Y_L X_{21} \qquad (4\text{-}165)$$

which may be verified by invoking Lemma 4-6 and an argument similar to that of Proposition 4-7. The reader is invited to fill in the details.

If the roles of impedance and admittance in Proposition 4-9 are interchanged, one may obtain an admittance matrix

$$Y = U_{11} - U_{12}Z_L(1 + U_{22}Z_L)^{-1}U_{21}$$
$$= U_{11} - U_{12}(1 + Z_L U_{22})^{-1} Z_L U_{21} \qquad (4\text{-}166)$$

for the n-port realized by the cascade loading configuration when the coupling network is characterized by the admittance matrix

$$U = \begin{bmatrix} U_{11} & U_{12} \\ \hline U_{21} & U_{22} \end{bmatrix} \qquad (4\text{-}167)$$

and the load by the impedance matrix

$$Z_L \qquad (4\text{-}168)$$

The details are, however, left to the reader as an exercise.

Of course, as with the scattering formulation, the cascade loading formula can be employed to "kill" undesired ports in the immittance matrices obtained by the topological analysis of Section 4-6. In this case, however, the short circuit has an impedance but no admittance matrix, while the open circuit has an admittance but no impedance matrix. Hence, the formula of Proposition 4-9 can be used only to "kill" parallel ports (by open circuiting them), whereas that of Equation (4-166) is applicable only to the problem of shorting undesired series ports.

4-8 Analysis of Elementary n-Ports

In the network synthesis techniques to be considered in the sequel, one is interested in decomposing a given network matrix into a network of elementary n-ports composed of memoryless devices and elementary components (such as capacitors, inductors, transmission lines, and so forth). In this section we, therefore, consider the characteristics of such n-ports. Consider first the n-port illustrated in Figure 4-14, which is composed of an ideal transformer loaded in an

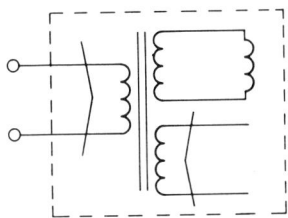

Figure 4-14 Elementary n-port.

inductor and $n - 1$ open circuits. Now, if one assumes that the ideal transformer has an orthogonal time-invariant turns ratio matrix T, then the cascade loading formula may be applied with coupling network

$$\Sigma(p) = \begin{bmatrix} 0 & T' \\ \hline T & 0 \end{bmatrix} \tag{4-169}$$

and load

$$S_L(p) = \begin{bmatrix} \dfrac{p-k}{p+k} & & & & \boxed{0} \\ & 1 & & & \\ & & 1 & & \\ & & & \ddots & \\ & \boxed{0} & & & 1 \end{bmatrix}$$

$$= \begin{bmatrix} 1 & & & & \boxed{0} \\ & 1 & & & \\ & & 1 & & \\ & & & \ddots & \\ & \boxed{0} & & & 1 \end{bmatrix} + \begin{bmatrix} \dfrac{-2k}{p+1} & & & & \boxed{0} \\ & 0 & & & \\ & & 0 & & \\ & & & \ddots & \\ & \boxed{0} & & & 0 \end{bmatrix} \tag{4-170}$$

Here, the second equality in Equation (4-170) results from the fact that

$$\frac{p-k}{p+k} = 1 - \frac{2k}{p+k} \qquad (4\text{-}171)$$

The cascade loading formula now yields the n-port scattering matrix

$$S = \Sigma_{11} + \Sigma_{12}S_L(1 - \Sigma_{22}S_L)^{-1}\Sigma_{21} = 0 + \Sigma_{12}S_L(1 - 0S_L)^{-1}\Sigma_{21}$$
$$= \Sigma_{12}S_L\Sigma_{21} \qquad (4\text{-}172)$$

$$= T' \begin{bmatrix} 1 & & & & \lceil 0 \\ & 1 & & & \\ & & 1 & & \\ & & & \cdot & \\ \overline{0\rceil} & & & & 1 \end{bmatrix} + \frac{-2k}{p+1} \begin{bmatrix} 1 & & & & \lceil 0 \\ & 0 & & & \\ & & 0 & & \\ & & & \cdot & \\ \overline{0\rceil} & & & & 0 \end{bmatrix} T$$

$$= (T'T) - \frac{2k}{p+k}(T'\Lambda\, T)$$

where

$$\Lambda = \begin{bmatrix} 1 & & & & \lceil 0 \\ & 0 & & & \\ & & \cdot & & \\ \overline{0\rceil} & & & & 0 \end{bmatrix} \qquad (4\text{-}173)$$

Now, since T is orthogonal,

$$T'T = 1 \qquad (4\text{-}174)$$

while we may define a vector X of norm 1 (that is, $X'X = 1$) as the first column of T',

$$X = [T_{11}\ T_{12} \cdots T_{1n}]' \qquad (4\text{-}175)$$

which yields the equality

$$T'\Lambda T = XX' \qquad (4\text{-}176)$$

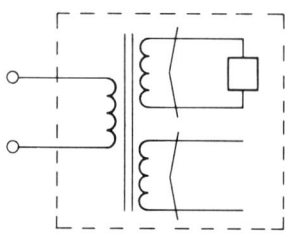

Figure 4-15 Elementary n-port loaded in m reactors.

Upon substituting Equations (4-174) and (4-176) into Equation (4-172), the scattering matrix for the n-port of Figure 4-14 is found to be

$$S = 1 - \frac{2kXX'}{p+k} \qquad (4\text{-}177)$$

Of course, a similar argument applies to the case of Figure 4-15, where the transformer is loaded by several inductors and X becomes an $n \times m$ matrix of norm 1 and rank m (where m is the number of inductors). We have thus proven the following proposition.

Proposition 4-10

A $2n$-port ideal transformer with orthogonal time-invariant–turns-ratio-matrix T and loaded with $m \leq n$ $1/k$ H inductors has scattering matrix

$$S = 1 - \frac{2kXX'}{p+k}$$

where X is an $n \times m$ matrix of norm 1 and rank m. □

For the particular application of network synthesis, one is interested in the inverse result of Proposition 4-10. That is, given an n-port scattering matrix

$$S = 1 - \frac{2kXX'}{p+k} \qquad (4\text{-}178)$$

can it be realized as an ideal transformer loaded in m inductors? Indeed, this is the case when X is an n vector of norm 1, for then it is always possible [132] to extend X to an orthogonal matrix such that

$$T' = [X \quad \tilde{X}] \qquad (4\text{-}179)$$

where $\tilde{X}$ is an $n \times n-1$ matrix, which can be used in the synthesis of S. In fact, it may be verified by analysis that if X is an $n \times m$ matrix of norm 1 and rank m, then the scattering matrix of Equation (4-178) may be realized by by the configuration of Figure 4-15 with m inductors. Here the turns ratio matrix for the transformer is the orthogonal extension of X with $\tilde{X}$ an $n \times n\text{-}m$ matrix,

$$T' = [X \quad \tilde{X}] \qquad (4\text{-}180)$$

We thus have the following proposition.

*Proposition 4-11

Proposition 4-10 is both necessary and sufficient. □

Besides the time-invariant case, one is also interested in the case where component time variations are allowed. Let us, therefore, consider the configuration of Figure 4-14 with possibly time-variable orthogonal turns ratio matrix $T(t)$, loaded by a (time-invariant) unit inductor. The coupling network is, therefore, characterized by

$$\Sigma(t,q) = \left[\begin{array}{c|c} 0 & T'(t)\delta(t-q) \\ \hline T(t)\delta(t-q) & 0 \end{array}\right] \quad \text{(4-181)}$$

while the load is of the form

$$S_L(t,q) = \left[\begin{array}{cccc} \delta(t-q) & -2e^{-(t-q)}U(t-q) & & \\ & \delta(t-q) & & \\ & & \delta(t-q) & \boxed{0} \\ & \boxed{0} & & \\ & & & \delta(t-q) \end{array}\right]$$

$$= \delta(t-q) \quad -2\Lambda e^{-(t-q)}U(t-q) \quad \text{(4-182)}$$

where Λ is as in Equation (4-173). Now, the cascade loading formula yields

$$S(t,q) = T'(t)\delta(t-q) \circ S_L(t,q) \circ T(t)\delta(t-q)$$
$$= \delta(t-q) - 2e^{-t}T'(t)\Lambda T(q)e^q U(t-q) \quad \text{(4-183)}$$

Defining $\phi(t)$ and $\phi'(q)$ as

$$\phi(t) = T'(t)\Lambda \quad \text{(4-184)}$$

and

$$\phi'(q) = \Lambda T(q) \quad \text{(4-185)}$$

this scattering matrix for the n-port becomes

$$S(t,q) = \delta(t-q) - 2\phi(t)\phi'(q)e^{(t-q)}U(t-q) \quad \text{(4-186)}$$

where $\phi(t)$ is of norm 1. As before, one can extend the result to the case where the transformer is loaded by several inductors or capacitors and, in fact, the result is necessary and sufficient.

*Proposition 4-12

A $2n$-port ideal transformer with orthogonal turns ratio matrix $T(t)$, loaded with $m \leq n$ inductors (capacitors) has the scattering matrix

$$S(t,q) = \delta(t-q) - 2\phi(t)\phi'(q)e^{-(t-q)}U(t-q)$$

with inductor load and

$$S(t,q) = -\delta(t-q) + 2\phi(t)\phi'(q)e^{-(t-q)}U(t-q)$$

with capacitor load where $\phi(t)$ is an $n \times m$ matrix of time functions which is of norm 1 and rank m for all t. Conversely, any such scattering matrix can be realized by such a configuration. □

It would be desirable to apply a similar class of analyses to elementary n-ports characterized by immittance matrices. Unfortunately, the ideal transformer has neither admittance nor impedance matrices; hence, the cascade loading formula is not applicable. In Chapter 2 we did, however, obtain the formula

$$Y(t,q) = T'(t)\delta(t-q) \text{ o } Y_L(t,q) \text{ o } T(t)\delta(t-q) \qquad \textbf{(4-187)}$$

for an arbitrary $n \times m$ ideal transformer with turns ratio matrix $T(t)$, loaded with an admittance, and a similar formula in the impedance case. Applying

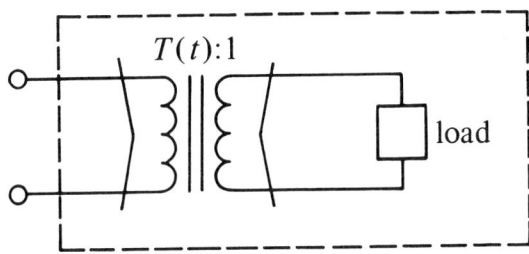

Figure 4-16 Loaded ideal transformer.

this to the network of Figure 4-16, where an ideal transformer is loaded in unit inductors, we have

$$\begin{aligned} Z(t,q) &= T'(t)\delta(t-q) \text{ o } \dot{\delta}(t-q) \text{ o } T(t)\delta(t-q) \\ &= T'(t)T(q)\dot{\delta}(t-q) \end{aligned} \qquad \textbf{(4-188)}$$

while, if the load is replaced by unit capacitors, this becomes

$$Z(t,q) = T'(t)T(q)U(t-q) \qquad \textbf{(4-189)}$$

Clearly, the converse also holds; hence, any impedance matrix of the form shown in Equations (4-188) and (4-189) is realizable by the configuration of Figure 4-16.

Of course, similar formulas apply to the admittance case and also to the time invariant case. The reader may verify the following.

*Proposition 4-13

A necessary and sufficient condition for a network to have the form of Figure 4-16 is that its impedance matrix be

$$Z(t,q) = T'(t)T(q)\dot{\delta}(t-q)$$

or

$$Z(t,q) = T'(t)T(q)U(t-q)$$

with inductor and capacitor load, respectively, or its admittance matrix be

$$Y(t,q) = T'(t)T(q)U(t - q)$$

or

$$Y(t,q) = T'(t)T(q)\dot{\delta}(t - q)$$

in the two cases. □

*Proposition 4-14

Under the hypotheses of Proposition 4-13, if the transformers are time-invariant, the frequency domain impedance matrices are

$$Z(p) = T'Tp$$

and

$$Z(p) = \frac{T'T}{p}$$

for inductor and capacitor loading, respectively, while the corresponding admittance matrices are

$$Y(p) = \frac{T'T}{p}$$

and

$$Y(p) = T'Tp$$ □

Although inductor and capacitor loads have been used almost exclusively in the preceding development, similar results could be obtained with essentially any loads. For example, if in the scattering development the load were simply characterized by a scattering matrix

$$S_1(t,q)$$

then the overall scattering matrix is

$$S(t,q) = \delta(t - q) - 2\phi(t)\phi'(q)[S_1(t,q) - \delta(t - q)] \qquad \textbf{(4-190)}$$

whereas in the immittance cases it is

$$Z(t,q) = T'(t)T(q)Z_1(t,q) \qquad \textbf{(4-191)}$$

or

$$Y(t,q) = T'(t)T(q)Y_1(t,q) \qquad \textbf{(4-192)}$$

Of course, similar formulas hold in the frequency domain for time-invariant networks. These the reader is invited to derive as an exercise.

4-9 Discussion

In the preceding, techniques have been formulated by which the component and connection constraints for a network are combined to yield a description of the n-port realized by the network. The main advantage of the topological approach to the problem is that the results are essentially algebraic, holding in both the time and frequency domain for any network whose components are characterized by the appropriate matrices. No assumption of component characteristics, form of the connection, or reduction of multiports to equivalent 1-ports is needed, or assumed. This is achieved via algebraic manipulation of the network and graph matrices as abstract linear operators. Since such manipulation is completely independent of the characteristics of these matrices, the process allows analysis of all linear networks by a single set of techniques.

In the final sections we considered the cascade loading formula and its application to the analysis of some elementary n-ports. Though this configuration does not possess the generality of the preceding techniques, the simplicity of the formula, together with the common occurrence of the cascade configuration, justifies the special attention given thereto.

CHAPTER

5
Realizability Theory

5-1 Introduction

In the preceding chapter, the topological analysis problem was approached from an essentially algebraic viewpoint, thereby allowing the results obtained to be applied to all linear networks (after making appropriate notational adjustments). In this way, a single unified theory for networks characterized in either the time or frequency domain is obtained. In the case of network synthesis, an equally general theory has yet to be obtained; however, it is possible to apply algebraic techniques to reduce the general synthesis problem to a (hopefully simpler) special case. It is such techniques which are considered here [149], [152].

The essence of the approach is to reduce the complexity of an n-port at the cost of increasing the number of its ports. This approach, illustrated in Figure 5-1, permits the realization of an n-port at the first n ports of an $(n + m)$-port coupling network appropriately loaded at the remaining m ports. Of course, for the technique to be of value, the $(n + m)$-port must be (in some sense) less complex than the original n-port. This technique, of synthesis by cascade loading, appears with numerous variations in this chapter and repeatedly in the remainder of the book. It is used to realize an active network as a passive network loaded in negative resistors, a passive device as a lossless network loaded in positive resistors, an RLC network as a memoryless network loaded in capacitors, and, finally, the technique is used to realize an n-variable network as an $(n - 1)$-variable network loaded in a 1-variable network.

Here we consider those loading techniques which hold for any linear network and can be implemented by wholly algebraic means. These include the reduction of a π and/or active scattering matrix to a passive network possessing a

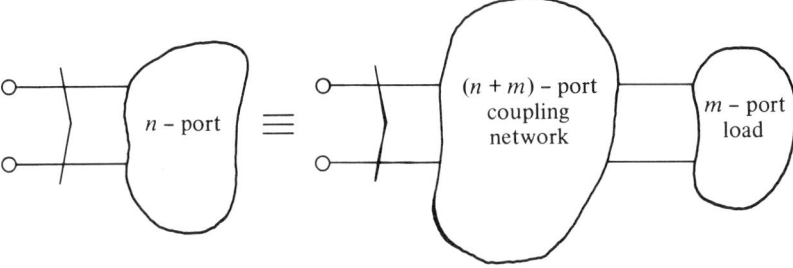

Figure 5-1 Synthesis by cascade loading.

scattering matrix and loaded in negative resistors, in the first section of the chapter. In the next three sections, the problem of reducing the realization of a passive n-port to that of a resistor-loaded lossless network is considered. In this latter case, a complete solution of the problem is not obtained. We do, however, obtain a partial solution which serves as a starting point for the passive synthesis procedures of later chapters. The final section is concerned with the memoryless synthesis problem, this being the one class of networks for which a complete synthesis is possible via the algebraic techniques of this chapter. Of course, the realization of the various special classes of network considered in the following chapters can often be reduced to a memoryless synthesis problem (by appropriate cascade loading techniques); hence, such results prove to be of considerable importance, even though memoryless networks themselves are of little interest.

The methods of this chapter are essentially algebraic, as were those of the previous chapter; hence, the operator notation is used exclusively. This allows the translation of the results to both the time and frequency domains, upon making the appropriate notational changes.

5-2 Realization of Active π and Scattering Matrices

In this section the problem of reducing the synthesis of a π matrix to that of a passive network characterized by a scattering matrix is considered. This is achieved via the configuration of Figure 5-2, where a $2n$-port loaded in negative resistors is used to realize an arbitrarily prescribed π matrix at the remaining ports. Upon identifying an active scattering matrix

$$b = Sa \qquad (5\text{-}1)$$

with the π matrix

$$0 = [\pi_a \mid \pi_b]\begin{bmatrix}a\\b\end{bmatrix} = [S \mid -1]\begin{bmatrix}a\\b\end{bmatrix} \qquad (5\text{-}2)$$

174 REALIZABILITY THEORY

the technique also allows the realization of such a matrix to be reduced to that of a passive scattering matrix.

As a first step in the procedure, zero rows or columns are added to the given π matrix to make its dimensions $k \times 2k$ where

$$k = \max\{m,n\} \tag{5-3}$$

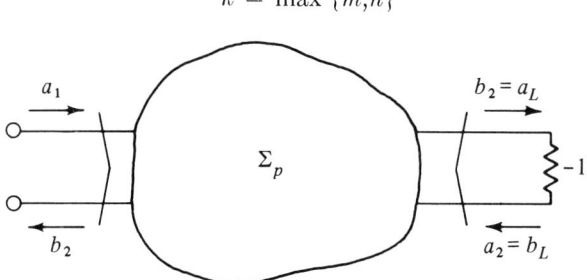

Figure 5-2 Realization of a π matrix by negative resistor loading.

is the maximum of the number of ports and the number of rows in π. Of course, for normal networks we may assume that

$$k = m = n \tag{5-4}$$

Hence, the step is not needed.

Lemma 5-1
Let

$$\pi = [\pi_a \mid \pi_b]$$

be an $m \times 2n$ π matrix of an n-port. Then

$$\tilde{\pi} = \begin{bmatrix} \pi_a & 0 & \pi_b & 0 \\ 0 & 0 & 0 & 0 \end{bmatrix} \begin{matrix} m \\ k-m \end{matrix}$$
$$\phantom{\tilde{\pi} = }\;\; n \;\; k-n \;\; n \;\; k-n$$

is the π matrix for a k-port,

$$k = \max\{m,n\}$$

which is identical to the n-port characterized by π at its "first" n ports, and has unconstrained variables at the remaining $k - n$ ports (if any). That is, a realization of $\tilde{\pi}$ realizes π at its first n ports.

Proof
Letting the variables measured at the first n ports of $\tilde{\pi}$ be denoted by a and b, and the variables measured at the extra $k - n$ ports (if any) be denoted by $\tilde{a}$ and $\tilde{b}$, the admissible pairs defined by $\tilde{\pi}$ are characterized by

$$0 = \begin{bmatrix} \pi_a & 0 & \pi_b & 0 \\ 0 & 0 & 0 & 0 \end{bmatrix} \begin{bmatrix} a \\ \tilde{a} \\ b \\ \tilde{b} \end{bmatrix} \tag{5-5}$$

Now, the bottom equation does not constrain the variables, whereas the top row is satisfied if and only if

$$0 = \pi_a a + 0\,\tilde{a} + \pi_b b + 0\,\tilde{b} = \pi_a a + \pi_b b = \pi \begin{bmatrix} a \\ b \end{bmatrix} \quad (5\text{-}6)$$

Hence, the variables at the first n ports are exactly those determined by π, whereas the extra variables are unconstrained. The first n ports of $\tilde{\pi}$ thus realize π. $\square$

Consider the over-normal 2-port characterized by the π matrix

$$\pi(t,q) = [\delta(t - 1 - q)\ \ e^t\dot{\delta}(t - q)\ |\ U(t - q)\ \ \delta(t - q)] \quad (5\text{-}7)$$

Here

$$n = 2 \quad (5\text{-}8)$$

and

$$m = 1 \quad (5\text{-}9)$$

Hence,

$$k = \max\{2,1\} = 2 \quad (5\text{-}10)$$

and

$$\tilde{\pi}(t,q) = \begin{bmatrix} \delta(t - 1 - q) & e^t\dot{\delta}(t - q) & | & U(t - q) & \delta(t - q) \\ \hline 0 & 0 & | & 0 & 0 \end{bmatrix} \quad (5\text{-}11)$$

Similarly, the under-normal network characterized by the π matrix

$$\pi(p) = \begin{bmatrix} 0 & | & \cosh^{-1}(p) \\ \dfrac{p^2}{1 + p^2} & | & e^{-p} \end{bmatrix} \quad (5\text{-}12)$$

which also has $k = 2$ corresponds to the $\tilde{\pi}$ matrix

$$\pi(p) = \begin{bmatrix} 0 & | & 0 & | & \cosh^{-1}(p) & | & 0 \\ \hline \dfrac{p^2}{1 + p^2} & | & 0 & | & e^{-p} & | & 0 \end{bmatrix} \quad (5\text{-}13)$$

obtained from π by adding two columns of zeros. It should be observed that the zero rows and zero columns never occur simultaneously in $\tilde{\pi}$, since either $k - n$ or $k - m$ is zero.

The utility of the lemma is that it allows us to consider only π matrices for which π_a and π_b are square. If this is not the case, we may, without loss of generality, consider $\tilde{\pi}$. Moreover, it can be assumed that π (or $\tilde{\pi}$) has finite norm and is causal, since if this were not the case, it could be induced by multiplying both π_a and π_b by an appropriate factor.

With these assumptions (restrictions?), the passive coupling network needed for the realization of Figure 5-2 may be obtained.

Theorem 5-1

Let a causal k-port be characterized by a $\tilde{\pi}$ matrix (as obtained from a π matrix via Lemma 5-1) with finite norm. Then a realization of the passive $2k$-port, with scattering matrix

$$\Sigma_p = \begin{bmatrix} 0 & | & \dfrac{1}{c} \\ -- & + & -- \\ \dfrac{\tilde{\pi}_a}{c} & | & \dfrac{\tilde{\pi}_b}{c^2} \end{bmatrix}$$

loaded at its last k ports with -1-ohm resistors, realizes $\tilde{\pi}$ at its first k ports (hence, π at its first n ports if n is less than k). Here c is a (yet to be determined) real constant.

Proof

It is first required to show that

$$\Sigma_p = \begin{bmatrix} 0 & | & \dfrac{1}{c} \\ -- & + & -- \\ \dfrac{\tilde{\pi}_a}{c} & | & \dfrac{\tilde{\pi}_b}{c^2} \end{bmatrix} \tag{5-14}$$

is passive for some c. By definition, S exists, whereas the causality of $\tilde{\pi}$ implies that

$$\Sigma_p(t,q) = 0 \quad \text{for } t < q \tag{5-15}$$

in the time domain, and that

$$\Sigma_p(p) \tag{5-16}$$

is analytic for RHP p in the frequency domain. Now, since $\tilde{\pi}$ has finite norm, it is always possible to choose a sufficiently large (but finite) c such that

$$\|\Sigma_p\| \leq 1 \tag{5-17}$$

Σ_p is thus passive (for an appropriate c) in both the time and frequency domain representations.

To show that $\tilde{\pi}$ is indeed realized at the first k ports of Σ_p when the last k ports are loaded in negative resistors, we must show that

$$0 = \tilde{\pi} \begin{bmatrix} a_1 \\ b_1 \end{bmatrix} = \tilde{\pi}_a a_1 + \tilde{\pi}_b b_1 \tag{5-18}$$

when the "2" variables are constrained by the negative resistor loading. In this case, we have

$$b_2 = a_L = v_L + i_L = (-1)i_L + i_L = 0 \tag{5-19}$$

while

$$a_2 = b_L = v_L - i_L = (-1)i_L + i_L = -2i_L \tag{5-20}$$

is arbitrary. Substituting Equations (5-19) and (5-20) into the equations defined by Σ_p yields

$$b_1 = (0)a_1 + \left(\frac{1}{c}\right)a_2 \qquad (5\text{-}21)$$

and

$$b_2 = \left(\frac{\pi_a}{c}\right)a_1 + \left(\frac{\pi_b}{c^2}\right)a_2 = 0 \qquad (5\text{-}22)$$

Hence,

$$b_1 = \frac{a_2}{c} \qquad (5\text{-}23)$$

and

$$0 = \left(\frac{\tilde{\pi}_a}{c}\right)a_1 + \left(\frac{\tilde{\pi}_b}{c^2}\right)a_2 \qquad (5\text{-}24)$$

Now, upon substituting Equation (5-23) into Equation (5-24) and multiplying the result by c, one obtains

$$\begin{aligned}0 &= c\left(\frac{\tilde{\pi}_a}{c}\right)a_1 + c\left(\frac{\tilde{\pi}_b}{c^2}\right)(cb_1) \\ &= \tilde{\pi}_a a_1 + \tilde{\pi}_b b_1 = \tilde{\pi}\begin{bmatrix}a_1\\b_1\end{bmatrix}\end{aligned} \qquad (5\text{-}25)$$

which completes the proof. □

The theorem has two immediate corollaries.

*Corollary 5-1

Under the hypothesis of Theorem 5-1, if c is chosen sufficiently large, then the network defined by Σ_p has admittance and impedance matrices.

*Corollary 5-2

Let a linear, nonenergetic n-port be characterized by a scattering matrix S. Then a realization of the passive $2n$-port, with scattering matrix

$$\Sigma_p = \left[\begin{array}{c|c} 0 & \frac{1}{c} \\ \hline \frac{S}{c} & \frac{-1}{c^2} \end{array}\right]$$

loaded at the last n ports with -1-ohm resistors, realizes S at its first n ports. □

Theorem 5-1 can be applied to the problem of realizing a norator with negative resistors and passive (normal) components, as is illustrated in Chapter 2. In that case,

$$\tilde{\pi} = [0 \mid 0] \tag{5-26}$$

Hence,

$$\tilde{\pi}_a = [0] \tag{5-27}$$

and

$$\tilde{\pi}_b = [0] \tag{5-28}$$

Choosing

$$c = 1 \tag{5-29}$$

then yields the passive coupling network with scattering matrix

$$\Sigma_p = \begin{bmatrix} 0 & | & 1 \\ & | & \\ 0 & | & 0 \end{bmatrix} \tag{5-30}$$

The norator may thus be realized as per Figure 5-3, where Σ_p is realized as a 3-port circulator loaded with a 1-ohm resistor. This circulator realization may

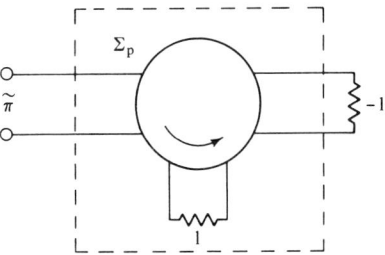

Figure 5-3 Circulator-negative resistor realization of a norator.

be obtained by inspection, or via the passive synthesis technique of the next section.

Recognizing that a 3-port circulator is equivalent to the gyrator configuration of Figure 5-4, an alternate gyrator-resistor realization of the norator is that of Figure 5-5.

Like the π matrix analysis of the previous chapter, the active synthesis procedure is carried out entirely by a process of substitution into an appropriate partitioned matrix; hence, the synthesis is as easily carried out in the time or

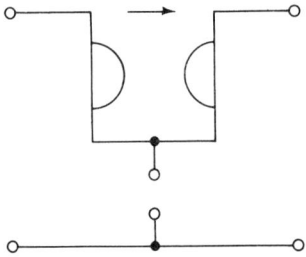

Figure 5-4 Gyrator realization of a circulator.

frequency domain as for the memoryless case just considered. For instance, to realize the π matrix

$$\pi(p) = \begin{bmatrix} 0 & | & \cosh^{-1}(p) \\ \dfrac{p^2}{1+p^2} & | & e^{-p} \end{bmatrix} \qquad (5\text{-}31)$$

[Figure showing gyrator-resistor circuit with resistors labeled 1 and -1]

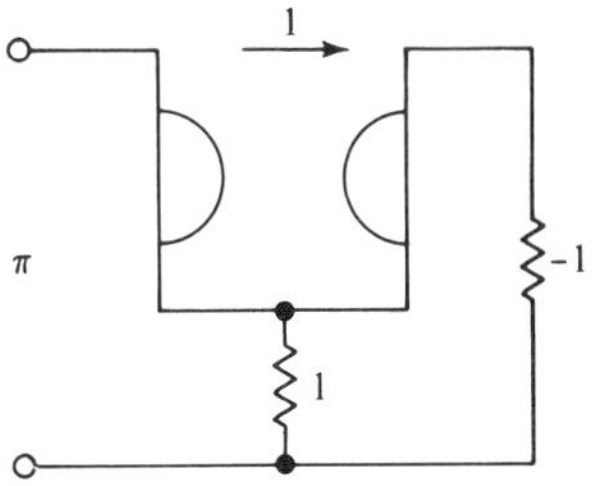

Figure 5-5 Gyrator-resistor realization of a norator.

one may apply the synthesis to the $\tilde{\pi}$ matrix derived therefrom [Equation (5-13)], yielding the passive coupling network characterized by the 4-port passive scattering matrix (for $c = 2$),

$$\Sigma_p = \left[\begin{array}{cc|cc} 0 & 0 & \dfrac{1}{2} & 0 \\ 0 & 0 & 0 & \dfrac{1}{2} \\ \hline 0 & 0 & \dfrac{\cosh^{-1}(p)}{4} & 0 \\ \dfrac{p^2}{2(1+p^2)} & 0 & \dfrac{e^{-p}}{4} & 0 \end{array} \right] \qquad (5\text{-}32)$$

Now, Lemma 5-1, together with the synthesis theorem, implies that if the last two ports of a realization of Σ_p are loaded in -1-ohm resistors, then the original π matrix of Equation (5-31) is realized at the first port of Σ_p (while the variables at the second port of Σ_p are unconstrained). This is illustrated in Figure 5-6.

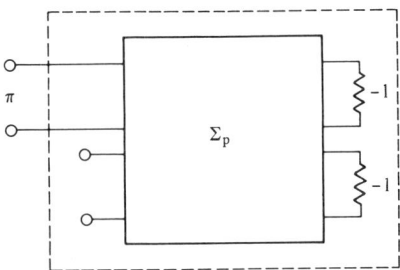

Figure 5-6 Realization of an under-normal π matrix via a passive coupling network loaded in negative resistors.

5-3 Synthesis of Passive Networks by Bordering

In this section we carry the preceding technique one step further by starting with a passive n-port characterized by a scattering matrix (possibly obtained by the method of the preceding section from a π matrix) and realizing it as a lossless (?) coupling network loaded in positive resistors. The loading scheme is illustrated in Figure 5-7. The problem, unfortunately, does not have a com-

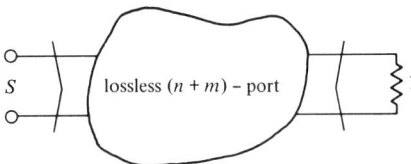

Figure 5-7 Realization of a passive n-port by positive resistor loading.

plete solution in the general case. Rather, it is possible to find a Σ_L which satisfies the existence and adjoint $(1 - S^*S = 0)$ conditions for losslessness, but which does not, in general, satisfy the causality (analyticity in the RHP for the frequency domain) condition required for losslessness. The technique thus obtained does, however, give a lossless coupling network in a number of special cases and, additionally, serves as a starting point (after appropriate modification) for the synthesis of larger classes of passive n-ports. The technique employed is a generalization of the bordering approach of Belevitch [29], [31], [123], [149].

The problem is to find a lossless $(n + m)$-port characterized by a scattering matrix Σ_L, such that when the last m ports are loaded in 1-ohm resistors, a given passive n-port S is seen at the first n ports of Σ_L. The condition on a coupling network, lossless or not, to realize S when loaded in 1-ohm resistors, is given by Proposition 5-1.

Proposition 5-1

Let an $(n + m)$-port have scattering matrix

$$\Sigma_L = \begin{bmatrix} \Sigma_{11} & \Sigma_{12} \\ \hline \Sigma_{21} & \Sigma_{22} \end{bmatrix}$$

Then, when a realization of Σ_L is loaded in 1-ohm resistors at its "2" ports, the scattering matrix observed at the "1" ports is Σ_{11}.

Proof

Applying the cascade loading formula,

$$S = \Sigma_{11} + \Sigma_{12} S_L (1 - \Sigma_{22} S_L)^{-1} \Sigma_{21} \tag{5-33}$$

for the scattering matrix seen at the first n ports under loading by S_L at the last m ports. In our case, the load is 1-ohm resistors; hence,

$$S_L = 0 \tag{5-34}$$

and the n-port realized by Σ_L under such conditions is just Σ_{11}. Thus,

$$S = \Sigma_{11} \tag{5-35}$$

is a necessary and sufficient condition for an $(n + m)$-port

$$\Sigma_L = \begin{bmatrix} \Sigma_{11} & \Sigma_{12} \\ \hline \Sigma_{21} & \Sigma_{22} \end{bmatrix} \tag{5-36}$$

to realize S at its first n ports. □

The proposition implies that if S is to be realized by the configuration of Figure 5-7, then

$$\Sigma_L = \begin{bmatrix} S & \Sigma_{12} \\ \hline \Sigma_{21} & \Sigma_{22} \end{bmatrix} \tag{5-37}$$

where Σ_{12}, Σ_{22}, and Σ_{21} are arbitrary. The synthesis problem is thus reduced to finding "bordering" matrices Σ_{12}, Σ_{21}, and Σ_{22} which render Σ_L lossless. Of course, a necessary condition for this is that S be passive, for, otherwise, an active network could be realized as the interconnection of passive components which would violate conservation of energy.

A partial solution to the problem is given by Proposition 5-2, where we obtain a sufficient condition for Σ_L [having the form of Equation (5-37)] to be an isometry, that is, to satisfy the adjoint condition for losslessness that

$$1 - \Sigma_L^* \Sigma_L = 0 \tag{5-38}$$

Proposition 5-2

Let

$$\Sigma_L = \begin{bmatrix} S & \Sigma_{12} \\ \hline \Sigma_{21} & \Sigma_{22} \end{bmatrix}$$

be the scattering matrix of an $(n + m)$-port for which

$$\Sigma_{21}^* \Sigma_{21} = (1 - S^*S)$$

$$\Sigma_{12} \Sigma_{12}^* = (1 - SS^*)$$

and

$$\Sigma_{22} = -\Sigma_{21} S^* \Sigma_{12}^{*-R}$$

Then

$$1 - \Sigma_L^* \Sigma_L = 0$$

Proof

It must be shown that the four entries in the matrix product

$$\Sigma_L^*\Sigma_L = \begin{bmatrix} S^* & | & \Sigma_{21}^* \\ \hline \Sigma_{12}^* & | & \Sigma_{22}^* \end{bmatrix} \begin{bmatrix} S & | & \Sigma_{12} \\ \hline \Sigma_{21} & | & \Sigma_{22} \end{bmatrix} \tag{5-39}$$

are those of the identity. The "1 − 1" entry is

$$S^*S + \Sigma_{21}^*\Sigma_{21} = S^*S + (1 - S^*S) = 1 \tag{5-40}$$

To show that the "1 − 2" entry is zero, we employ the identity

$$S^*(1 - SS^*) = (1 - S^*S)S^* \tag{5-41}$$

and have

$$\begin{aligned}
S^*\Sigma_{12} + \Sigma_{21}^*\Sigma_{22} &= S^*\Sigma_{12} - \Sigma_{21}^*\Sigma_{21}S^*\Sigma_{12}^{*-R} \\
&= S^*\Sigma_{12} - (1 - S^*S)S_{12}^*\Sigma^{-R} \\
&= S^*\Sigma_{12} - S^*(1 - SS^*)\Sigma_{12}^{*-R} \\
&= S^*\Sigma_{12} - S^*\Sigma_{12}\Sigma_{12}^*\Sigma_{12}^{*-R} \\
&= S^*\Sigma_{12} - S^*\Sigma_{i2} = 0
\end{aligned} \tag{5-42}$$

The "2-1" entry in the matrix is the adjoint of the "1-2" entry (since the product of a matrix and its adjoint is always hermitian); hence, this is also zero. Finally, the "2-2" entry is found to be

$$\begin{aligned}
\Sigma_{12}^*\Sigma_{12} + \Sigma_{22}^*\Sigma_{22} &= \Sigma_{12}^*\Sigma_{12} + (\Sigma_{21}S^*\Sigma_{12}^{*-R})^*(\Sigma_{21}S^*\Sigma_{12}^{*-R}) \\
&= \Sigma_{12}^*\Sigma_{12} + \Sigma_{12}^{-L}S\Sigma_{21}^*\Sigma_{21}S^*\Sigma_{12}^{*-R} \\
&= \Sigma_{12}^*\Sigma_{12} + \Sigma_{12}^{-L}S(1 - S^*S)S^*\Sigma_{12}^{*-R} \\
&= \Sigma_{12}^*\Sigma_{12} + \Sigma_{12}^{-L}SS^*(1 - SS^*)\Sigma_{12}^{*-R} \\
&= \Sigma_{12}^*\Sigma_{12} + \Sigma_{12}^{-L}SS^*\Sigma_{12}\Sigma_{12}^*\Sigma_{12}^{*-R} \\
&= \Sigma_{12}^*\Sigma_{12} + \Sigma_{12}^{-L}SS^*\Sigma_{12} \\
&= \Sigma_{12}^*\Sigma_{12} + \Sigma_{12}^{-L}(1 - \Sigma_{12}\Sigma_{12}^*)_{12} \\
&= \Sigma_{12}^*\Sigma_{12} + \Sigma_{12}^{-L}\Sigma_{12} = \Sigma_{12}^{-L}\Sigma_{12}\Sigma_{12}^*\Sigma_{12} \\
&= \Sigma_{12}^*\Sigma_{12} + 1 - \Sigma_{12}^*\Sigma_{12} = 1
\end{aligned} \tag{5-43}$$

On an entry-by-entry basis, it is thus found that the product

$$\Sigma_L^*\Sigma_L = 1 \tag{5-44}$$

when the hypotheses of the Proposition are satisfied, which completes the proof. □

Although passivity is not used in the proof of the Proposition, it is necessary if the Proposition is to be applied successfully. In particular, if factorizations

$$\Sigma_{21}^* \Sigma_{21} = (1 - S^*S) \tag{5-45}$$

and

$$\Sigma_{12} \Sigma_{12}^* = (1 - SS^*) \tag{5-46}$$

exist, then the resistivity operators

$$(1 - S^*S) \tag{5-47}$$

and

$$(1 - SS^*) \tag{5-48}$$

must be positive (since they are decomposed as the product of an operator and its adjoint, which is always positive). It now follows from the development of Chapter 1 that S must have norm less than or equal to one. This condition for passivity is, therefore, necessary if the hypotheses of the Proposition are to be satisfied. Furthermore, if Σ_L is to be causal, its "1-1" entry S must also be causal; hence, if the Proposition is to be applied successfully to achieve a synthesis, S must satisfy the norm, causality, and (by hypothesis) the existence conditions for passivity, and is, therefore, passive. Of greater interest, possibly, is the converse problem. That is, if S is passive, does a Σ_L exist? Although no answer to this question is known, in general, it may be answered affirmatively in a number of special cases considered in the following chapters. In these instances, the Proposition yields a starting point from which passive synthesis may be carried out. It is, however, necessary to exercise great care in choosing the factorization of $(1 - SS^*)$ and $(1 - S^*S)$ to assure that Σ_L will be causal and no more complex than S (that is, we do not want to start with a passive S which is realizable with resistors and capacitors, and end up with a lossless Σ_L which requires transmission lines in its realization). In the following chapters, the Proposition is employed in a number of different contexts, where the required factorizations can be found. In fact, if appropriate care is exercised, we can achieve minimal resistor synthesis via the bordering approach.

The causal factorizations of the proposition are by no means unique to network synthesis and, in fact, in one equivalent form or another, such factorizations have proven to be the primary tool of linear system theory. Factorizations of the type required by n-ports represented in the frequency domain were probably first encountered in the theory of stochastic filtering, their existence properties having been studied by Weiner [184]. The special case of these frequency domain factorizations for the rational matrices encountered in Chapter 6 has been studied by numerous individuals [20], [27], [98], [123]. The appropriate factorization theorem is given (without proof) in Chapter 6. In fact, the "so called" matrix Ricatti equation techniques which are common in system theory, being used to solve optimal control, stability, and filtering problems [13], in addition to the passive synthesis problem as formulated in Chapter 9, are mathematically equivalent to this classical special case of the frequency domain factorization problem [20]. In the time domain, where the analyticity of the frequency domain factorization is replaced by a causality constraint, considerably less is known about the existence of factorizations. In this instance, the results of Krein [91] are probably most complete. The factorization problem has also been considered from an abstract point of view [70], [152], wherein one attempts to verify the existence of the required factorization without specifying whether time or frequency domain is being used. Although this approach is more closely allied with the philosophy of the present chapter, the theory requires some rather deep function analytic arguments, and hence will not be considered here.

One restricted, but significant, class of networks (including all memoryless networks) to which the proposition can be applied directly are those for which S^* represents the scattering matrix of a passive network.

*Proposition 5-3

Under the hypotheses of Proposition 5-2, if both S and S^* are the scattering matrices of passive networks, then Σ_L is lossless. □

The passive synthesis procedure may be illustrated by the memoryless coupling network obtained in the active synthesis procedure of Section 5-2. Here

$$S = \begin{bmatrix} 0 & 1 \\ 0 & 0 \end{bmatrix} \tag{5-49}$$

$$S^* = S' = \begin{bmatrix} 0 & 0 \\ 1 & 0 \end{bmatrix} \tag{5-50}$$

$$1 - SS^* = \begin{bmatrix} 0 & 0 \\ 0 & 1 \end{bmatrix} = \begin{bmatrix} 0 \\ 1 \end{bmatrix} [0 \quad 1] = \Sigma_{12}\Sigma_{12}^* \tag{5-51}$$

$$1 - S^*S = \begin{bmatrix} 1 & 0 \\ 1 & 0 \end{bmatrix} \begin{bmatrix} 1 \\ 0 \end{bmatrix} [1 \quad 0] = \Sigma_{21}^*\Sigma_{21} \tag{5-52}$$

Hence,

$$\Sigma_{12} = \begin{bmatrix} 0 \\ 1 \end{bmatrix} \tag{5-53}$$

$$\Sigma_{21} = [1 \quad 0] \tag{5-54}$$

and

$$\Sigma_{22} = -\Sigma_{21}S^*\Sigma_{12}^{*-R}$$

$$= -[1 \quad 0]\begin{bmatrix} 0 & 0 \\ 1 & 0 \end{bmatrix}\begin{bmatrix} 0 \\ 1 \end{bmatrix} = [0] \tag{5-55}$$

Combining these matrices yields the lossless coupling network

$$\Sigma_L = \begin{bmatrix} S & | & \Sigma_{12} \\ \hline \Sigma_{21} & | & \Sigma_{22} \end{bmatrix} = \begin{bmatrix} 0 & 1 & | & 0 \\ 0 & 0 & | & 1 \\ \hline 1 & 0 & | & 0 \end{bmatrix} \tag{5-56}$$

which is exactly the circulator used in the realization of the passive coupling network employed in Section 5-2.

$$S = \begin{bmatrix} 0 & 1 \\ 0 & 0 \end{bmatrix} \tag{5-57}$$

is thus realized as a circulator loaded (at port 3) with a 1-ohm resistor, as shown in Figure 5-3.

Unlike the active synthesis, the passive realization is considerably more complex when applied to nonmemoryless networks in either the time or frequency domain than for the memoryless case. Consider, for example, the lossy delay characterized in the time domain by the 1-port scattering matrix

$$S(t,q) = \left(\frac{1}{2}\right)\delta(t - T - q) \tag{5-58}$$

Now, for this matrix,

$$S(t,q)^* = \left(\frac{1}{2}\right)\delta(q - T - t) \tag{5-59}$$

and

$$\begin{aligned} 1 - S^*S &= \delta(t - q) - S(t,q)^* \circ S(t,q) \\ &= \delta(t - q) - \left(\tfrac{1}{4}\right)\int_{-\infty}^{\infty} \delta(u - T - t)\delta(u - T - q) \\ &= \delta(t - q) - \left(\tfrac{1}{4}\right)\delta(t - q) = \left(\tfrac{3}{4}\right)\delta(t - q) \end{aligned} \tag{5-60}$$

Similarly,

$$\begin{aligned} 1 - SS^* &= \delta(t - q) - S(t,q) \circ S(t,q)^* \\ &= \left(\tfrac{3}{4}\right)\delta(t - q) \end{aligned} \tag{5-61}$$

Now, one factorization of these resistivity operators is

$$\Sigma_{12}(t,q) = \Sigma_{21}(t,q) = \left(\frac{\sqrt{3}}{2}\right)\delta(t - q) \tag{5-62}$$

(since $\delta(t - q)$ is the identity, it factors into two copies of itself, which, being self-adjoint, yield the desired factorization). Using this factorization, $\Sigma_{22}(t,q)$ is found, as per the Proposition, to be

$$\begin{aligned} \Sigma_{22}(t,q) &= -\Sigma_{21}(t,q) \circ S^*(t,q) \circ \Sigma_{12}^*(t,q)^{-R} \\ &= \left(\sqrt{\tfrac{3}{2}}\right)\delta(t - q) \circ \left(\tfrac{1}{2}\right)\delta(q - T - t) \circ \left(\sqrt{\tfrac{3}{2}}\right)\delta(t - q)^{-R} \\ &= -\left(\tfrac{1}{2}\right)\delta(q - T - t) \end{aligned} \tag{5-63}$$

and the 2-port coupling network, as per the Proposition, is

$$\Sigma_L(t,q) = \left[\begin{array}{c|c} \left(\tfrac{1}{2}\right)\delta(t - T - q) & \left(\sqrt{\tfrac{3}{2}}\right)\delta(t - q) \\ \hline \left(\sqrt{\tfrac{3}{2}}\right)\delta(t - q) & -\left(\tfrac{1}{2}\right)\delta(q - T - t) \end{array}\right] \tag{5-64}$$

which satisfies the adjoint condition for losslessness, but is clearly noncausal (since $\Sigma_{22}(t,q)$ is noncausal). Note here that $\Sigma_L(t,q)$ is noncausal, even though the two factorizations yield causal operators. Thus it is not sufficient for the factorizations to be causal; rather, they must be causal and, moreover, chosen to assure that $\Sigma_{22}(t,q)$ will also be causal (even though it contains $S(t,q)^*$, which is, in general, noncausal, as a factor). Let us consider the synthesis of the lossy delay once again, still factoring $1 - S^*S$ as per Equation (5-62), but with $1 - SS^*$ now factored as

$$\Sigma_{21}(t,q) \circ \Sigma_{21}^*(t,q) = \left(\sqrt{\frac{3}{2}}\right)\delta(t - T - q) \circ \left(\sqrt{\frac{3}{2}}\right)\delta(q - T - t)$$
$$= \left(\frac{3}{4}\right)\delta(t - q) \tag{5-65}$$

Here, $\Sigma_{21}(t,q)$ is a delay (rather than being memoryless, as before), but the product of the delay and its adjoint, a predictor, is still the memoryless resistivity operator. Calculating $\Sigma_{22}(t,q)$ with this factorization yields

$$\Sigma_{22}(t,q) = -\left(\sqrt{\frac{3}{2}}\right)\delta(t - T - q) \circ \left(\frac{1}{2}\right)\delta(q - T - t) \circ \left(\sqrt{\frac{3}{2}}\right)\delta(t - q)^{-R}$$
$$= -\left(\frac{1}{2}\right)\delta(t - q) \tag{5-66}$$

and the coupling network

$$\Sigma_L(t,q) = \left[\begin{array}{c|c} \left(\frac{1}{2}\right)\delta(t - T - q) & \left(\sqrt{\frac{3}{2}}\right)\delta(t - q) \\ \hline \left(\sqrt{\frac{3}{2}}\right)\delta(t - T - q) & -\left(\frac{1}{2}\right)\delta(t - q) \end{array}\right] \tag{5-67}$$

which not only satisfies the adjoint condition for losslessness, but is also causal, and hence lossless. To obtain the complete realization, let us factor $\Sigma_L(t,q)$ as

$$\Sigma_L(t,q) = \left[\begin{array}{cc} \left(\frac{1}{2}\right)\delta(t - q) & \left(\sqrt{\frac{3}{2}}\right)\delta(t - q) \\ \left(\sqrt{\frac{3}{2}}\right)\delta(t - q) & -\left(\frac{1}{2}\right)\delta(t - q) \end{array}\right] \circ \left[\begin{array}{cc} \delta(t - T - q) & 0 \\ 0 & \delta(t - q) \end{array}\right] \tag{5-68}$$

and observe that the first factor is the scattering matrix of an ideal transformer of turns ratio $\sqrt{3}$ (see Table 2-1), while the second is the scattering matrix of an uncoupled 2-port for which one port is a delay and the other is an open circuit. Now, the product may be realized by coupling two such 2-ports through a circulator (to realize the product). Loading the "back" port of this coupling network with a 1-ohm resistor then gives the desired realization of the lossy delay, as illustrated in Figure 5-8.

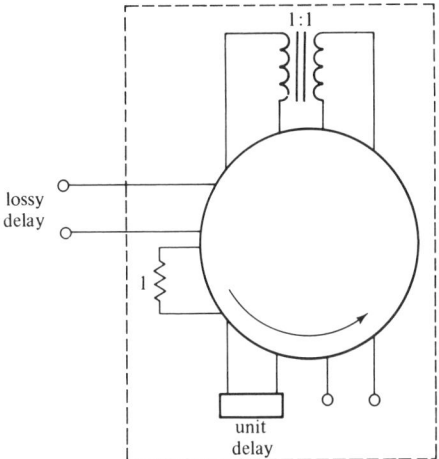

Figure 5-8 Realization of a lossy delay.

From the example, it is clear that the synthesis of the Proposition taken alone is not sufficient; rather, this result must be coupled with a means for obtaining a factorization of the resistivity operators which will assure that the coupling network is causal. In general, no such technique for achieving this goal is known (though an existence theorem has been proven [152]). In the following chapters, however, factorizations which assure the causality (hence, losslessness) of the synthesis when applied to certain special cases are developed.

5-4 Equivalent Realizations

One approach to the problem of finding a causal coupling network via the technique of the preceding section is to start with any coupling network which satisfies the adjoint condition for losslessness, as per Proposition 5-2, and then apply an equivalence transformation to obtain alternative coupling networks which also satisfy the adjoint condition and (hopefully) the causality condition. The transformation to be employed is due to Oono and Yassura [110], [123] and is stated in the following theorem, the proof of which is left to the reader as an exercise.

*Theorem 5-2

Let S be the scattering matrix of a passive n-port and

$$\Sigma_L = \begin{bmatrix} S & \Sigma_{12} \\ \Sigma_{21} & \Sigma_{22} \end{bmatrix}$$

188 REALIZABILITY THEORY

be a lossless $(n + m)$-port (which realizes S at its first n ports when loaded in 1-ohm resistors at its last m ports). Then the $(n + m + k)$-port

$$\tilde{\Sigma}_L = \begin{bmatrix} 1 & 0 \\ \hline & \\ 0 & \Theta \end{bmatrix} \begin{bmatrix} S & \Sigma_{12} & 0 \\ \hline \Sigma_{21} & \Sigma_{22} & 0 \\ \hline 0 & 0 & \sigma \end{bmatrix} \begin{bmatrix} 1 & 0 \\ \hline & \\ 0 & \phi \end{bmatrix}$$

where Θ and ϕ are lossless $(m + k)$-port scattering matrices and σ is a lossless k-port scattering matrix, is lossless and realizes S at its first n ports when its last $(m + k)$-ports are loaded in 1-ohm resistors. □

The theorem never yields a realization containing less resistors than the "initial realization" Σ_L. However, it yields realizations containing any number of resistors greater than or equal to m. Now, if Σ_L is a minimal resistor realization (that is, m is minimal), and we consider a class of networks for which losslessness implies the existence of a scattering matrix (for instance, normal time-invariant networks), the transformations of Theorem 5-2 yield all possible realizations [123].

Applying the theorem to the lossless coupling network

$$\Sigma_L = \begin{bmatrix} 0 & 1 & 0 \\ 0 & 0 & 1 \\ \hline 1 & 0 & 0 \end{bmatrix} \quad (5\text{-}69)$$

obtained in the synthesis of

$$S = \begin{bmatrix} 0 & 1 \\ 0 & 0 \end{bmatrix} \quad (5\text{-}70)$$

the 4-port coupling network

$$\tilde{\Sigma}_L = \begin{bmatrix} 0 & 1 & 0 & 0 \\ 0 & 0 & -1 & 0 \\ \hline 0 & 0 & 0 & -1 \\ 1 & 0 & 0 & 0 \end{bmatrix}$$

$$= \begin{bmatrix} 1 & 0 \\ \hline & 0 & 1 \\ 0 & & \\ & 1 & 0 \end{bmatrix} \begin{bmatrix} \Sigma_L & 0 \\ \hline 0 & 1 \end{bmatrix} \begin{bmatrix} 1 & 0 \\ \hline & -1 & 0 \\ 0 & & \\ & 0 & -1 \end{bmatrix} \quad (5\text{-}71)$$

is obtained. Of course, there is no reason to restrict our consideration to memoryless networks (except for computational convenience). For instance, if in the previous example we let σ be the lossless operator

$$\sigma(p) = \frac{e^{-p}(p - 1)}{p + 1} \quad (5\text{-}72)$$

then the lossless coupling network

$$\Sigma_L(p) = \begin{bmatrix} 0 & 1 & 0 & 0 \\ 0 & 0 & -1 & 0 \\ \hline 0 & 0 & 0 & \dfrac{e^{-p}(1-p)}{p+1} \\ 1 & 0 & 0 & 0 \end{bmatrix} \quad (5\text{-}73)$$

is obtained. Note also that in the realization of the lossy delay considered in the preceding section, the causal coupling network could have been obtained from the noncausal one via the equivalence transformation with

$$\Theta(t,q) = \delta(t - T - q) \quad (5\text{-}74)$$

$$\phi(t,q) = \delta(t - q) \quad (5\text{-}75)$$

and no $\sigma(t,q)$, that is, no additional load resistors. In fact, Theorem 5-2 may always be applied to coupling networks, such as those constructed in the preceding section, which satisfy the adjoint condition for losslessness, but are not necessarily causal. In this case, the equivalent coupling networks also satisfy the adjoint condition for losslessness, and may or may not be causal. It is, however, possible, such as in the example, for a noncausal coupling network to be converted to a causal (hence, lossless) one. The justification of this contention is left to the reader as an exercise.

5-5 Realization of Immittance Matrices

As with most scattering matrix results, the algebraic synthesis techniques of this chapter can be modified to realize networks characterized by immittance matrices. In the case of the π matrix results, this is unnecessary since

$$[\pi_v \mid \pi_i] \quad (5\text{-}76)$$

can readily be converted to

$$[\pi_a \mid \pi_b] = [\pi_v + \pi_i \mid \pi_v - \pi_i] \quad (5\text{-}77)$$

to which the scattering matrix techniques may be applied. Furthermore, if c is chosen sufficiently large, the passive coupling network which results from the procedure will be assured of possessing an immittance matrix. In this section we thus restrict our consideration to the passive synthesis of a network characterized by an impedance matrix (though a dual result holds for admittance matrices).

As has been the practice in the preceding sections, a passive n-port, characterized by Z, is realized at the first n ports of a lossless $(n + m)$-port, characterized by X_L, when loaded in 1-ohm resistors at its last m ports. Also, as in Section 5-3, only a partial solution is obtained, X_L satisfying all the conditions for losslessness except causality (analyticity in the RHP in the frequency domain).

The impedance analog of the cascade loading proposition of the preceding section is obtained from the immittance form of the cascade loading theorem. If one lets a realization of the $(m + n)$-port with impedance matrix

$$X_L = \begin{bmatrix} X_{11} & X_{12} \\ \hline X_{21} & X_{22} \end{bmatrix}$$

be loaded at the "2" ports with an m-port characterized by the admittance matrix

$$Y_L$$

then the n-port "seen" at the "1" ports of X_L is characterized by the impedance

$$Z = X_{11} - X_{12}Y(1 + X_{22}Y_L)^{-1}X_{21} \qquad (5\text{-}78)$$

If it is assumed that the load is composed of 1-ohm resistors,

$$Y_L = 1 \qquad (5\text{-}79)$$

and

$$Z = X_{11} - X_{12}(1 + X_{22})^{-1}X_{21} \qquad (5\text{-}80)$$

The passive synthesis problem thus requires that we find a lossless $(n + m)$-port

$$X_L = \begin{bmatrix} X_{11} & X_{12} \\ \hline X_{21} & X_{22} \end{bmatrix} \qquad (5\text{-}81)$$

such that Equation (5-80) is satisfied for a specified passive n-port Z.

Though one cannot, in the general case, find a "truly lossless" X which satisfies this condition, it is possible to identify a class of $(n + m)$-ports which satisfy the existence and adjoint (skew hermitian) conditions for losslessness (and can be used as a starting point for a complete lossless synthesis). For X_L to satisfy the adjoint condition, we have the following lemma.

*Lemma 5-2

$$X_L = \begin{bmatrix} X_{11} & X_{12} \\ \hline X_{21} & X_{22} \end{bmatrix}$$

is skew hermitian (that is, $X_L + X_L^* = 0$) if and only if

$$X_{11} + X_{11}^* = 0$$

$$X_{22} + X_{22}^* = 0$$

and

$$X_{21} = -X_{12}^*$$

□

Substituting the result of Lemma 5-2 into Equation (5-80) leaves

$$Z = X_{11} + X_{12}(1 + X_{22})^{-1}X_{12}^* \qquad (5\text{-}82)$$

which must be satisfied with

$$X_{11} + X_{11}^* = 0 \qquad (5\text{-}83)$$

and

$$X_{22} + X_{22}^* = 0 \qquad (5\text{-}84)$$

to complete the realization. This is, in turn, achieved most readily by splitting both sides of Equation (5-82) into their hermitian and skew hermitian parts.

*Lemma 5-3

Let

$$Z = X_{11} + X_{12}(1 + X_{22})^{-1}X_{12}^*$$

where

$$X_{11} + X_{11}^* = 0$$

and

$$X_{22} + X_{22}^* = 0$$

Then the hermitian and skew hermitian parts of Z are

$$Z_h = X_{12}(1 + X_{22})^{-1}(1 + X_{22}^*)^{-1}X_{12}^*$$

and

$$Z_s = X_{11} - X_{12}(1 + X_{22})^{-1}X_{22}(1 + X_{22}^*)^{-1}X_{12}^* \qquad \square$$

With these preliminaries now completed, the realization may be obtained by splitting Z into its hermitian and skew hermitian parts, solving the first equation of Lemma 5-3 for X_{12} (after arbitrarily specifying X_{22}), substituting this into the second equation, and solving for X_{11}. The method, derived from a result of Bayard [42], [110], [116], is formalized in the following theorem.

*Theorem 5-3

Let Z be a specified n-port impedance matrix with

$$Z = Z_h + Z_s$$

and let

$$NN^* = Z_h$$

Then

$$X_L = \begin{bmatrix} X_{11} & X_{12} \\ \hline X_{21} & X_{22} \end{bmatrix}$$

satisfies the existence and adjoint conditions for losslessness if

$$X_{22} + X_{22}^* = 0$$

but it is otherwise arbitrary,

$$X_{12} = N(1 + X_{22})$$

$$X_{21} = X_{12}^*$$

and

$$X_{11} = Z_s + NX_{22}N^*$$

Furthermore, any realization of X loaded at its "2" ports with 1-ohm resistors realizes Z at its "1" ports. □

The passive synthesis for impedance matrices is thus similar to the analogous result for scattering matrices (though complicated by the "weaker" cascade loading formula), and, as in that case, is predicated on our ability to find a factorization

$$NN^* = Z_h \tag{5-85}$$

of the hermitian operator Z_h. The existence of such a factorization implies that Z_h is positive; hence, the procedure is only applicable to passive n-ports (assuring that conservation of energy is not violated). The choice of this factorization and X_{22} are the determining factors in whether or not X is causal; hence, great care must be exercised in their formulation.

Consider the time-variable resistor characterized by the impedance

$$Z(t,q) = Z_h(t,q) = t^2 \delta(t - q) \tag{5-86}$$

One factorization for $Z_h(t,q)$ is

$$N(t,q) = t\delta(t - q) \tag{5-87}$$

which yields the coupling network

$$X_L(t,q) = \begin{bmatrix} 0 & -t\delta(t - q) \\ t\delta(t - q) & 0 \end{bmatrix} \tag{5-88}$$

for

$$X_{22}(t,q) = 0 \tag{5-89}$$

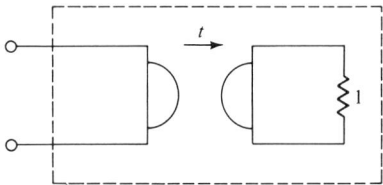

Figure 5-9 Realization of memoryless time-variable impedance.

Since this coupling network is just a time-variable gyrator with

$$\gamma(t) = t \tag{5-90}$$

the given impedance of Equation (5-86) may be realized as such a gyrator loaded in a 1-ohm resistor, as shown in Figure 5-9.

5-6 Synthesis of Memoryless n-Ports

The synthesis procedures thus far considered have been entirely algebraic in nature and, therefore, applicable to any network in either the time or frequency domain. Such procedures, though powerful, cannot be used to obtain a complete realization, since one must invoke the characteristics of the particular elementary components with which an n-port is to be realized before obtaining a final realization in terms of such. The one exception is the case of a memoryless n-port, being in some sense universal, since it corresponds to the "algebraic" operators which appear naturally within any operator space [140]. In fact, the synthesis procedures to follow reduce a general n-port to a network of elementary n-ports (considered in Chapter 4) and memoryless n-ports, the latter being independent of the particular elementary n-ports employed. This section is, therefore, devoted to the problem of realizing a memoryless n-port.

Let us begin with a lossless n-port characterized by a scattering matrix

$$S(t,q) = A(t)\delta(t-q) \tag{5-91}$$

Since $S(t,q)$ is lossless, one is assured that $A(t)$ is orthogonal. That is,

$$A'(t)A(t) = A(t)A'(t) = 1 \tag{5-92}$$

Clearly, it suffices to realize the $2n$-port with scattering matrix

$$\tilde{S}(t,q) = \begin{bmatrix} A(t)\delta(t-q) & 0 \\ \hline 0 & A'(t)\delta(t-q) \end{bmatrix} \tag{5-93}$$

which is also lossless and equivalent to $S(t,q)$ at its first n ports (since the "front" and "back" ports are uncoupled). Now, $S(t,q)$ can be factored as

$$\tilde{S}(t,q) = S_1(t,q) \circ S_2(t,q)$$

$$= \begin{bmatrix} 0 & 1 \\ \hline 1 & 0 \end{bmatrix} \circ \begin{bmatrix} 0 & A'(t)\delta(t-q) \\ \hline A(t)\delta(t-q) & 0 \end{bmatrix} \tag{5-94}$$

whence a realization of $\tilde{\tilde{S}}(t,q)$ [thus, $S(t,q)$] can be obtained by loading a $6n$-port circulator with realizations of $S_1(t,q)$ and $S_2(t,q)$, as per Figure 5-10(a). Since $A(t)$ is orthogonal, $S_2(t,q)$ represents a $2n$-port ideal transformer with turns ratio

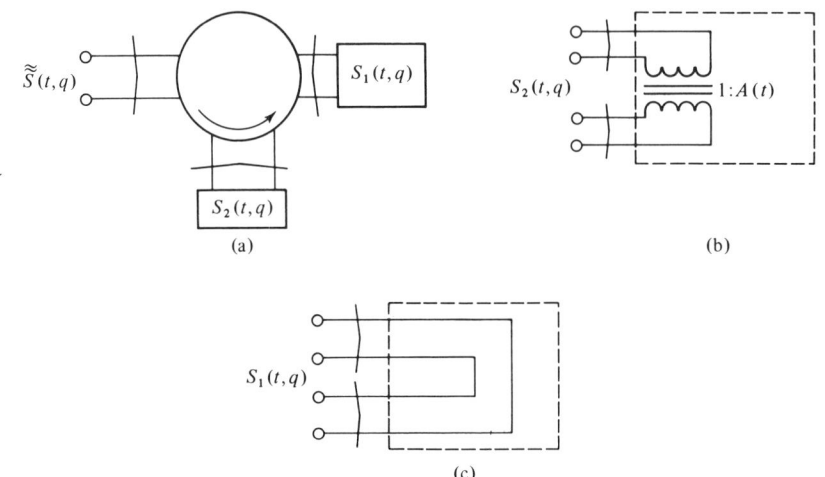

Figure 5-10 Component parts for realization of lossless memoryless n-ports.

matrix $A(t)$, while the reader may verify that $S_1(t,q)$ corresponds to a $2n$-port composed of a pair of wires between the ith port and the $(n+i)$th port, as illustrated in Figure 5-10(b). The desired realization is thus that of Figure 5-11, which we summarize as follows.

*Proposition 5-4

Every lossless memoryless n-port characterized by a scattering matrix

$$S(t,q) = A(t)\delta(t-q)$$

can be realized as per the configuration of Figure 5-11. Moreover, if the n-port is time-invariant, so is the realization. ☐

Of course, in the time-invariant case, one can work in the frequency domain rather than the time domain simply by dropping the $\delta(t-q)$ from the previous development.

The active and passive syntheses of the preceding sections always yield a "truly lossless" coupling network when one is dealing with memoryless n-ports (use the square root of the resistivity matrix [132] as the factorization). The general memoryless synthesis problem can, therefore, always be reduced to the lossless one of Proposition 5-4 via these techniques. Alternatively, one might desire a direct realization, especially in the passive case. As a preliminary to formulating such, we require the following lemma, the proof of which is left to the reader.

*Lemma 5-4

Let $A(t)$ be any $n \times n$ matrix for which

$$\|A(t)\| \leq 1$$

Then there exists an $n \times n$ matrix $T(t)$ such that

$$2T(t)(1 + T(t)'T(t))^{-1} = A(t) \qquad \square$$

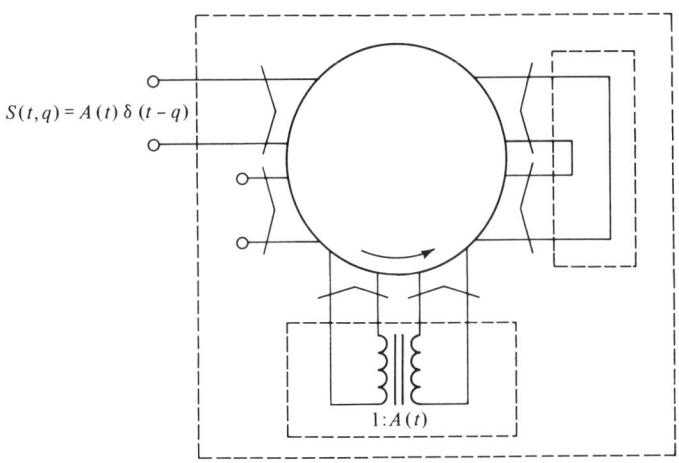

Figure 5-11 Realization of a lossless memoryless n-port.

To realize the n-port scattering matrix

$$S(t,q) = A(t)\delta(t - q)$$

for which the norm of $A(t)$ is bounded by one if the network is passive, consider the $2n$-port

$$\tilde{\tilde{S}}(t,q) = S_1(t,q) \circ S_2(t,q) \qquad (5\text{-}95)$$

$$= \begin{bmatrix} 0 & \delta(t-q) \\ \hline 0 & 0 \end{bmatrix}$$

$$\circ \begin{bmatrix} (1 + T'T)^{-1}(t'T - 1)\delta(t-q) & 2(1 + T'T)^{-1}T'\delta(t-q) \\ \hline 2T(1 + T'T)^{-1}\delta(t-q) & (1 + TT')^{-1}(1 - TT')\delta(t-q) \end{bmatrix}$$

Here $S_1(t,q)$ corresponds to the gyrator resistor network of Figure 5-12, while $S_2(t,q)$ is an ideal transformer with turns ratio matrix $T(t)$ (the explicit dependance of T on t has been left out of Equation (5-95) for the sake of brevity). Hence, $\tilde{\tilde{S}}(t,q)$ can be realized as a circulator loaded in a realization of the network of Figure 5-12, and an ideal transformer. It remains to be shown that a realization of $\tilde{\tilde{S}}(t,q)$ will lead to a realization of the given passive scattering matrix

$$S(t,q) = A(t)\delta(t - q) \tag{5-96}$$

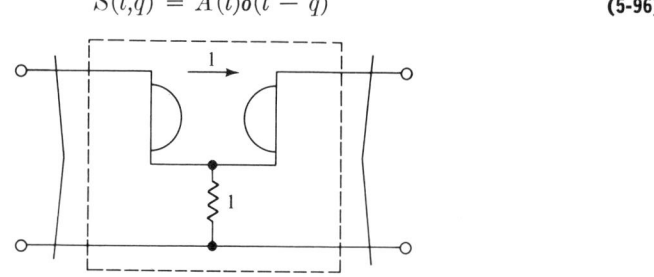

Figure 5-12 Realization of $S_1(t,q)$.

If, however, we choose the turns ratio matrix for the ideal transformer as the $T(t)$ which satisfies the equality

$$2T(t)(1 + T'(t)T(t))^{-1} = A(t) \tag{5-97}$$

(and is assured to exist by Lemma 5-4), then $\tilde{\tilde{S}}(t,q)$ has the form

$$\tilde{\tilde{S}}(t,q) = \begin{bmatrix} 2T(1 + T'T)^{-1}\delta(t - q) & (1 + TT')^{-1}(1 - TT')\delta(t - q) \\ \hline 0 & 0 \end{bmatrix} \tag{5-98}$$

$$= \begin{bmatrix} A(t)\delta(t - q) & (1 + TT')^{-1}(1 - TT')\delta(t - q) \\ \hline 0 & 0 \end{bmatrix}$$

from which it may be shown, upon invoking the cascade loading theorem, that the scattering matrix

$$S(t,q) = A(t)\delta(t - q) \tag{5-99}$$

is realized at the first n ports of $\tilde{\tilde{S}}(t,q)$ when the last n ports are unloaded (that is,

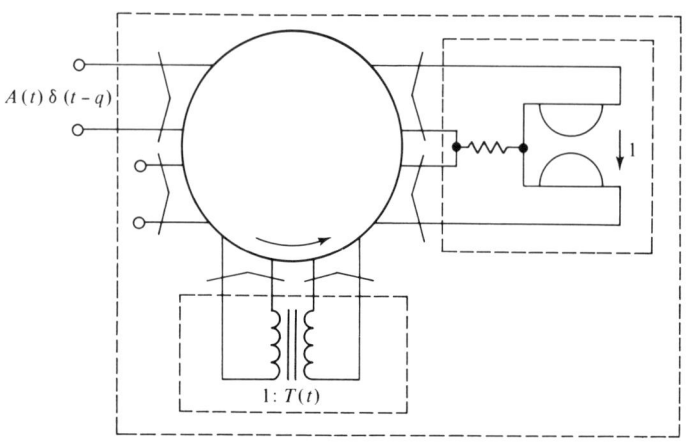

Figure 5-13 Realization of a passive memoryless scattering matrix.

loaded in open circuits). The required realization for $\tilde{S}(t,q)$, and hence $S(t,q)$, is, therefore, that shown in Figure 5-13. In summary, we have the following proposition.

*Proposition 5-5

Every passive, memoryless n-port characterized by a scattering matrix

$$S(t,q) = A(t)\delta(t - q)$$

can be realized as per the configuration of Figure 5-13. Moreover, if the n-port is time-invariant, so is the realization. □

As with the lossless synthesis, the same development holds in the frequency domain once the $\delta(t - q)$ have been dropped and the n-port is assumed to be time-invariant.

The preceding realizations, although conveniently carried out, often require a very large number of gyrators, these being used both in the realization of $S_1(t,q)$ and in the circulator. Alternative realizations are possible [3], which reduce the number of gyrators. These, however, require that one carry out a rather complex decomposition on the matrix $A(t)$, and will not be considered. One could also directly realize a memoryless immittance by such decompositions. In this case, one begins with an impedance (or admittance)

$$Z(t,q) = Z(t)\delta(t - q) \qquad \textbf{(5-100)}$$

and decomposes $Z(t)$ into symmetric and skew symmetric parts,

$$Z(t) = Z_h(t) + Z_s(t) \qquad \textbf{(5-101)}$$

these being further decomposed by congruence transformation [132] to the forms

$$Z_h(t) = T'_h(t) \begin{bmatrix} r_1 & & & & | & 0 \\ & r_2 & & & | & \\ & & \cdot & & | & \\ & & & \cdot & | & \\ & \overline{0|} & & & | & \\ & & & & | & r_k \end{bmatrix} T_h(t) \qquad \textbf{(5-102)}$$

and

$$Z_s(t) = T'_s(t) \begin{bmatrix} 0 & -1 & | & & & | & \\ 1 & 0 & | & & & | & 0 \\ \hline & & | & 0 & -1 & | & \\ & & | & 1 & 0 & | & \\ \hline & & | & & & | & \\ & \overline{0|} & | & & & | & \\ & & | & & & | & 0 & -1 \\ & & | & & & | & 1 & 0 \end{bmatrix} T_s(t) \qquad \textbf{(5-103)}$$

Now, from this decomposition, $Z_h(t)$ can readily be realized as a transformer with turns ratio matrix $T_h(t)$ loaded in k resistors, $r_1, r_2, \cdots, r_k$, while the decomposition allows $Z_s(t)$ to be realized as a transformer of turns ratio matrix $T_s(t)$ loaded in unit gyrators. These realizations are shown in Figure 5-14,

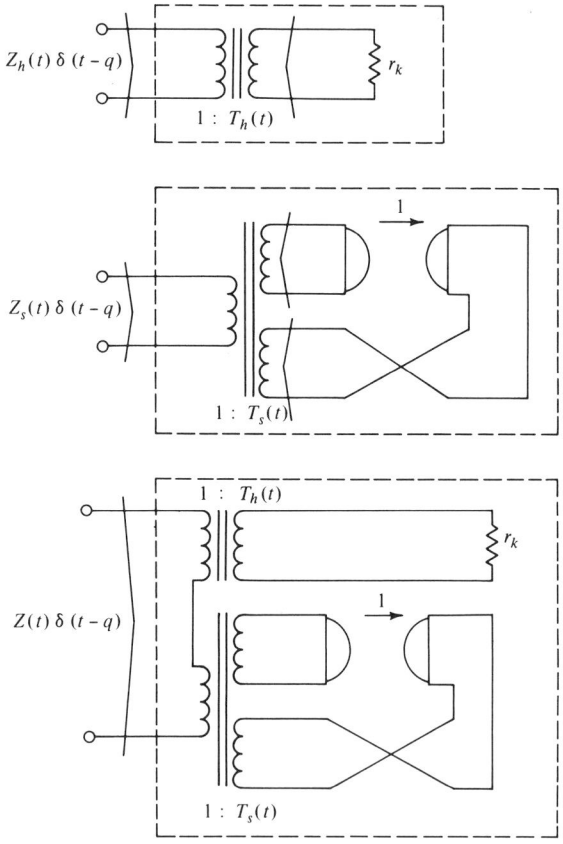

Figure 5-14 Component parts and realization of memoryless immittance.

together with the overall realization for $Z(t)$ obtained from a series connection of those for $Z_h(t)$ and $Z_s(t)$. Note that if $Z(t)\delta(t - q)$ is passive, its hermitian part is positive; hence, the resistors obtained by the similarity transformation will all be positive, thereby yielding a realization composed entirely of passive components. Similarly, if $Z(t)\delta(t - q)$ is lossless, $Z_h(t)$ is zero and the realization is entirely of lossless components, time-variable ideal transformers, and unit gyrators. In summary, we have the following proposition.

*Proposition 5-6

Any memoryless impedance (admittance)
$$Z(t,q) = Z(t)\delta(t - q)$$

can be realized as a series (parallel) combination of a resistor-loaded ideal transformer and a gyrator-loaded ideal transformer. Moreover, the components in the realization are passive, lossless, or time invariant if $Z(t,q)$ satisfies the corresponding properties. □

One application of Proposition 5-6 is the realization of an $(m+n)$-port gyrator characterized by an impedance matrix

$$Z(t,q) = \begin{bmatrix} 0 & -\gamma'(t) \\ \gamma(t) & 0 \end{bmatrix} \quad (5\text{-}104)$$

Here,

$$Z(t) = Z_s(t) = \begin{bmatrix} 0 & -\gamma'(t) \\ \gamma(t) & 0 \end{bmatrix} \quad (5\text{-}105)$$

while

$$Z_h(t) = 0 \quad (5\text{-}106)$$

Now,

$$Z_s(t) = \begin{bmatrix} \gamma'(t) & 0 \\ 0 & 1 \end{bmatrix} \begin{bmatrix} 0 & -1 \\ 1 & 0 \end{bmatrix} \begin{bmatrix} \gamma(t) & 0 \\ 0 & 1 \end{bmatrix} \quad (5\text{-}107)$$

which is the desired decomposition of $Z_s(t)$, except for a rearrangement of the rows and columns in the "gyrator" matrix. Letting Q be the orthogonal matrix which achieves such a rearrangement, that is,

$$\begin{bmatrix} \begin{matrix} 0 & -1 \\ 1 & 0 \end{matrix} & & \\ & \begin{matrix} 0 & -1 \\ 1 & 0 \end{matrix} & \bigg|0 \\ & & \cdot \\ \overline{0}\bigg| & & \begin{matrix} \cdot & \\ & 0 & -1 \\ & 0 & 1 \end{matrix} \end{bmatrix} = Q \begin{bmatrix} 0 & -1 \\ 1 & 0 \end{bmatrix} Q' \quad (5\text{-}108)$$

then if

$$T(t) = Q \begin{bmatrix} \gamma(t) & 0 \\ 0 & 1 \end{bmatrix} \quad (5\text{-}109)$$

$$Z_s(t) = T'(t) \begin{bmatrix} \begin{array}{cc|ccc|c} 0 & -1 & & & & \\ 1 & 0 & & & & \\ \hline & & 0 & -1 & & 0 \\ & & 1 & 0 & & \\ \hline & & & & \ddots & \\ \hline \overline{0} & & & & & \\ & & & & & \begin{matrix} 0 & -1 \\ 1 & 0 \end{matrix} \end{array} \end{bmatrix} T(t) \quad \textbf{(5-110)}$$

which is the required decomposition. We have thus shown the following.

*Proposition 5-7

A time-variable $(m + n)$-port gyrator with gyration matrix $\gamma(t)$ can be realized as an ideal transformer with the turns ratio matrix of Equation (5-109) loaded n unit 2-port gyrators.

5-7 Discussion

The developments of this chapter, although not solving the synthesis problem, yield a starting point from which such procedures may begin. In particular, the techniques developed allow us to consider only passive networks which possess scattering matrices, since the general case can always be reduced to this special case. The results of the sections on passive synthesis, though not self-contained, are indicative of the procedures employed to completely solve the passive synthesis problem for various special cases in the sequel. In these cases, the special properties of $(1 - S^*S)$ or Z_h are used to obtain a factorization which yields a causal coupling network, this completing the solution of the passive synthesis problem begun in Sections 5-3 and 5-5.

The essence of the techniques employed in this chapter (and the last) is algebraic, depending only on the algebraic input-output relation imposed by the network, and not on the particular form of the operator representation. For this reason, the results hold in both the time and frequency domains with appropriate notational modifications. It is this class of algebraic results which are ideally suited to the generalized network problem [148], [149], [152], since they are general and also readily applicable to the various special cases which may arise. A yet to be achieved, but potentially fruitful, goal is the "algebrazation" of the entire synthesis process.

Finally, we observe that the results of the chapter are all conceptually similar in that a given n-port was realized at the first n-ports of a "simpler" $(n + m)$-port appropriately loaded at its last m ports. Thus far, we have employed positive and negative resistors for this purpose, though in the remaining chapters similar techniques will be found, where inductors, capacitors, transmission lines, and variable parameter devices assume the role of load.

CHAPTER

6
RLC Networks

6-1 Introduction

In this chapter, as the first of four special cases, we undertake the study of RLC networks. In dealing with any special case, one faces the mixed blessing of having an extra structure which does not appear in the general case. While this structure yields an additional tool with which to manipulate the network representation, the operations performed must preserve this structure. In the RLC case, this additional structure takes the form of a real rational system function which allows the properties of the network to be completely determined by a finite number of parameters (the much maligned poles and zeros). To our detriment, however, every operation performed on the system function of an RLC network must preserve its real rational character. In particular, this limits the applications of the adjoint, since the adjoint of a real rational function is not rational (for example, $p^* = \bar{p}$, which is not rational). Rather, one must employ the Hurwitz conjugate

$$H(p)_* = H'(-p) \tag{6-1}$$

which preserves rationality and coincides with the adjoint on the $j\omega$ axis.

The RLC networks may be approached via either their rational system function [31], [92], [110] or the corresponding differential equation [117]. In this chapter, attention is restricted to the more classical frequency domain approach, leaving the differential operator and its corresponding impulse response to Chapter 9, where they may be studied as the time-invariant special case of the RLC time-variable n-ports. Although we take the classical approach, our viewpoint is that of showing how the general theory thus far developed can be applied to the special case. The development thus differs considerably from what it might have been had our consideration been restricted entirely to the RLC case. In particular, scattering techniques take precedence, whereas immittance techniques usually dominate the classical development [183]. Actually, several of the results (the Darlington passive synthesis, for example) are simplified by the scattering approach, while others (lossless synthesis), though made more complicated via scattering methods, are indicative of methods applicable to more general classes of networks (that is, transmission line and multivariable networks).

In the special case of RLC networks, it is often possible to obtain results for 1-ports which do not hold for n-ports (since the system functions for two RLC 1-ports always commute, which is not true of the n-port case). This special case is thus distinguished by referring to the corresponding 1×1 matrices as parameters rather than matrices, and denoting them by small letters rather than capitals.

In the first section, the underlying definitions of an RLC network and their implications are considered. In particular, the real rationality of the resultant system functions is exhibited, and simplified versions of the passivity and losslessness conditions are obtained by invoking the specific properties of rational functions.

The next three sections are devoted to 1-port synthesis. In these sections, the immittance and scattering techniques are used interchangeably, since the RLC 1-ports which do not possess all three of these parameters are trivially realized. After deriving Richards' theorem in Section 6-3, various realizations of an RLC lossless 1-port are obtained in Section 6-4. These include the gyrator-inductor realization of an LBR rational 1-port, our first (of many) encounters with synthesis by scattering matrix factorization. The final section on 1-port synthesis is devoted to the Darlington-type realizations, these being achieved by application of the general passive synthesis results of the preceding chapter. To obtain a coupling network which is truly lossless and also rational, the Hurwitz conjugate is employed rather than the adjoint of the general results, and a careful accounting of the poles is kept to assure RHP analyticity.

The final two sections are concerned with n-port synthesis. Since the results obtained are natural generalizations of the 1-port case and the results of Chapter 5, the proofs of most of the theorems are left as exercises.

Through the entirety of the chapter, only passing consideration is given to analysis and active (π matrix) synthesis, because the results of Chapters 4 and 5 in these areas are complete in themselves. In fact, the rational properties of RLC networks buy little in the way of improvement for these results.

6-2 Properties of RLC Networks

Intuitively, an RLC network should include the three basic components: the capacitor, resistor, and inductor; in addition, ideal transformers and gyrators need to be included if one is to achieve intercomponent coupling and non-reciprocity. In fact, an inductor can always be realized as a gyrator loaded with a capacitor, and an ideal transformer as cascaded gyrators; hence, the RCG networks are equivalent to the RLC networks.

Definition 6-1

An *RLC n-port* is an n-port which can be realized by a network of time-invariant resistors, capacitors, and gyrators. □

Since inductors, transformers, circulators, nullators, and norators can all be realized from the RCG components, they will be used freely in the development.

The fundamental property which distinguishes the RLC networks from the general case is the real rationality of their system functions.

Proposition 6-1

Let $H(p)$ be the system function of an RLC n-port. Then $H(p)$ is a matrix of real rational functions (that is, rational functions with real coefficients).

Proof

Since the gyrators and resistors are memoryless, while the capacitors are rational in the p variable, Proposition 4-8 may be invoked to show that the system function of an RLC (or RCG) n-port is indeed real rational in p. □

System functions which satisfy the property of Proposition 6-1 are said to be real rational.

The real rationality of the RLC system functions is, at the same time, an advantage and a disadvantage. Since every rational function is completely determined by its poles (zeros of its denominator polynomial) and zeros (zeros of its numerator polynomial), together with a multiplicative constant, an RLC n-port is completely described by a finite number of parameters. Possibly of equal significance is the fact that the only singularities of a rational function are its poles; hence, the RHP analyticity (that is, causality) conditions for RLC networks are readily obtained. To our detriment is the fact that the adjoint of a rational function is not rational, necessitating its replacement by the Hurwitz conjugate.

Upon invoking these special properties of real rational system functions, the BR conditions for an RLC network may be reformulated as follows.

Theorem 6-1

If $S(p)$ is the bounded real scattering matrix of a passive RLC n-port, it satisfies the following three conditions:

1. the entries in $S(p)$ are real rational functions
2. the poles of $S(p)$ are in the LHP (that is, $S(p)$ is analytic in the RHP and on the $j\omega$ axis)
3. $1 - S(j\omega)^* S(j\omega) \geq 0$ for all imaginary axis $j\omega$.

Moreover, any scattering matrix satisfying these three conditions is BR.

Proof

Clearly, the scattering matrix of a passive network is BR and, by Proposition 6-1, is real rational; hence condition 1. Now, since the only singularities of a real rational function are poles [144], they are all either in the LHP or on the $j\omega$ axis, since a BR function is analytic in the RHP. In fact, it is impossible for a pole to lie on the $j\omega$ axis, for such a function could not possibly have norm bounded by one in the RHP (at least, for that part of the RHP which is "close" to the $j\omega$ axis pole). Condition 2 is, therefore, a necessary condition. Since, by condition, 2 $S(p)$ is not only analytic in the RHP but also on the $j\omega$ axis, the RHP condition

$$1 - S(p)^*S(p) \geq 0 \qquad (6\text{-}2)$$

for bounded reality may be extended to the $j\omega$ axis (that is, the boundary of the RHP) by analytic continuation [144]; hence condition 3. Any RLC passive system function therefore satisfies conditions 1 through 3. Conversely, if a given $S(p)$ satisfies conditions 1 through 3, it is analytic in the RHP, since the only singularities are the poles, and these lie in the LHP, and it exists by hypothesis. It, therefore, remains only to show that the three conditions imply that

$$1 - S(p)^*S(p) \geq 0 \qquad (6\text{-}3)$$

for all RHP p, or equivalently, that

$$\|S(p)\| \leq 1 \qquad (6\text{-}4)$$

in the RHP. Now, conditions 1 and 2 imply that $S(p)$ is analytic in the RHP, and also on its boundary, the $j\omega$ axis. Hence, we may invoke the maximum modulus theorem [144], which assures that in any region of analyticity the norm attains its maximum on the boundary. In our case, condition 3 implies that

$$\|S(j\omega)\| \leq 1 \qquad (6\text{-}5)$$

on the $j\omega$ axis; hence, the maximum modulus theorem assures that the norm condition [of Equation (6-4)] holds for the entire RHP. Conditions 1, 2, and 3 (plus an *a priori* assumption of existence) thus imply that $S(p)$ is bounded real, completing the proof.

It should be noted that the theorem is really not yet complete, in that we have shown that the three conditions imply bounded reality, but we have not shown that they imply that the n-port is RLC. In fact, this is the case, every such scattering matrix being realizable as an RLC network. This last part of the sufficiency proof results from the construction of the synthesis to follow, and will therefore be deferred.

The theorem permits a considerable simplification of the passivity test for RLC networks over that required in the general case. Condition 1 may be checked by inspection, whereas condition 2 requires that the denominator of $S(p)$ (least common denominator in the matrix case) be Hurwitz, for which a number of tests are available [78]. Finally, because $S(p)$ is analytic in the closed RHP, it suffices to carry out the adjoint test for passivity only on the $j\omega$ axis.

It is significant that the effect of rationality (together with RHP boundedness) assures analyticity in the entire closed RHP (rather than the open RHP of the general BR condition), which, in turn, allows the

$$1 - S(p)*S(p) \geq 0 \tag{6-6}$$

condition for passivity to be transferred from the RHP to its boundary, the $j\omega$ axis. Since

$$S(j\omega)_* = S(j\omega)^* \tag{6-7}$$

this allows the nonrational adjoint to be replaced by the rational Hurwitz conjugate in the theorem.

Corollary 6-1

Under the hypothesis of Theorem 6-1, condition 3 is equivalent to:

3*. $1 - S(j\omega)_*S(j\omega) \geq 0$ for all imaginary axis $j\omega$. □

Consider the 2-port characterized by the scattering matrix

$$S(p) = \begin{bmatrix} 0 & \dfrac{p-1}{p+1} \\ 1 & 0 \end{bmatrix} \tag{6-8}$$

This is clearly real rational and its only pole is at

$$p = -1 \tag{6-9}$$

which is in the LHP, and

$$1 = S(j\omega)_*S(j\omega) = \begin{bmatrix} 1 & 0 \\ 0 & 1 \end{bmatrix} - \begin{bmatrix} 0 & 1 \\ \dfrac{-j\omega-1}{-j\omega+1} & 0 \end{bmatrix} \begin{bmatrix} 0 & \dfrac{j\omega-1}{j\omega+1} \\ 1 & 0 \end{bmatrix} \tag{6-10}$$

$$= \begin{bmatrix} 0 & 0 \\ 0 & 0 \end{bmatrix}$$

Hence, the 2-port is passive.

In the case of a lossless RLC network, the analogs of Theorem 6-1 and Corollary 6-1 are as follows.

*Corollary 6-2

If $S(p)$ is the LBR scattering matrix of a lossless RLC n-port, it satisfies the following three conditions:

1. the entries in $S(p)$ are real rational functions
2. the poles of $S(p)$ are in the LHP
3. $1 - S(j\omega)^*S(j\omega) = 0$ for all imaginary axis $j\omega$. ☐

It is noteworthy that in the lossless case, it is not possible to apply an argument similar to that used in Theorem 6-1 to extend condition 3 to the entire RHP (wherein it is, in fact, false), since the maximum modulus theorem yields only an inequality for the interior of a region, even though equality may hold on the boundary. It is, however, possible to extend the equivalent Hurwitz conjugate condition to the entire RHP, since a rational function has at most a finite number of zeros unless it is identically zero. Thus, the adjoint condition on the $j\omega$ axis implies that $1 - S(p)_*S(p)$ is identically zero.

*Corollary 6-3

Under the hypotheses of Corollary 6-2, condition 3 is equivalent to both of the following:

3'. $1 - S(j\omega)_*S(j\omega) = 0$ for all imaginary axis $j\omega$
3''. $1 - S(p)_*S(p) = 0$ for all RHP p. ☐

Upon invoking the corollary, it is clear that the 2-port of Equation (6-8) was not only passive, but also lossless. As in the general case, the passivity condition

$$1 - S(j\omega)^*S(j\omega) = 1 - S(j\omega)_*S(j\omega) \geq 0 \qquad (6\text{-}11)$$

may be replaced by

$$\|S(j\omega)\| \leq 1 \qquad (6\text{-}12)$$

which is also independent of the adjoint.

The passivity conditions for RLC immittance matrices are the following.

*Theorem 6-2

If $Z(p)$ [$Y(p)$] is the positive real immittance matrix of a passive RLC network, it satisfies the following conditions:

1. $Z(p)$ [$Y(p)$] has real rational entries
2. $Z(p)$ [$Y(p)$] has no poles in the RHP

3. the $j\omega$ axis poles of $Z(p)$ $[Y(p)]$ are simple and have positive hermitian residue matrices

4. $Z(j\omega) + Z(j\omega)^* \geq 0$ $[Y(j\omega) + Y(j\omega)^* \geq 0]$ for all imaginary axis $j\omega$ where it is defined.

Moreover, any system function satisfying these conditions is PR. ☐

As with the scattering case, the theorem is not complete, for we have not yet shown that the four conditions imply that the n-port is RLC (though we have shown that an RLC n-port satisfies the conditions). This final portion of the theorem will be proven by the construction inherent in the synthesis to follow.

*Corollary 6-4

If $Z(p)$ $[Y(p)]$ is the LPR immittance matrix of a lossless RLC network, it satisfies conditions 1 through 3 of Theorem 6-2, and satisfies condition 4 with equality. Moreover, any $Z(p)$ $[Y(p)]$ satisfying these conditions is LPR. ☐

As in the passive case, we will show later, by synthesis, that any function satisfying the four conditions of the theorem is not only LPR, but also RLC (in fact, LC).

As in the scattering case, the adjoint in condition 4 of the above theorem and corollary can be replaced by the Hurwitz conjugate.

A further simplification of the LPR conditions for an RLC network is the following.

*Corollary 6-5

If $Z(p)$ $[Y(p)]$ is the LPR immittance matrix of a lossless RLC n-port, it satisfies the following conditions:

1. $Z(p)$ $[Y(p)]$ has real rational entries
2. the poles of $Z(p)$ $[Y(p)]$ are simple, on the $j\omega$ axis, and have positive hermitian residues.

Moreover, any function satisfying these two conditions is LPR. ☐

The admittance parameter

$$y(p) = \frac{p^3 + p^2 + p + 1}{p^3 + p} = \frac{(p+1)(p^2+1)}{p(p^2+1)} = \frac{(p+1)}{p} \quad \text{(6-13)}$$

is real rational and has a simple pole at zero. Now,

$$y(j\omega) + y(-j\omega) = \frac{j\omega + 1}{j\omega} + \frac{-j\omega + 1}{-j\omega} = \frac{j\omega + 1 + j\omega - 1}{j\omega}$$
$$= \frac{2j\omega}{j\omega} = 2 \quad \text{(6-14)}$$

is non-negative for all ω; hence, the admittance is indeed PR. It is, however, not LPR, since Equation (6-14) is not identically zero.

The effect of the preceding theory is to yield simplified passivity (and analyticity) tests on which the various synthesis procedures can be based. In fact, the existence of synthesis procedures employing only the hypotheses of the above theorems serves as a proof that these results are sufficient as well as necessary.

The poles and zeros of a rational system function serve two purposes. First, as has just been achieved, the characteristics of various classes of RLC networks can be determined by a study of their corresponding pole-zero patterns, and second, a degree of complexity can be attributed to an RLC network by counting its poles and zeros.

For a scaler function, the number of poles and zeros coincide (if those at infinity are included); hence, the degree of such a function (and its corresponding n-port) may be taken as the number of poles. In the matrix case, a pole may appear in one entry and not another, or it may have different multiplicities in different entries; hence, a more complex definition is required [31], [86], [110], [199].

*Definition 6-2

Let $H(p)$ be a matrix of real rational functions. Then the *degree of $H(p)$*, $\delta H(p)$, is

$$\delta H(p) = \sum_{p_i} m_i$$

where m_i is the maximum multiplicity that the pole at p_i assumes in any minor of $H(p)$, and the summation is taken over all poles (including those at infinity).
□

The definition coincides with the degree of McMillan [106]; however, the definition employed is that of Youla and Tissi [199]. Since the number of finite poles of a rational function is equal to the order of its denominator, while the number of poles at infinity is the difference between the numerator and denominator orders (if the numerator order exceeds that of the denominator, and zero otherwise), the degree of a rational function is just the greater of its numerator and denominator orders. McMillan's degree for a matrix thus coincides with the usual degree of a rational function.

For the matrix

$$H(p) = \begin{bmatrix} \dfrac{p+1}{p(p+2)} & \dfrac{1}{p(p+2)} \\ \dfrac{p+1}{p(p+2)} & \dfrac{1}{p(p+2)} \end{bmatrix} \tag{6-15}$$

the determinant is zero, but any of the minors have two poles, one at 0 and one at -2; hence,

$$\delta H(p) = 2 \tag{6-16}$$

We note from the example that, in general, it is not sufficient to test the determinant alone in determining the degree of a matrix; rather, it is necessary to check each minor. Alternatively, consider the matrix

$$H(p) = \begin{bmatrix} \dfrac{p+1}{p(p+2)} & \dfrac{-1}{p(p+2)} \\ \dfrac{p+1}{p(p+2)} & \dfrac{1}{p(p+2)} \end{bmatrix} \quad (6\text{-}17)$$

for which the determinant has four poles (two at 0 and two at -2), while the minors have only two poles (at 0 and -2); hence, the degree in this case is 4.

For Definition 6-2 to be meaningful, the degree must be a function of the n-port represented by a matrix, and not the matrix itself. To this end we have the following proposition.

*Proposition 6-2

Let an n-port be represented by real rational scattering and/or immittance matrices $S(p)$, $Z(p)$, and $Y(p)$. Then whenever they exist,

$$\delta S(p) = \delta Z(p) = \delta Y(p) \qquad \square$$

Physically, the degree of an n-port (that is, its matrix representation) is the minimum number of reactors (inductors and capacitors) with which it can be realized [106]. The fact that the degree is a lower bound on the number of reactors needed in a realization is due to Proposition 6-3 [86]. The fact that this lower bound can actually be reached will be shown (by construction) in the sequel.

*Proposition 6-3

Let an RLC n-port be characterized by scattering (immittance) matrices $S(p)$ [$Z(p)$, $Y(p)$]. Then every RLC realization of the n-port has at least

$$\delta S(p) = \delta Z(p) = \delta Y(p)$$

reactors. $\qquad \square$

Because of this property, the degree of an n-port proves to be of fundamental importance in the synthesis of RLC networks, this because a sequential procedure which reduces the degree of the network at each step is assured to converge (to a zero-degree n-port, which is easily realized).

Although we are primarily interested in the properties of passive n-ports, since the techniques of Chapter 5 allow the general synthesis problem to be reduced to a passive one, it is interesting to note that the nonenergetic n-ports have a very simple characterization in the RLC case [150]. From the results of Chapter 3, it is clear that $S(p)$ represents a nonenergetic RLC n-port if and only if Theorem 6-1 is satisfied with condition 3 replaced by

$$\|S(j\omega)\| < \infty \qquad (6\text{-}18)$$

But this is automatically satisfied by the requirement that all the poles of $S(p)$ be in the LHP; hence, the following proposition.

*Proposition 6-4

If $S(p)$ is the bounded real scattering matrix of a nonenergetic RLC n-port, its entries are real rational functions whose poles are in the LHP. Moreover, any such scattering matrix is nonenergetic. □

6-3 Richards' Theorem

In Chapter 2 it was found that the product of two passive (lossless) scattering matrices was also passive (lossless). Moreover, a method of realizing such a product via cascaded circulators was obtained. It is thus reasonable to consider the possibility of realizing a scattering matrix or parameter by factoring it into simpler, easily realized, parts. The basic theorem required to obtain such a factorization is a result of Richards [136]. For our purposes the 1-port form of the result is sufficient.

Theorem 6-3

Let an RLC 1-port be characterized by a BR scattering parameter $s(p)$ and an impedance parameter $z(p)$ (which exist for nontrivial RLC 1-ports), and let

$$s_{z(k)}(p)$$

be the scattering parameter of the 1-port when normalized to $z(k)$ for some (strictly) positive real k [hence, $z(k)$ is positive and real by the passivity of $z(p)$]. Then the scattering parameter

$$s_{z(k)}{}^r(p) = \frac{(p+k)}{(p-k)} s_{z(k)}(p)$$

is BR (LBR if $s(p)$ is LBR) and of degree no greater than $s(p)$. Furthermore, if

$$z(k) = -z(-k)$$

the degree of $s_{z(k)}{}^r(p)$ is strictly less than that of $s(p)$.

Proof

Writing $s_{z(k)}{}^r(p)$ in terms of $z(p)$ yields

$$s_{z(k)}{}^r(p) = \frac{(p+k)}{(p-k)}[z(p) + z(k)]^{-1}[z(p) - z(k)] \tag{6-19}$$

At

$$p = k \tag{6-20}$$

$$[z(p) - z(k)]\big|_{p=k} = [z(k) - z(k)] = 0 \tag{6-21}$$

Thus the apparent pole of $s_{z(k)}{}^r(p)$ in the RHP (at k) does not exist, since it is cancelled by a zero of $s_{z(k)}(p)$. Moreover, the degree of $s_{z(k)}{}^r(p)$ is no greater than that of $s(p)$ (since the extra pole is cancelled). Since $s(p)$ is BR, its poles are in the LHP (Theorem 6-1); hence, so are the poles of $s_{z(k)}{}^r(p)$ (which coincide with those of $s_{z(k)}(p)$ after the cancellation of the pole at $p = k$). Now the factor

$$\frac{(p+k)}{(p-k)} \tag{6-22}$$

is an "all pass factor" [183]; hence, the $j\omega$ axis properties of $s_{z(k)}{}^r(p)$ and $s_{z(k)}(p)$ coincide, whereas the rationality of $s_{z(k)}(p)$ (if it represents an RLC network) is preserved. $s_{z(k)}{}^r(p)$ thus satisfies the three conditions for an RLC scattering parameter to be BR (or LBR). Finally, if

$$z(k) = -z(-k) \tag{6-23}$$

the pole of $s_{z(k)}(p)$ at $-k$ cancels with $(p + k)$ so that the degree of $s_{z(k)}{}^r(p)$ is strictly less than that of $s(p)$. □

The theorem allows the factorization of $s_{z(k)}(p)$ as

$$s_{z(k)}(p) = \frac{(p-k)}{(p+k)} s_{z(k)}{}^r(p) \tag{6-24}$$

where $s_{z(k)}{}^r(p)$ is no more complex than $s(p)$. For synthesis, it is, however, necessary to decrease the complexity of $s_{z(k)}{}^r(p)$, this being accomplished by an opportune choice of k. In the lossless case this is easily achieved.

Corollary 6-6

Under the hypotheses of Theorem 6-3, if $s(p)$ is LBR, strict degree reduction is achieved for any k.

Proof

For losslessness,

$$z(p) + z(p)_* = 0 \tag{6-25}$$

for all p in the RHP; hence, for real strictly positive k,

$$0 = z(k) + z(k)_* = z(k) + z(-k) \tag{6-26}$$

or, equivalently,

$$z(k) = -z(-k) \tag{6-27}$$

so that by the theorem strict degree reduction is assured. □

The above theorem can be extended to the n-port case if the corresponding impedance matrix exists [110] and can be formulated in the nonrational case, but without degree reduction (which is undefined in that case). For our application the 1-port version is sufficient, since it can be applied to an n-port by factoring its determinant rather than the matrix itself [27].

Consider the scattering parameter (with unit normalization)

$$s(p) = \frac{(p-1+j)(p-1-j)}{(p+1-j)(p+1+j)} = \frac{p^2 - 2p + 2}{p^2 + 2p + 2} \quad (6\text{-}28)$$

with corresponding impedance

$$z(p) = \frac{p^2 + 2}{2p} \quad (6\text{-}29)$$

Since the poles and zeros of $s(p)$ are complex, it cannot be factored with its present normalization. If, however, we let

$$k = \sqrt{2} \quad (6\text{-}30)$$

in which case

$$z(\sqrt{2}) = \frac{\sqrt{2}^2 + 2}{2\sqrt{2}} = \sqrt{2} \quad (6\text{-}31)$$

$$z(p)_{z(k)} = z(p)_{\sqrt{2}} = \frac{z(p)}{\sqrt{2}}$$

$$= \frac{p^2 + 2}{2\sqrt{2}p} \quad (6\text{-}32)$$

and

$$s_{\sqrt{2}}(p) = \frac{p^2 - 2\sqrt{2}p + 2}{p^2 + 2\sqrt{2}p + 2} = \frac{(p - \sqrt{2})^2}{(p + \sqrt{2})^2} \quad (6\text{-}33)$$

The normalized scattering parameter clearly has the desired factorization.

The example illustrates the fundamental characteristic of the theorem; that is, even though a scattering parameter may not have a specified factorization in its given normalization, the factorization can always be obtained after an appropriate change of normalization.

6-4 Synthesis of Lossless 1-Ports

Richards' theorem allows one to readily establish synthesis procedures for lossless 1-ports. This is achieved first by employing the scattering parameter, and then via Foster's approach [67], where an immittance parameter is employed. Although the latter is simpler, the former serves as a stepping stone to more general scattering techniques to which the immittance approaches are not applicable.

The lossless synthesis of an LBR rational scattering parameter follows readily upon repeated applications of Richards' theorem, realizing each factorization by a circulator (or its gyrator equivalent) and each change of normalization by an ideal transformer.

Theorem 6-4

Let $s(p)$ be a real rational LBR scattering parameter. Then it can be realized by the structure of Figure 6-1, where $s_{z(k)}{}^1(p)$ is LBR rational and has degree strictly less than that of $s(p)$ and is normalized to $z(k)$ [where $z(p)$ is the impedance corresponding to $s(p)$], and k is a strictly positive real constant.

Figure 6-1 Step of lossless synthesis by factorization.

Proof

By Richards' theorem

$$s_{z(k)}(p) = \frac{(p-k)}{(p+k)} s_{z(k)}{}^1(p) \tag{6-34}$$

where

$$\frac{p-k}{p+k} \tag{6-35}$$

is the scattering parameter [normalized to $z(k)$] of a $1/k$ henry inductor. Hence, Proposition 2-6 allows $s_{z(k)}(p)$ to be realized by a circulator loaded with realizations of

$$\frac{p-k}{p+k} \tag{6-36}$$

and

$$s_{z(k)}{}^1(p) \tag{6-37}$$

To realize $s(p)$ it now suffices to insert an ideal transformer at the first port of the circulator, the turns ratio of which is such as to scale the voltage and current in correspondence to the required normalization. $s(p)$ is thus realized by the configuration of Figure 6-1. Finally, the corollary to Richards' theorem assures that the degree of $s_{z(k)}{}^1(p)$ is strictly less than that of $s(p)$ (since $s(p)$ is LBR). □

Since the degree of $s_{z(k)}{}^1(p)$ is strictly less than that of $s(p)$, repeated application of the theorem will yield, after $\delta s(p)$ steps, a zero-degree remainder

$$s_{z^{\delta-1}(k)}{}^{\delta}(p) \tag{6-38}$$

which is realizable by an open or short circuit (these being the only lossless RLC 1-ports of degree zero). We thus obtain a realization as per Figure 6-2.

Corollary 6-7

Every LBR rational scattering parameter $s(p)$ can be realized by the configuration of Figure 6-2 using $\delta s(p)$ inductors. Furthermore, this is the minimum number of inductors with which $s(p)$ can be realized. □

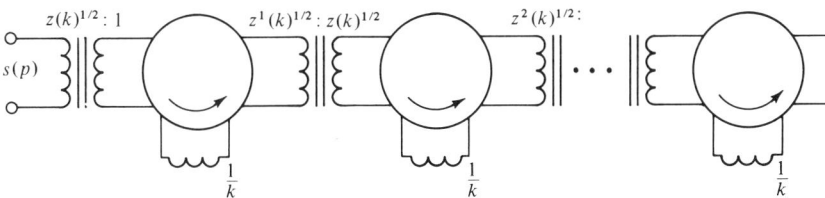

Figure 6-2 Realization of an LBR rational scattering parameter.

Combining the above corollary with the necessary condition of Corollary 6-2 yields the following necessary and sufficient condition.

Corollary 6-8

A necessary and sufficient condition for $s(p)$ to be the scattering parameter of a lossless RLC 1-port (actually LC, since resistors are not lossless) is that it be LBR rational. □

The realization of Figure 6-2 can be greatly simplified if one replaces the circulators by their gyrator equivalents and then embeds the ideal transformer turns ratios into the gyration constants. In this case, the realization of Figure 6-3 is obtained.

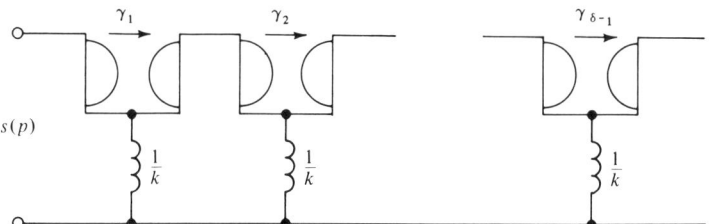

Figure 6-3 Alternate realization of an LBR rational scattering parameter.

Applying the realization to the unnormalized scattering parameter of the preceding section,

$$s(p) = \frac{p^2 - 2p + 2}{p^2 + 2p + 2} \qquad (6\text{-}39)$$

yields, after normalization to $\sqrt{2}$,

$$s_{\sqrt{2}}(p) = \frac{(p - \sqrt{2})^2}{(p + \sqrt{2})^2} \tag{6-40}$$

Hence, the two alternative realizations of Figure 6-4 are obtained.

Although, in general, the existence characteristics for scattering matrices are much more powerful than those of immittances, in the 1-port case the

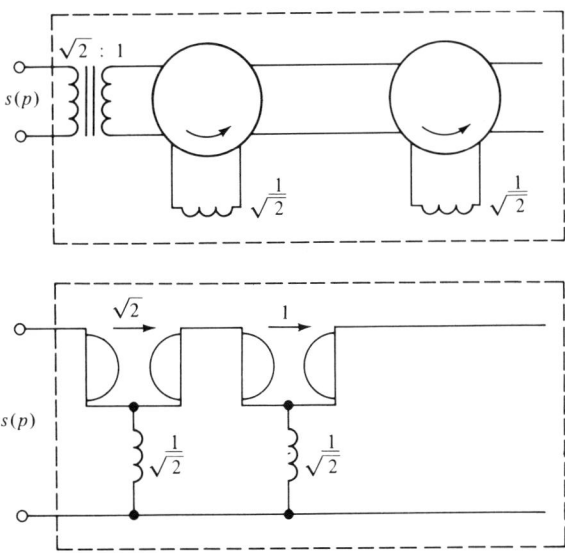

Figure 6-4 Two realizations of a lossless scattering parameter.

existence of one implies that of the other (except for some trivially realized networks, such as the open circuit). In this special case, it is, therefore, reasonable to take an immittance approach to the synthesis problem, rather than the scattering approach hitherto employed. In fact, the immittance techniques yield several straightforward lossless syntheses, which are not only transformerless, but, in fact, use the minimum possible number of components. Unlike the previous case, where the natural decomposition of a scattering parameter is as a product (which may be realized by a circulator), in the immittance synthesis one would like to decompose an immittance parameter into the sum of an elementary 1-port and a remainder (which can be realized by a series or parallel configuration). Our basic decomposition theorem is, therefore, the following.

Theorem 6-5

Let $z(p)$ [$y(p)$] be a rational immittance parameter with a pair of complex conjugate $j\omega$ axis poles

$$\frac{pA}{p^2 + \omega_0^2}$$

Then
$$z^1(p) = z(p) - \frac{pA}{p^2 + \omega_0{}^i}$$

is PR or LPR if $z(p)$ is PR or LPR [and similarly for $y(p)$].

Note that if ω_0 is either zero or infinite, the theorem still holds. However, the complex conjugate pair of poles reduces to a simple pole. Furthermore, the degree of $z^1(p)$ is strictly less than that of $z(p)$.

Proof

Since $z(p)$ is rational, we may employ the PR and LPR characterizations of Theorem 6-2 and its corollaries, rather than the more general description of Chapter 3. Now, clearly, if $z(p)$ is rational, so is $z^1(p)$, and since we are subtracting a pole of $z(p)$ to form $z^1(p)$, its poles have the same characteristics and residues as those of $z(p)$, except for the deletion of the pair at ω_0. In particular, the poles of $z^1(p)$ are in the LHP and/or on the $j\omega$ axis, in which case they are simple and have positive and real residues. Moreover, since we have removed a pole, the degree of $z^1(p)$ is strictly less than that of $z(p)$. To complete the proof, it remains to show that

$$z^1(j\omega) + z^1(-j\omega) \geq (= 0) \tag{6-41}$$

in the PR and LPR cases, respectively. Now,

$$z^1(j\omega) + z^1(-j\omega) = z(j\omega) - \frac{j\omega A}{-\omega^2 + \omega_0{}^2}$$
$$+ z(-j\omega) - \frac{-j\omega A}{-\omega^2 + \omega_0{}^2}$$
$$= z(j\omega) + z(-j\omega) \tag{6-42}$$

which is positive if $z(p)$ is PR, and zero if $z(p)$ is LPR, thereby completing the proof. □

Although the theorem plays essentially the same role in immittance synthesis as does Richards' theorem in the scattering case, it is actually a weaker result. This is because normalization of an immittance does not move its poles and zeros; hence, the theorem only permits the extraction of a pole that is initially on the $j\omega$ axis, but does not give one a means of moving a pole to such a location, as does Richards' theorem.

The theorem may be implemented in a number of ways to obtain a transformerless realization for an LPR rational 1-port. Possibly most direct is the synthesis of Foster [67]. Since an LPR function has all of its poles on the $j\omega$ axis, and these are simple with positive real residues, one may start with an arbitrary LPR rational $z(p)$ and extract a pair of poles (or a single pole at zero or infinity) yielding the decomposition

$$z(p) = \frac{pA_1}{p^2 + \omega_1{}^2} + z^1(p) \tag{6-43}$$

where, by the theorem, $z^1(p)$ is still LPR and of degree strictly less than that of $z(p)$. Now, extracting another pair of poles from $z^1(p)$ yields

$$z(p) = \frac{pA_1}{p^2 + \omega_1{}^2} + \frac{pA_2}{p^2 + \omega_2{}^2} + z^2(p) \tag{6-44}$$

Repeating the process of pole extraction until a zero remainder is reached (after $\delta z(p)$ poles have been extracted), we obtain the partial fraction expansion

$$z(p) = A_\infty p + \frac{A_0}{p} + \sum_i \frac{pA_i}{p^2 + \omega_i{}^2} \tag{6-45}$$

for $z(p)$. Here we have included possible poles at zero and infinity, as well as pairs of poles on the $j\omega$ axis. Note also that all of the A's are positive and real, since they are the residues of $j\omega$ axis poles.

To realize $z(p)$ it remains to synthesize each term of the partial fraction expansion and realize $z(p)$ as the series connection of such. Now, clearly the impedance

$$A_\infty p \tag{6-46}$$

corresponds to an inductor, whereas the impedance

$$\frac{A_0}{p} \tag{6-47}$$

corresponds to a capacitor. Finally, the impedance

$$\frac{pA_i}{p^2 + \omega_i{}^2} \tag{6-48}$$

may be realized via the "tank" circuit of Figure 6-5(c). The complete realization is thus the series configuration of Figure 6-6, for which we have the following proposition.

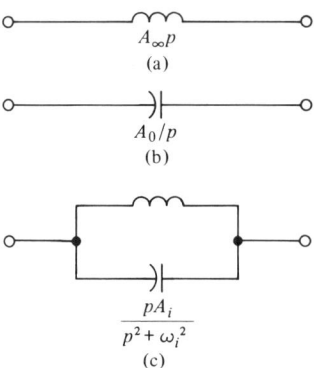

Figure 6-5 Components for Foster's LC synthesis.

Proposition 6-5

Every LPR rational impedance has a transformerless realization in the form of Figure 6-6 using $\delta z(p)$ reactors. □

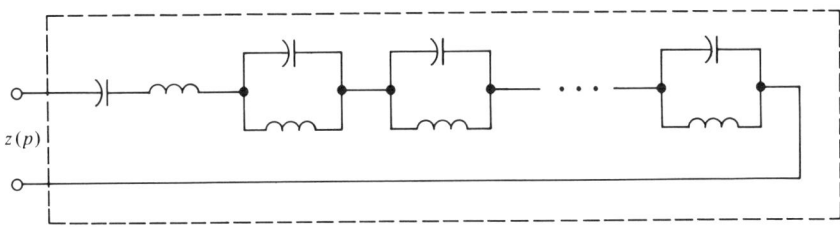

Figure 6-6 Foster form of a general LPR rational impedance.

Since it has already been shown that $z(p)$ could not be realized with less than $\delta z(p)$ reactors, Foster's synthesis is minimal.

Upon converting the scattering parameter

$$s(p) = \frac{p^2 - 2p + 2}{p^2 + 2p + 2} \qquad (6\text{-}49)$$

realized in the preceding section to an impedance,

$$z(p) = \frac{p^2 + 2}{2p} \qquad (6\text{-}50)$$

is obtained. Now, the partial fraction expansion of $z(p)$ is

$$z(p) = \frac{p}{2} + \frac{1}{p} \qquad (6\text{-}51)$$

Hence, the realization of Figure 6-7 is readily obtained.

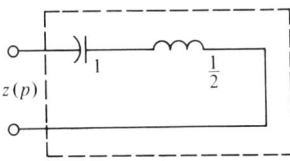

Figure 6-7 Foster form of an LPR rational scattering parameter.

If, rather than starting with an impedance, one starts with an LPR rational admittance and carries out a process of pole extraction similar to that employed for the impedance, one obtains the partial fraction expansion

$$y(p) = B_\infty p + \frac{B_0}{p} + \sum_i \frac{p B_i}{p^2 + \omega_i^2} \qquad (6\text{-}52)$$

which leads to a realization of parallel connected series "tanks," as shown in Figure 6-8.

If we invert the impedance of Equation (6-51) and carry out the indicated partial fraction expansion, we obtain

$$y(p) = \frac{2p}{p^2 + 2} \tag{6-53}$$

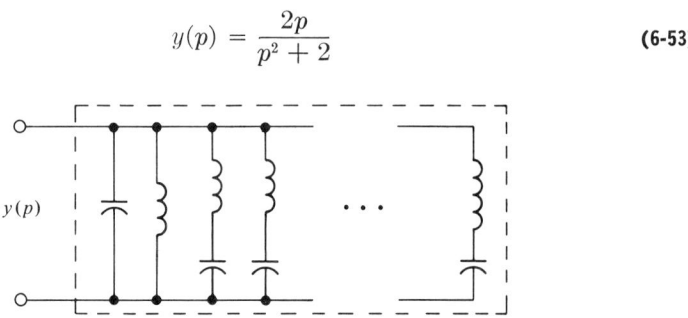

Figure 6-8 Foster form for a general LPR rational admittance.

and the realization shown in Figure 6-9, which, in this special case, happens to coincide with that obtained by the impedance approach.

In general, the poles of a PR function are not restricted to the $j\omega$ axis; hence, the extraction techniques indicated by Theorem 6-4 will not, in general, yield

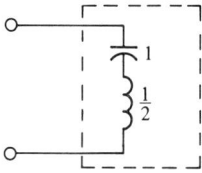

Figure 6-9 Foster form for an LPR rational admittance.

a complete synthesis of a given PR rational function. It is, however, possible to obtain a complete, and, in fact, transformerless, realization of an arbitrary PR rational function by combining the Foster and Richards theorem techniques to obtain the Bott–Duffin realization [38], [183]. This result is a well-known part of classical network theory, and will not be considered here.

6-5 Darlington-Type Realizations

In this section we consider two realizations for a passive RLC 1-port obtained by specializing the general techniques of Chapter 5. Both realizations take on the Darlington form [56] of Figure 6-10. Although they are not transformerless, the number of components is quite small. In fact, if one is willing to allow a nonreciprocal coupling network, the realization may be chosen so as to simultaneously use a minimum of reactors and resistors. Moreover, the existence of such a realization is a useful tool in itself. Here we consider only the determination of the necessary lossless coupling networks, with their realization being deferred until the final section of the chapter.

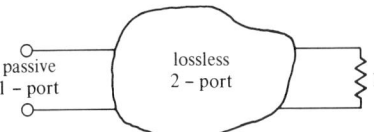

Figure 6-10 Darlington-type realization of a 1-port.

Upon invoking the general results of Proposition 5-2, it is found that a BR rational scattering parameter $S(p)$ can be realized at the first port of the Darlington configuration by the coupling network

$$\Sigma(p) = \left[\begin{array}{c|c} s(p) & \sigma_{12}(p) \\ \hline \sigma_{21}(p) & \sigma_{22}(p) \end{array}\right] \quad (6\text{-}54)$$

which satisfies the existence and adjoint condition

$$1 - \Sigma(j\omega)^*\Sigma(j\omega) = 0 \quad (6\text{-}55)$$

for losslessness if

$$\sigma_{12}(p)\sigma_{12}(p)^* = 1 - s(p)s(p)^* \quad (6\text{-}56)$$

$$\sigma_{21}(p)^*\sigma_{21}(p) = 1 - s(p)^*s(p) \quad (6\text{-}57)$$

and

$$\sigma_{22}(p) = -\sigma_{21}(p)s(p)^*\sigma_{12}(p)^{*-1} \quad (6\text{-}58)$$

The synthesis problem is thus reduced to choosing $\sigma_{12}(p)$ and $\sigma_{21}(p)$ as factorizations of

$$1 - s(p)s(p)^* = 1 - s(p)^*s(p) \quad (6\text{-}59)$$

so as to assure the $\Sigma(p)$ is real rational and analytic in the RHP (the remaining condition needed to render $\Sigma(p)$ lossless). Achieving the first of these conditions greatly simplifies the second, since the only singularities of a rational function are its (easily located) poles.

To achieve rationality, one must replace the adjoint in Equations (6-55) through (6-58) by the Hurwitz conjugate, since the adjoint is not real rational. This is done without loss of the $j\omega$ axis adjoint condition of Equation (6-55), since the adjoint and the Hurwitz conjugate coincide on the $j\omega$ axis, while it achieves the required rationality because

$$H(p)_* = H(-p)' \quad (6\text{-}60)$$

is rational if $H(p)$ is rational. We then have the coupling network

$$\Sigma(p) = \left[\begin{array}{c|c} s(p) & \sigma_{12}(p) \\ \hline \sigma_{21}(p) & \sigma_{22}(p) \end{array}\right] \quad (6\text{-}61)$$

where

$$\sigma_{12}(p)\sigma_{12}(-p) = 1 - s(p)s(-p) \tag{6-62}$$

$$\sigma_{21}(-p)\sigma_{21}(p) = 1 = s(p)s(-p) \tag{6-63}$$

and

$$\sigma_{22}(p) = -\sigma_{21}(p)s(-p)\sigma_{12}(-p)^{-1} \tag{6-64}$$

The choice of a factorization of $1 - s(p)s(-p)$, which yields the required RHP analyticity, is aided by the following lemma.

Lemma 6-1

Let $s(p)$ be a BR rational scattering parameter. Then

$$1 - s(p)s(-p) = \frac{k(p)k(-p)}{m(p)m(-p)}$$

where $k(p)$ is Hurwitz and $m(p)$ is the strictly Hurwitz denominator of $s(p)$, that is,

$$s(p) = \frac{n(p)}{m(p)}$$

Proof

Let

$$s(p) = \frac{n(p)}{m(p)} \tag{6-65}$$

where $m(p)$ is (strictly) Hurwitz, since the poles of a BR rational scattering parameter are (by Theorem 6-1) in the LHP. Now,

$$\begin{aligned} 1 - s(p)s(-p) &= 1 - \frac{n(p)n(-p)}{m(p)m(-p)} \\ &= \frac{m(p)m(-p) - n(p)n(-p)}{m(p)m(-p)} \end{aligned} \tag{6-66}$$

and, since $s(p)$ is BR,

$$1 - s(j\omega)s(-j\omega) \geq 0 \tag{6-67}$$

or, equivalently,

$$m(j\omega)m(-j\omega) - n(j\omega)n(-j\omega) \geq 0 \tag{6-68}$$

Since Equation (6-68) is positive and even, it has a rational square root $k(j\omega)$ such that

$$k(j\omega)k(j\omega)^* = k(j\omega)k(-j\omega) = m(j\omega)m(-j\omega) - n(j\omega)n(-j\omega) \tag{6-69}$$

and $k(j\omega)$ is Hurwitz. Upon setting

$$k(p) = k(j\omega)\,|_{j\omega=p} \tag{6-70}$$

one has, by analytic continuation, that

$$k(p)k(-p) = m(p)m(-p) - n(p)n(-p) \qquad (6\text{-}71)$$

and

$$1 - s(p)s(-p) = \frac{k(p)k(-p)}{m(p)m(-p)} \qquad (6\text{-}72)$$

where $m(p)$ is the strictly Hurwitz denominator of $s(p)$. □

A convenient graphical interpretation of the lemma is that the poles and zeros of a positive, even, rational function [such as $1 - s(p)_*s(p)$] are distributed

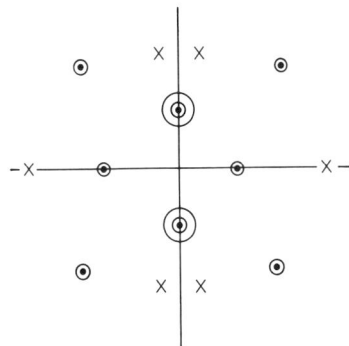

Figure 6-11 Quadrilateral symmetry for the poles and zeros of an even function.

on the complex plane with quadrilateral symmetry, as illustrated in Figure 6-11. By this it is meant that the poles and zeros appear symmetrically about the origin in fours as

$$a + jb$$
$$a - jb$$
$$-a + jb$$

and

$$-a - jb$$

as illustrated. Of course, if $b = 0$, only a pair of poles or zeros need appear.

With this interpretation, the factorization of the lemma simply corresponds to taking the LHP zeros together with half of the $j\omega$ axis zeros to form $k(p)$, in which case the remainder of the zeros form $k(-p)$.

With the establishment of the lemma, the 1-port Darlington synthesis via Belevitch's approach [27], [29] may be derived in either of two forms, one with a minimal degree $\Sigma(p)$, and the other with a reciprocal $\Sigma(p)$.

Theorem 6-6

Let $s(p)$ be a BR rational scattering parameter, and let

$$1 - s(p)s(-p) = \frac{k(p)k(-p)}{m(p)m(-p)}$$

where

$$s(p) = \frac{n(p)}{m(p)}$$

are as in Lemma 6-1. Then $s(p)$ can be realized by the configuration of Figure 6-10, with the coupling network

$$\Sigma(p) = \begin{bmatrix} s(p) & \sigma_{12}(p) \\ \hline \sigma_{21}(p) & \sigma_{22}(p) \end{bmatrix} = \begin{bmatrix} \dfrac{n(p)}{m(p)} & \dfrac{k(-p)}{m(p)} \\ \hline \dfrac{k(p)}{m(p)} & \dfrac{-n(-p)}{m(p)} \end{bmatrix}$$

which is lossless and satisfies

$$\delta\Sigma(p) = \delta s(p)$$

Proof

$\Sigma(p)$ is clearly real rational, since $s(p)$ is real rational and has the same poles as $s(p)$; hence, its poles are in the LHP (since those of a BR rational $s(p)$ are in the LHP). Moreover, the poles of the minors and the determinant are the same as those of $s(p)$; hence,

$$\delta\Sigma(p) = \delta s(p) \qquad (6\text{-}73)$$

To complete the proof, it must be shown that

$$1 - \Sigma(j\omega)_* \Sigma(j\omega) = 0 \qquad (6\text{-}74)$$

Now,

$$\sigma_{12}(p)\sigma_{12}(-p) = \frac{k(-p)}{m(p)}\frac{k(p)}{m(-p)} = \frac{k(p)k(-p)}{m(p)m(-p)} \qquad (6\text{-}75)$$

$$= 1 - s(p)s(-p)$$

$$\sigma_{21}(-p)\sigma_{21}(p) = \left(\frac{k(p)}{m(p)}\right)\left(\frac{k(-p)}{m(-p)}\right) = \frac{k(p)k(-p)}{m(p)m(-p)} \qquad (6\text{-}76)$$

and

$$= 1 - s(p)s(-p) - 1 - s(-p)s(p)$$

$$\sigma_{22}(p) = \frac{-n(-p)}{m(p)} = -\frac{k(p)n(-p)m(-p)}{m(p)m(-p)k(p)} \qquad (6\text{-}77)$$

$$= -\sigma_{21}(p)s(-p)\sigma_{12}(-p)^{-1}$$

which are exactly the conditions required by Proposition 5-2 for Equation (6-74) to hold. $\Sigma(p)$ is thus real rational with poles in the LHP, and satisfies the condition of Equation (6-74); hence, by Corollary 6-1, it is lossless. □

Consider the BR rational scattering parameter

$$s(p) = \frac{p^2 - 2\sqrt{2}p + 2}{2p^2 + 4\sqrt{2}p + 4} \qquad (6\text{-}78)$$

$$= \frac{(p - \sqrt{2})^2}{2(p + \sqrt{2})^2}$$

In this case,

$$1 - s(p)s(-p) = \frac{2(p + \sqrt{2})^2(-p + \sqrt{2})^2 - (p - \sqrt{2})^2(-p - \sqrt{2})^2}{2(p + \sqrt{2})^2(-p + \sqrt{2})^2} \qquad (6\text{-}79)$$

$$= \left(\frac{(p + \sqrt{2})^2}{\sqrt{2}(p + 2)^2}\right)\left(\frac{(-p + \sqrt{2})^2}{\sqrt{2}(-p + \sqrt{2})^2}\right) = \frac{k(p)k(-p)}{m(p)m(-p)}$$

where

$$k(p) = \frac{(p + \sqrt{2})^2}{\sqrt{2}} \qquad (6\text{-}80)$$

and

$$m(p) = (p + \sqrt{2})^2 \qquad (6\text{-}81)$$

The lossless coupling network obtained by the Belevitch approach of the theorem is thus

$$\Sigma(p) = \begin{bmatrix} \dfrac{n(p)}{m(p)} & \dfrac{k(-p)}{m(p)} \\ \dfrac{k(p)}{m(p)} & \dfrac{-n(-p)}{m(p)} \end{bmatrix}$$

$$= \begin{bmatrix} \dfrac{(p - \sqrt{2})^2}{2(p + \sqrt{2})^2} & \dfrac{(p + \sqrt{2})^2}{2(p + \sqrt{2})^2} \\ \dfrac{1}{\sqrt{2}} & \dfrac{-(p + \sqrt{2})^2}{2(p + \sqrt{2})^2} \end{bmatrix} \qquad (6\text{-}82)$$

To complete the realization of $s(p)$, it is still necessary to realize $\Sigma(p)$. In Section 6-7 this will be accomplished using

$$\delta\Sigma(p) = \delta s(p) \qquad (6\text{-}83)$$

reactors. Since this is the minimum number of reactors with which $s(p)$ could have been realized, the complete realization will be minimal reactor and minimal resistor (since every passive network which is not lossless requires at least one resistor in any of its realizations).

Since the 1-port being realized is known to be reciprocal (every 1-port with a scattering matrix is reciprocal), it would be desirable to obtain a $\Sigma(p)$ which was reciprocal, so that the complete realization will need only reciprocal components. This is, indeed, possible, but at the cost of increasing the degree of $\Sigma(p)$. To this end, we have the following theorem, the proof of which is essentially the same as in the preceding case.

*Theorem 6-7

Let $s(p)$ be a BR rational scattering parameter, and let

$$1 - s(p)s(-p) = \frac{k(p)k(-p)}{m(p)m(-p)}$$

where

$$s(p) = \frac{n(p)}{m(p)}$$

are as in Lemma 6-1. Then $s(p)$ can be realized by the configuration of Figure 6-10, with the coupling network

$$\Sigma(p) = \begin{bmatrix} s(p) & | & \sigma_{12}(p) \\ ---- & + & ---- \\ \sigma_{21}(p) & | & \sigma_{22}(p) \end{bmatrix}$$

$$= \begin{bmatrix} \dfrac{n(p)}{m(p)} & | & \dfrac{k(-p)}{m(p)} \\ ---- & + & ---------- \\ \dfrac{k(-p)}{m(p)} & | & -\dfrac{k(-p)n(-p)}{m(p)k(p)} \end{bmatrix}$$

which is lossless and reciprocal. □

Note that the $j\omega$ axis zeros of $k(p)$ and $k(-p)$ coincide; hence, in $\sigma_{22}(p)$, such zeros cancel and do not induce $j\omega$ axis poles in $\Sigma(p)$.

If the reciprocal version of the Belevitch realization is employed for the realization of the scattering parameter

$$s(p) = \frac{(p - \sqrt{2})^2}{2(p + \sqrt{2})^2} \tag{6-84}$$

the factorization of Equation (6-79) now yields

$$\Sigma(p) = \begin{bmatrix} \dfrac{(p - \sqrt{2})^2}{2(p + \sqrt{2})^2} & | & \dfrac{(-p + \sqrt{2})^2}{\sqrt{2}(p + \sqrt{2})^2} \\ ---------- & | & ---------- \\ \dfrac{(-p + \sqrt{2})^2}{\sqrt{2}(p + \sqrt{2})^2} & | & \dfrac{-(-p - \sqrt{2})^2(-p + \sqrt{2})^2}{2(p + \sqrt{2})^4} \end{bmatrix} \tag{6-85}$$

Although this coupling network is reciprocal, its degree is greater than that of the nonreciprocal coupling network of Equation (6-82).

Alternatively, the Darlington-type realization of a passive RLC 1-port can be obtained via the Bayard approach, the general form of which is given in

Section 5-5. In this case, it follows from Theorem 5-4 that if a PR rational impedance parameter

$$z(p) = z_h(p) + z_s(p) \tag{6-86}$$

is written as the sum of its hermitian and skew hermitian parts, it can be realized as per Figure 6-10 if the coupling network has impedance matrix

$$x(p) = \begin{bmatrix} x_{11}(p) & | & x_{12}(p) \\ \hline x_{21}(p) & | & x_{22}(p) \end{bmatrix} \tag{6-87}$$

where

$$x_{22}(p) + x_{22}(p)^* = 0 \tag{6-88}$$

$$x_{11}(p) = z_s(p) + n(p)x_{22}(p)n(p)^* \tag{6-89}$$

and

$$x_{12}(p) = -x_{21}(p)^* = n(p)(1 + x_{22}(p)) \tag{6-90}$$

with

$$n(p)n(p)^* = z_h(p) \tag{6-91}$$

As in the previous case, so as to maintain rationality, the adjoint in Equations (6-87) through (6-91) is replaced by the Hurwitz conjugate. This does not affect the $j\omega$ axis behavior of $X(p)$, while assuring its rationality; hence,

$$X(p) = \begin{bmatrix} x_{11}(p) & | & x_{12}(p) \\ \hline x_{21}(p) & | & x_{22}(p) \end{bmatrix} \tag{6-92}$$

satisfying

$$x_{22}(p) + x_{22}(-p) = 0 \tag{6-93}$$

$$x_{11}(p) = \text{Od } z(p) + n(p)x_{22}(p)n(-p) \tag{6-94}$$

and

$$x_{12}(p) = -x_{21}(-p) = n(p)(1 + x_{22}(p)) \tag{6-95}$$

with

$$n(p)n(-p) = \text{Ev } z(p) \tag{6-96}$$

is rational and satisfying the $j\omega$ axis condition for losslessness that

$$X(j\omega) + X(j\omega)^* = X(j\omega) + X(j\omega)_* = 0 \tag{6-97}$$

The synthesis problem is thus reduced to choosing $x_{22}(p)$ and $n(p)$ so as to place the poles of $X(p)$ on the $j\omega$ axis (where the poles of any lossless impedance must lie, by Corollary 6-5). Here,

$$\text{Od } z(p) = \frac{1}{2}[z(p) - z(-p)] \tag{6-98}$$

and

$$\text{Ev } z(p) = \frac{1}{2}[z(p) + z(-p)] \tag{6-99}$$

are the odd and even parts of $z(p)$, which are the analogs of the skew hermitian and hermitian parts of $z(p)$ when the Hurwitz conjugate replaces the adjoint.

It is convenient to deal with the special case where the poles of the impedance to be realized are in the (strict) LHP. This can be done without loss of generality, since any PR rational impedance can be decomposed, via a process of pole extraction (Theorem 6-5), as

$$z(p) = z_L(p) + z_H(p) \tag{6-100}$$

where $z_L(p)$ is lossless and $z_H(p)$ is PR rational with all poles in the LHP. Since $z_L(p)$ can be realized by the methods of Section 6-4, the general synthesis problem is reduced to the required special case.

The immittance analog of Lemma 6-1 is Lemma 6-2, which may also be given a graphical interpretation in terms of quadrilateral symmetry.

*Lemma 6-2

Let $z(p)$ be a PR rational impedance parameter with poles in the LHP. Then

$$\text{Ev } z(p) = \frac{z(p) + z(-p)}{2} = \frac{k(p)k(-p)}{m(p)m(-p)}$$

where $k(p)$ is a Hurwitz polynomial and $m(p)$ is the strictly Hurwitz denominator of $z(p)$. □

An additional lemma needed for the synthesis is the following classical result of Hurwitz [78], [183].

*Lemma 6-3

$m(p)$ is a (strictly) Hurwitz polynomial if and only if

$$\frac{\text{Ev } m(p)}{\text{Od } m(p)}$$

is an LPR rational impedance parameter. □

Lemma 6-3 is the network theoretic interpretation of the Hurwitz test [78] of classical analysis, the continued fraction expansion portion of which corresponds to the Cauer synthesis [183] of

$$\frac{\text{Ev } m(p)}{\text{Od } m(p)} \tag{6-101}$$

(that is, $m(p)$ is Hurwitz if and only if the Cauer synthesis [78] of Equation (6-101) succeeds in realizing a lossless 1-port).

The lossless coupling network needed to realize a PR rational impedance with LHP poles may now be obtained via the following theorem.

*Theorem 6-8

Let $z(p)$ be a PR rational impedance parameter with LHP poles and

$$\text{Ev } z(p) = \frac{k(p)k(-p)}{m(p)m(-p)}$$

where

$$z(p) = \frac{n(p)}{m(p)}$$

are as in Lemma 6-2. Then $z(p)$ is realized by the configuration of Figure 6-10, with the coupling network

$$x(p) = \left[\begin{array}{c|c} \text{Od } z(p) + \dfrac{\text{Ev } z(p) \text{Ev } m(p)}{\text{Od } m(p)} & \dfrac{k(p)}{\text{Od } m(p)} \\ \hline \dfrac{k(-p)}{\text{Od } m(p)} & \dfrac{\text{Ev } m(p)}{\text{Od } m(p)} \end{array} \right]$$

which is lossless. □

Of course, in the above theorem, impedance has no special significance and a dual theorem may be obtained using admittances.

Consider the impedance

$$z(p) = \frac{p^2 + p + 1}{p^2 + p + 4} \tag{6-102}$$

which has no $j\omega$ axis poles or zeros. Here,

$$\text{Od } z(p) = \frac{3p}{p^4 + 4p^2 + 16} = \frac{3p}{m(p)m(-p)} \tag{6-103}$$

and

$$\text{Ev } z(p) = \frac{p^4 + 4p^2 + 4}{p^4 + 7p^2 + 16} = \frac{k(p)k(-p)}{m(p)m(-p)} \tag{6-104}$$

where

$$k(p) = p^2 + 2 \tag{6-105}$$

and

$$m(p) = p^2 + p + 4 \tag{6-106}$$

is the denominator of the original function. Now, upon invoking Theorem 6-8, the desired lossless coupling network with impedance matrix

$$X(p) = \left[\begin{array}{c|c} \dfrac{p^2 + 1}{p} & \dfrac{p^2 + 2}{p} \\ \hline \dfrac{p^2 + 2}{2} & \dfrac{p^2 + 4}{p} \end{array} \right] \tag{6-107}$$

is obtained. Decomposing $X(p)$ as

$$X(p) = X_1(p) + X_2(p) = \begin{bmatrix} 1 & 1 \\ 1 & 1 \end{bmatrix} p + \begin{bmatrix} 1 & 2 \\ 2 & 4 \end{bmatrix} \frac{1}{p}$$

$$= \begin{bmatrix} 1 \\ 1 \end{bmatrix} p [1 \quad 1] + \begin{bmatrix} 1 \\ 2 \end{bmatrix} \frac{1}{p} [1 \quad 2]$$

(6-108)

and recognizing, from Chapter 4, that the congruence transformation

$$Z(p) = T'Z_L(p)T \qquad (6\text{-}109)$$

indicates the realization of $Z(p)$ as an ideal transformer with turns ratio matrix T loaded by a realization of $Z_L(p)$, one may realize $X(p)$ by the series configuration shown in Figure 6-12, and hence $z(p)$, upon loading the second port of $X(p)$ with a 1-ohm resistor.

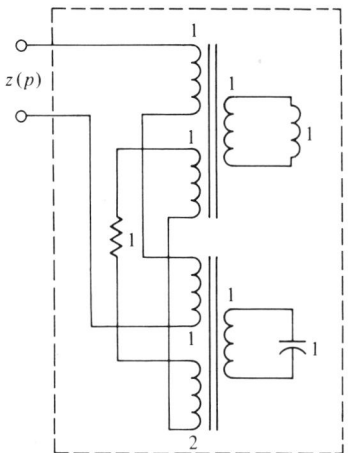

Figure 6-12 Realization of a PR rational impedance.

6-6 Synthesis of Passive n-Ports

The techniques of the preceding section can be readily extended to include the n-port case. As in the special case, one starts with the general results of Chapter 5, replacing the adjoint by the Hurwitz conjugate to preserve rationality. The synthesis problem is, as in the previous section, reduced to one of factoring the appropriate positive parahermitian (that is, even) matrix, so as to assure that $\Sigma(p)$ or $X(p)$ will be analytic in the RHP. In the n-port case, the factorization Lemmas 6-1 and 6-2 are replaced by the Gauss factorization [110] or a similar polynomial factorization [27], [31], [98], [123], [192]. In any case, a factorization theorem, such as that of Theorem 6-9, is obtained. The reader is referred to the references cited above for the rather tedious proof, which will not be given here.

*Theorem 6-9

Let $R(p)$ be an even $[R(p) = R(p)_* = R(-p)']$ matrix of real polynomials with (normal) rank r such that
$$R(j\omega) \geq 0$$
for all imaginary axis $j\omega$. Then
$$R(p) = K(p)K(p)_* = K(p)K(-p)'$$
where $K(p)$ is an $n \times r$ matrix of real polynomials such that $K(p)_*$ has a right inverse which is analytic in the RHP. Here r is the rank (a.e) of $R(p)$. □

The theorem is readily modified to cover the case of real rational matrices by treating such a matrix as a polynomial matrix divided by its least common denominator, a polynomial.

With the factorization established, we may proceed to the actual synthesis. From the general results of Chapter 5, after substitution of the Hurwitz conjugate for the adjoint, one may realize a BR rational scattering matrix
$$S(p) = \frac{N(p)}{m(p)} \qquad (6\text{-}110)$$
where $m(p)$ is the least common denominator of $S(p)$ and $N(p)$ is a polynomial matrix by loading a realization of
$$\Sigma(p) = \begin{bmatrix} S(p) & \Sigma_{12}(p) \\ \hline \Sigma_{21}(p) & \Sigma_{22}(p) \end{bmatrix} \qquad (6\text{-}111)$$

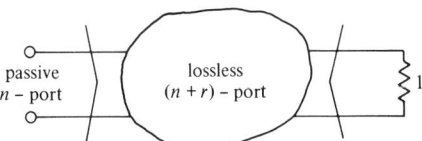

Figure 6-13 Configuration of passive n-port synthesis.

in 1-ohm resistors, as shown in Figure 6-13. Here, if
$$\Sigma_{12}(p)\Sigma_{12}(p)_* = 1 - S(p)S(p)_* \qquad (6\text{-}112)$$
$$\Sigma_{21}(p)_*\Sigma_{21}(p) = 1 - S(p)_*S(p) \qquad (6\text{-}113)$$
and
$$\Sigma_{22}(p) = -\Sigma_{21}(p)S(p)_*\Sigma_{12}(p)_*^{-1} \qquad (6\text{-}114)$$
$\Sigma(p)$ is always real rational (if $S(p)$ is real rational) and
$$1 - \Sigma(j\omega)_*\Sigma(j\omega) = 0 \qquad (6\text{-}115)$$

Hence, it is lossless if its poles are in the LHP. Unlike the 1-port case,

$$1 - S(p)S(p)_* \neq 1 - S(p)_*S(p) \qquad (6\text{-}116)$$

However, the required factorizations can still be obtained upon invoking Theorem 6-9.

*Lemma 6-4

Let $S(p)$ be an $n \times n$ BR rational scattering matrix. Then

$$1 - S(p)S(p)_* = \frac{K(p)K(p)_*}{m(p)m(p)_*}$$

and

$$1 - S(p)_*S(p) = \frac{L(p)L(p)_*}{m(p)m(p)_*}$$

Here $m(p)$ is the strictly Hurwitz least common denominator of $S(p)$, while $K(p)$ and $L(p)$ are $n \times r$ [r, the normal rank of $1 - S(p)S(p)_*$] polynomial matrices such that $K(p)_*$ and $L(p)_*$ have right inverses which are analytic in the RHP. □

The lemma follows readily from the theorem, upon writing

$$1 - S(p)S(p)_* = 1 - \frac{N(p)N(p)_*}{m(p)m(p)_*}$$

$$= \frac{m(p)m(p)_* 1 - N(p)N(p)_*}{m(p)m(p)_*} \qquad (6\text{-}117)$$

and applying the theorem to the numerator. With these factorizations, the remainder of the realization follows from essentially the same arguments as the 1-port realization of the previous section.

*Theorem 6-10

Let $S(p)$ be an $n \times n$ BR rational scattering matrix with

$$1 - S(p)S(p)_* = \frac{K(p)K(p)_*}{m(p)m(p)_*}$$

and

$$1 - S(p)_*S(p) = \frac{L(p)L(p)_*}{m(p)m(p)_*}$$

as in Lemma 6-4, where

$$S(p) = \frac{N(p)}{m(p)}$$

SYNTHESIS OF PASSIVE n-PORTS 233

Then $S(p)$ can be realized at the first n ports of the lossless $(n + r)$-port

$$\Sigma(p) = \left[\begin{array}{c|c} \dfrac{N(p)}{m(p)} & \dfrac{K(p)}{m(p)} \\ \hline \dfrac{L(p)_*}{m(p)} & \dfrac{-L(p)_*N(p)_*}{m(p)}K(p)_*^{-R} \end{array}\right]$$

when the last r ports are loaded in 1-ohm resistors (as per Figure 6-13). □

Note that $L(p)$ and $K(p)$ have the same $j\omega$ axis zeros; hence, the apparent $j\omega$ axis poles in $\Sigma_{22}(p)$ are canceled.

Consider the one-way coupler characterized by the scattering matrix

$$S(p) = \begin{bmatrix} 0 & \dfrac{p-1}{p+1} \\ 0 & 0 \end{bmatrix} \tag{6-118}$$

for which

$$1 - S_*(p)S(p) = \begin{bmatrix} 1 & 0 \\ 0 & 0 \end{bmatrix} \tag{6-119}$$

and

$$1 - S(p)S_*(p) = \begin{bmatrix} 0 & 0 \\ 0 & 1 \end{bmatrix} \tag{6-120}$$

Now, upon factoring these matrices as per the theorem, one obtains

$$\Sigma_{21}(p)_*\Sigma_{21}(p) = \begin{bmatrix} \dfrac{p+1}{p-1} \\ 0 \end{bmatrix} \begin{bmatrix} \dfrac{p-1}{p+1} & 0 \end{bmatrix} = \begin{bmatrix} 1 & 0 \\ 0 & 0 \end{bmatrix} \tag{6-121}$$

$$\Sigma_{12}(p)\Sigma_{12}(p)_* = \begin{bmatrix} 0 \\ \dfrac{p-1}{p+1} \end{bmatrix} \begin{bmatrix} 0 & \dfrac{p+1}{p-1} \end{bmatrix} = \begin{bmatrix} 0 & 0 \\ 0 & 1 \end{bmatrix} \tag{6-122}$$

and

$$\Sigma_{22}(p) = -\Sigma_{21}(p)S_*(p)\Sigma_{12}(p)_*^{-R} = [0] \tag{6-123}$$

where $\Sigma_{22}(p)$ is analytic in the RHP, as predicted by the theorem. Combining these matrices, one obtains the coupling network

$$\Sigma(p) = \left[\begin{array}{cc|c} 0 & \dfrac{p-1}{p+1} & 0 \\ 0 & 0 & \dfrac{p-1}{p+1} \\ \hline \dfrac{p-1}{p+1} & 0 & 0 \end{array}\right] \tag{6-124}$$

which realizes the given scattering matrix when its last port is loaded with a 1-ohm resistor. Using a 6-port circulator to implement the realization of $\Sigma(p)$, one obtains the overall realization of the given one-way coupler shown in Figure 6-14.

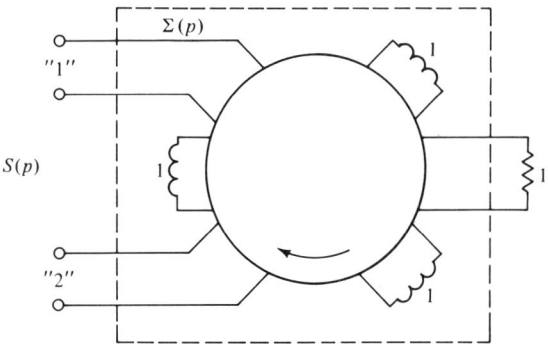

Figure 6-14 Realization of a one-way coupler.

Although the technique of Theorem 6-10 is assured to yield a valid realization, it does not necessarily give a realization employing the minimum number of reactors. In fact, the coupling network $\Sigma(p)$ obtained by the technique had degree 3; hence, 3 reactors were used in the realization of Figure 6-14. If, however, we had chosen the factorizations

$$\Sigma_{21}(p)_*\Sigma_{21}(p)\Sigma_{21}(p) = \begin{bmatrix} 1 \\ 0 \end{bmatrix} [1 \ 0] = \begin{bmatrix} 1 & 0 \\ 0 & 0 \end{bmatrix} \quad \text{(6-125)}$$

and

$$\Sigma_{12}(p)\Sigma_{12}(p)_* = \begin{bmatrix} 0 \\ 1 \end{bmatrix} [0 \ 1] = \begin{bmatrix} 0 & 0 \\ 0 & 1 \end{bmatrix} \quad \text{(6-126)}$$

the 3-port coupling network characterized by

$$\Sigma(p) = \left[\begin{array}{cc|c} 0 & \dfrac{p-1}{p+1} & 0 \\ 0 & 0 & 1 \\ \hline 1 & 0 & 0 \end{array} \right] \quad \text{(6-127)}$$

which is also analytic in the RHP is obtained. Here, since

$$\delta\Sigma(p) = 1 \quad \text{(6-128)}$$

a realization using only one reactor is possible, this being illustrated in Figure 6-15. One reactor and a 4-port circulator are used to realize the lossless $\Sigma(p)$, which, in turn, realizes the given one-way coupler when the last port is loaded in a 1-ohm resistor. Since this realization has only one resistor and one reactor, it is clearly minimal in both respects.

Unlike the 1-port case, Belevitch's n-port synthesis yields a $\Sigma(p)$ which may have higher degree than $S(p)$. This is due to the fact that the poles (in the LHP) of $K(p)_*^{-R}$ may not cancel with the zeros of $L(p)_*$, as they did in the special case. The realization thus usually requires more than $\delta S(p)$ reactors, and is nonminimal in this sense. It does, however, give a minimal resistor realization.

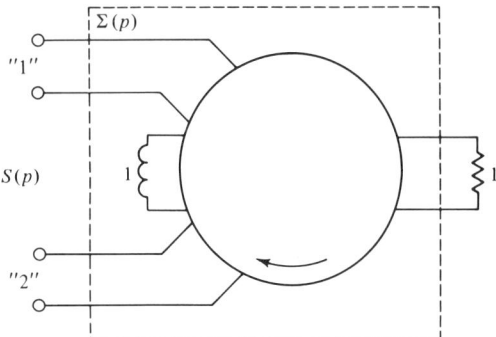

Figure 6-15 Minimal reactor realization of a one-way coupler.

*Corollary 6-9

The minimal number of resistors with which a BR rational scattering matrix $S(p)$ can be realized is r [the normal rank of $1 - S(p)_* S(p)$]. Furthermore, this minimum is achieved by the realization of Theorem 6-10. □

Whereas the 1-port Belevitch synthesis yields an easily implemented synthesis procedure, the complexity of factoring a polynomial matrix renders the n-port version of the result quite impractical. It serves, however, as a useful theoretical tool.

The n-port Bayard synthesis for an immittance, like its scattering counterpart, follows naturally from the 1-port result once the n-port factorization has replaced the 1-port result. As in the 1-port case, it is desirable to deal with impedances having no $j\omega$ axis poles; hence, an arbitrary PR rational impedance matrix is first decomposed as

$$Z(p) = Z_L(p) + Z_H(p) \qquad (6\text{-}129)$$

where $Z_L(p)$ is lossless and $Z_H(p)$ has no $j\omega$ axis poles. As in the 1-port case, this decomposition may be carried out via a process of pole extraction, which is justified by an n-port generalization of Theorem 6-5.

*Theorem 6-11

Let $Z(p)$ be a rational impedance matrix with a pair of complex conjugate $j\omega$ axis poles,

$$\frac{pA + B}{p^2 + \omega_0^2}$$

where A is positive and hermitian, and B is skew hermitian. Then

$$Z^1(p) = Z(p) - \frac{pA + B}{p^2 + \omega_0^2}$$

is PR or LPR rational if $Z(p)$ is PR or LPR. □

Repeated application of the theorem clearly yields the required decomposition; hence, we may proceed with the remaining development under the assumption that the given impedance matrix is PR and has no $j\omega$ axis poles.

Given such a matrix, one must find a lossless

$$X(p) = \begin{bmatrix} X_{11}(p) & | & X_{12}(p) \\ \hline X_{21}(p) & | & X_{22}(p) \end{bmatrix}$$

such that

$$X_{22}(p) + X_{22}(p)_* = 0 \tag{6-130}$$

$$X_{11}(p) = \text{Od } Z(p) + N(p)X_{22}(p)N(p)_* \tag{6-131}$$

and

$$X_{12}(p) = -X_{21}(p)_* = N(p)(1 + X_{22}(p)) \tag{6-132}$$

where

$$N(p)N(p)_* = \text{Ev } Z(p) \tag{6-133}$$

The required factorization is achieved by the following lemma.

*Lemma 6-5

Let $Z(p)$ be a PR rational $n \times n$ impedance matrix with LHP poles. Then

$$\text{Ev } Z(p) = \frac{K(p)K(p)_*}{m(p)m(p)_*}$$

Here $m(p)$ is the strictly Hurwitz least common denominator of $Z(p)$ and $K(p)$ is an $n \times r$ polynomial matrix [r, the normal rank of Ev $Z(p)$].

To obtain the n-port Bayard synthesis one now proves the following.

*Theorem 6-12

Let $Z(p)$ be a PR rational $n \times n$ impedance matrix with LHP poles, and let

$$\text{Ev } Z(p) = \frac{K(p)K(p)_*}{m(p)m(p)_*}$$

where

$$Z(p) = \frac{N(p)}{m(p)}$$

are as in Lemma 6-5. Then $Z(p)$ can be realized at the first n ports of the lossless $(n+r)$-port

$$X(p) = \left[\begin{array}{c|c} \text{Od } Z(p) + \text{Ev } Z(p)\dfrac{\text{Ev } m(p)}{\text{Od } m(p)} & \dfrac{K(p)}{\text{Od } m(p)} \\ \hline \dfrac{K(p)_*}{\text{Od } m(p)} & \dfrac{\text{Ev } m(p)}{\text{Od } m(p)}1 \end{array} \right]$$

with the last r ports loaded in 1-ohm resistors (as per Figure 6-13). □

As in the scattering case, the difficulty in factoring Ev $Z(p)$ renders the result of little practical value. Also, as in that case, the number of resistors needed in the realization is minimal.

6-7 Synthesis of Lossless n-Ports

The results of the preceding sections reduce the general passive synthesis problem to the realization of a lossless multiport. It still remains to exhibit a realization for this remainder to complete the realization. Here, this end is achieved by extending the Richards theorem factorization synthesis of a lossless 1-port to cover the n-port case. This is carried out by applying Richards' theorem to the determinant of $S(p)$ rather than the matrix itself. The justification for such an approach is due to the following lemma.

*Lemma 6-6

If $S(p)$ is an LBR rational scattering matrix then

$$\det S(p)$$

is an LBR rational scattering parameter, and furthermore,

$$\delta \det S(p) = \delta S(p)$$ □

With this preliminary completed, one may proceed with the realization using the technique of Belevitch [27], [31], [110]. For notational convenience, let

$$s(p) = \det S(p) \qquad (6\text{-}134)$$

and let

$$z(p) = [s(p) - 1]^{-1}[s(p) + 1] \qquad (6\text{-}135)$$

be the corresponding impedance. An application of Richards' theorem to $s(p)$ now yields

$$s_{z(k)}(p) = \frac{(p-k)}{(p+k)} s_{z(k)}{}^r(p) \qquad (6\text{-}136)$$

where the degree of $s_{z(k)}{}^r(p)$ is strictly less than that of $s(p)$ (hence, $S(p)$, by the lemma). Equation (6-136) shows that $s_{z(k)}(p)$ has a zero at

$$p = k \qquad (6\text{-}137)$$

or, equivalently, the matrix $S_{z(k)}(p)$ is singular at $p = k$. Now, for any singular matrix, there exists an x such that

$$\|x\| = 1 \qquad (6\text{-}138)$$

and

$$S_{z(k)}(k)x = 0 \qquad (6\text{-}139)$$

Once k and x have been found, we may now obtain a step of the factorization synthesis via the following theorem.

*Theorem 6-13

Let $S(p)$ be an LBR rational scattering matrix and k and x be as above. Then

$$S_{z(k)}(p) = \bar{S}_{z(k)}(p) S_{z(k)}{}^1(p)$$

where

$$S_{z(k)}{}^1(p) = 1 - \frac{2kxx'}{p+k}$$

and

$$\bar{S}_{z(k)}(p) = S_{z(k)}(p) S_{z(k)}{}^1(p)_*$$

are LBR rational scattering matrices with

$$\delta \bar{S}_{z(k)}(p) = \delta S_{z(k)}(p) - 1 \qquad \square$$

Consistent with the theorem and our previous developments, $S(p)$ may now be realized by the circulator network of Figure 6-16, thereby reducing the

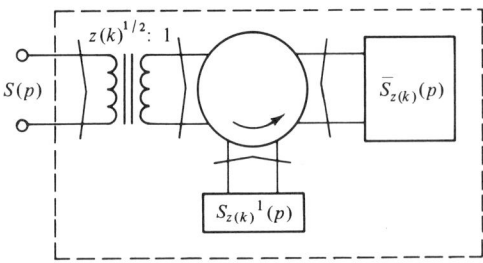

Figure 6-16 Step of factorization synthesis for LBR rational scattering matrices.

synthesis problem for a degree n matrix to that of a degree 1 matrix and a degree $n - 1$ matrix. Upon repeated application of the theorem to $\bar{S}_{z(k)}(p)$, one obtains the realization of Figure 6-17 in terms of the degree 1 n-ports,

$$S_{z(k)}{}^i(p) = 1 - \frac{2kxx'}{p+k} \qquad (6\text{-}140)$$

and the degree zero remainder,

$$S_{z(k)}{}^0 \qquad (6\text{-}141)$$

The former is precisely the elementary n-port considered in Chapter 4, and is realizable as a $2n$-port ideal transformer loaded in an inductor, whereas the latter is memoryless and, thereby, realizable by the gyrator-transformer con-

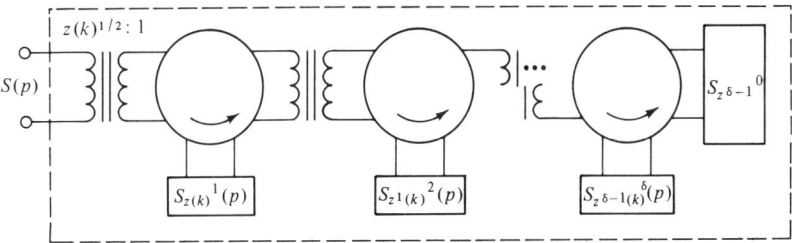

Figure 6-17 Realization of an LBR rational scattering matrix by factorization.

figuration considered in Chapter 5. These realizations are illustrated, once again, in Figure 6-18.

Consider the nonreciprocal degree 1 scattering matrix

$$S(p) = \begin{bmatrix} 0 & \dfrac{p-1}{p+1} \\ 1 & 0 \end{bmatrix} \tag{6-142}$$

for which

$$s(p) = \frac{1-p}{1+p} \tag{6-143}$$

has a zero at $k = 1$ and

$$z(1) = z(p)\Big|_{p=1} = \frac{1}{p}\Big|_{p=1} = 1 \tag{6-144}$$

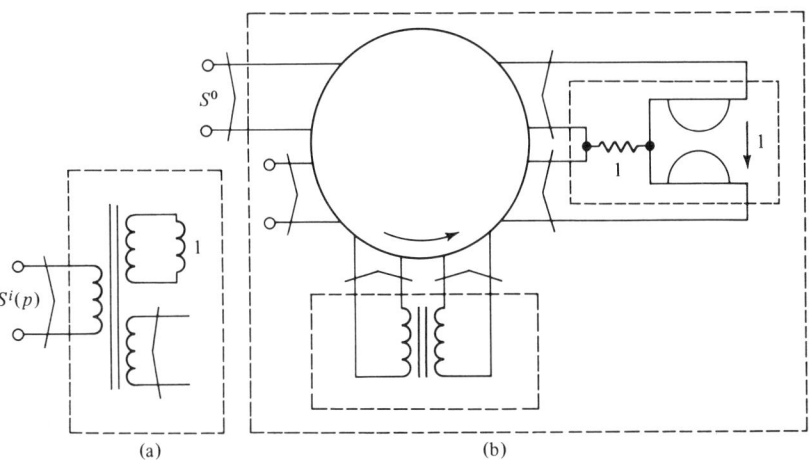

Figure 6-18 Realization of degree-one and degree-zero networks.

Now,
$$S(1) = \begin{bmatrix} 0 & 0 \\ 1 & 0 \end{bmatrix} \tag{6-145}$$

Hence, if one chooses for x the vector
$$x = \begin{bmatrix} 0 \\ 1 \end{bmatrix} \tag{6-146}$$

we have
$$S(1)x = 0 \tag{6-147}$$

By the theorem,
$$\begin{aligned} S^1(p) &= \begin{bmatrix} 1 & 0 \\ 0 & 1 \end{bmatrix} - \frac{2}{p+1} \begin{bmatrix} 0 & 0 \\ 0 & 1 \end{bmatrix} \\ &= \begin{bmatrix} 1 & 0 \\ 0 & \dfrac{p-1}{p+1} \end{bmatrix} \end{aligned} \tag{6-148}$$

$$\bar{S}(p) = S(p)S^1(p)_* = \begin{bmatrix} 0 & \dfrac{p-1}{p+1} \\ 1 & 0 \end{bmatrix} \begin{bmatrix} 1 & 0 \\ 0 & \dfrac{p+1}{p-1} \end{bmatrix} \tag{6-149}$$

$$= \begin{bmatrix} 0 & 1 \\ 1 & 0 \end{bmatrix}$$

is the desired lossless degree zero remainder. Realizing $S^1(p)$ as per Figure 6-18, and observing that the degree zero remainder is the scattering matrix of a pair of wires, the desired realization, as shown in Figure 6-19, is obtained.

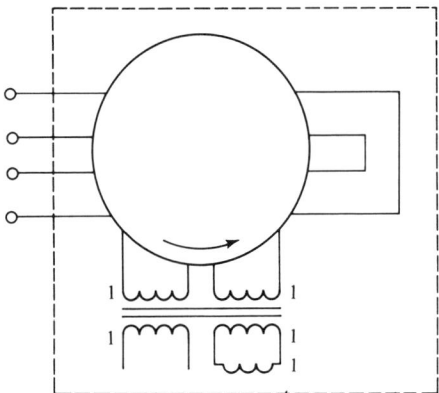

Figure 6-19 Realization of a degree-one lossless 2-port.

Finally, the possibility of realizing an LPR rational impedance matrix (possibly obtained from the Bayard realizations for passive RLC n-ports) must be considered. This is possible as an n-port form of the n-port Foster synthesis, employing Theorem 6-11, the n-port case differing from the original result in that ideal transformers are required to realize the terms of the partial fraction expansion, in addition to the gyrators needed in the nonreciprocal case [31], [110], [178].

6-8 Discussion

Our primary goal in the preceding development of RLC network theory was to exhibit, by example, methods by which the general techniques already developed could be applied to a special case of interest. The formulation was thus quite nonclassical, concentrating on techniques which are applicable to the general case, rather than those (perfectly valid but restricted) techniques which hold in the RLC case, but have no further implications. For this reason, great emphasis was placed on synthesis by scattering matrix factorization. Though this technique is clearly more complex than the Foster (and Cauer) forms of classical lossless synthesis [183], the techniques developed have further applications. Similarly, certain classical results, such as the Bott–Duffin synthesis [38], which have no application in the general theory, were not considered at all.

It is apparent from the development that the most significant aspect of the RLC networks is their poles (and zeros). These permit the definition of degree, a necessary tool in showing that the synthesis procedures converge after a (prespecified) finite number of steps. Moreover, they give us a "handle" on the singularities of the RLC system function, which, in turn, allows the analyticity (causality) questions of the general passive synthesis procedures of Chapter 5 to be resolved.

The extra structure of the special case thus allows solution of problems for which general techniques are not (yet) available, but at the same time, induces a collection of details (those poles and zeros, again) which might obscure the main ideas under consideration.

CHAPTER

7
Transmission Line Networks

7-1 Introduction

In the preceding chapter we considered various classes of RLC or "finite" networks, so termed because they are characterized by a finite number of parameters in the frequency plane. Here the transmission line networks are considered, which, by analogy with the RLC case, may be termed "periodic" networks; that is, although they have an infinite number of parameters, these repeat periodically. Since any one period (in frequency) contains only finitely many parameters, these networks, like their RLC counterparts, are still characterized by a finite amount of information.

The transmission line networks may be studied by considering only a single (frequency) period on the p plane (which contains all relevant information), or by transforming the p plane into a λ plane via

$$\lambda = \coth \tau p \qquad (7\text{-}1)$$

and studying the new λ plane. This transformation, due to Richards [136], [137], amounts essentially to taking a single period of the p plane and stretching it out to cover the entire λ plane, as illustrated in Figure 7-1. Richards' transformation can be viewed as converting a p plane transmission line network into a λ plane RLC network, the latter of which may be studied by the methods of Chapter 6.

INTRODUCTION   243

Historically, the scattering approach was developed as an extension of classical transmission line theory; hence, it might be expected, as is indeed the case, that the scattering matrix is ideally suited for the study of such networks.

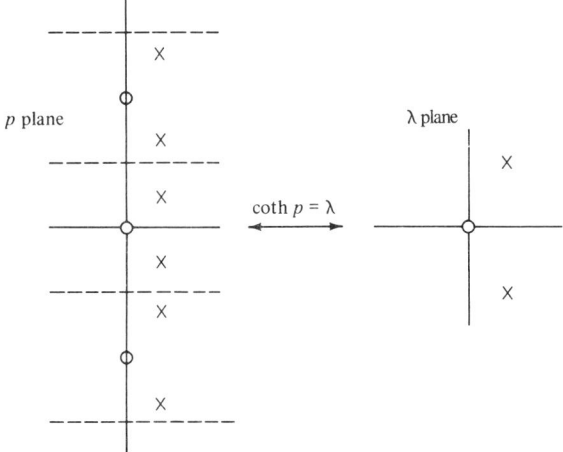

Figure 7-1 Frequency domain interpretation of Richards' transformation.

We have already seen in Chapter 4 that the definition of a uniform transmission line is greatly simplified by the use of an appropriately normalized scattering matrix, this being

$$S_R(p) = \begin{bmatrix} 0 & e^{-\tau p} \\ e^{-\tau p} & 0 \end{bmatrix} \tag{7-2}$$

where R is the characteristic impedance of the line and τ is its (electrical) length. Here the characteristic impedance does not appear at all in the scattering matrix (though it is implicit in its definition). Moreover, under this normalization the scattering matrix itself exhibits explicitly the two-way delay character of such a line. Possibly of even greater significance is the Richards theorem approach to synthesis [48], [137]. This technique, which, like the scattering matrix itself, was developed originally for application to transmission line networks, yields physically desirable cascade realizations (without circulators or gyrators), whereas the standard immittance techniques yield realizations containing open and short circuit matching stubs which cannot be conveniently implemented.

In the first section, the fundamental properties of the transmission line networks are investigated, with special emphasis on the characteristics of cascaded lines, this being the only physically acceptable structure for a number of practical types of transmission line (waveguides, fiber optic devices, acoustical delay lines, and so forth). In the second section, Richards' transformation is defined and the relation between the p and λ domain representation of a network is investigated. In particular, the effect of the transformation on bounded and positive reality is considered. Section 7-4 is devoted to an investigation of the cascade structure and the development of synthesis procedures for such, while the final section is devoted to the study of a class of frequency dependent lines. These latter, more general, lines include most commonly encountered LC, RC, and lossy LC lines, but like the usual (LC) case, they can be studied by transforming them to an equivalent λ domain. Once such a transformation is achieved, the theory in the frequency dependent case is exactly the same as for the uniform LC line.

7-2 Transmission Line Characteristics

In defining the transmission line n-ports, one takes an approach similar to that used in the RLC case, with the uniform transmission line characterized by the normalized scattering matrix

$$S_R = \begin{bmatrix} 0 & e^{-\tau p} \\ e^{-\tau p} & 0 \end{bmatrix} \qquad (7\text{-}3)$$

replacing the capacitor. Of course, a transmission line is a 2-port, while a capacitor is a 1-port; hence, the analogy between the two cases is not complete.

Definition 7-1

A *transmission line n-port* is an n-port which can be realized (per Chapter 4) by a network composed of a finite number of resistors, uniform transmission lines, and gyrators. □

Of course, as in the RLC case, the resistors and gyrators combine to allow the realization of transformers, circulators, and other memoryless devices which are thus valid components in a transmission line network.

Unlike the capacitor, which was characterized by a single parameter, a transmission line has two defining parameters, its characteristic impedance and length. The latter can, however, be eliminated if one deals with networks of commensurate lines.

Definition 7-2

The lines of a transmission line n-port are said to be *commensurate* if their lengths have a nonzero common devisor. The *unit length* for a network of such commensurate lines, each of length τ_i, $i = 1, 2, \cdots, m$, is

$$\tau = (\tfrac{1}{2}) \, \text{GCD} \, \{\tau_1, \tau_2, \cdots, \tau_m\}$$

where GCD $\{\cdot\}$ is the greatest common devisor of the lengths $\tau_1, \tau_2, \cdots, \tau_m$. □

In essence, a network of commensurate lines can, without loss of generality, be taken as having all lines of unit length τ, the "physical" line of length τ_i being (mathematically) replaced by a cascade of lines with unit length τ. This, in effect, eliminates the length as a parameter of the transmission line, because it is then a fixed parameter of the entire network, thereby greatly simplifying the required calculations. The convenience inherent in this process is such as to justify consideration only of networks whose transmission lines are of commensurable lengths. Of course, a network of arbitrary transmission lines can be approximated by a network of commensurate lines; hence, such a restriction induces little "real" loss in generality [48], [88]. In the remainder of the chapter, we thus assume that all transmission line n-ports are composed of lines with commensurable lengths and have a specified unit element length τ.

Although the length of both the transmission lines for the network of Figure 7-2 are irrational, they have a nonzero common devisor, 2π, and as such are

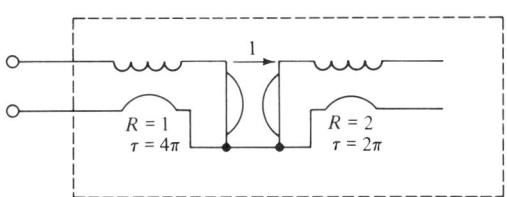

Figure 7-2 A network of commensurate transmission lines.

commensurate with unit element length π. If, however, one of the lines had length one, no nonzero common devisor would exist, and the lines are not commensurable. In general, the commensurable assumption is a property of the relative lengths of the lines in a network, but is independent of their absolute lengths, which may be rational or irrational.

Upon analyzing the network of Figure 7-2, the scattering parameter observed at the input port is found to be

$$s(p) = \frac{e^{-8\pi p} - 3e^{-12\pi p}}{3 - e^{-4\pi p}} \tag{7-4}$$

which is rational in the variable $e^{-4\pi p} = e^{-4\tau p}$. Although this does not hold, in general, as is indicated by the scattering matrix of a single transmission line, one can show that the system function for a network of commensurate transmission lines with unit element length τ is rational in the variable $e^{-2\tau p}$ and, in general, one has the following theorem.

Theorem 7-1

The system function of a network of commensurable transmission lines with unit element length τ, resistors, and gyrators is real rational in the variable

$$e^{-2\tau p}$$

Proof

Since

$$\tau = (\tfrac{1}{2})\,\mathrm{GCD}\,\{\tau_1, \tau_2, \cdots, \tau_m\} \tag{7-5}$$

for each line there exists an n_i such that

$$\tau_i = 2n_i\tau \tag{7-6}$$

Hence,

$$e^{-\tau_i p} = e^{-2n_i \tau p} = (e^{-2\tau p})^{n_i} \tag{7-7}$$

Each line is therefore rational in $e^{-2\tau p}$ and, upon combining the various lines with the memoryless components (which are not functions of p), a system function rational in $e^{-2\tau p}$ is found. □

Intuitively, one might expect that the scattering parameter of a 1-port would be rational in $e^{-4\tau p}$, as is the case with the example shown in Figure 7-2, since in a 1-port the signal "should be" delayed 2τ going down a line, and another 2τ coming back up; hence, a total delay of 4τ. Although this is true in most practical cases, it does not hold in general, as is illustrated by the 1-port of Figure 7-3, the scattering parameter of which was found (in Chapter 4) to be

$$s(p) = e^{-2\tau p} \tag{7-8}$$

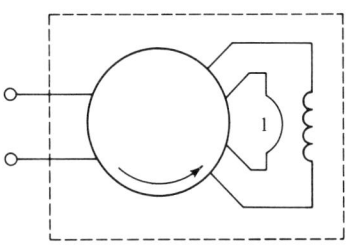

Figure 7-3 1-port with scattering parameter $e^{-2\tau p}$.

Clearly, the effect of the nonreciprocal device in this network is to allow the incident wave to return to the input port after only one transmission through the line. In fact, completely reciprocal, though more complex, realizations of this same scattering parameter are possible; hence, nonreciprocity is not necessary for this condition [200]. Although the scattering parameter of a 1-port is not, in general, rational in $e^{-4\tau p}$, as we have shown, when this is the case (most of the time), one may choose

$$\tau = \text{GCD } \{\tau_1, \tau_2, \cdots, \tau_m\} \tag{7-9}$$

with Theorem 7-1 still valid. Although such a choice of τ is of no theoretical value, it halves the number of lines needed for the realization of a scattering parameter, and hence is of considerable practical value.

Theorem 7-1 allows one to justify the contention of the introduction that the poles and zeros of a transmission line system function repeat periodically in frequency. For either the poles or zeros, one is interested in the zeros of a polynomial in $e^{-2\tau p}$,

$$Q(e^{-2\tau p}) = \sum_{i=0}^{n} a_i (e^{-2\tau p})^i \tag{7-10}$$

Noting that
$$e^{-2\tau p} = e^{-2\tau(p+jn\pi/\tau)} \tag{7-11}$$
we see that if
$$Q(e^{-2\tau p_0}) = 0 \tag{7-12}$$
then
$$Q(e^{-2\tau(p_0+jn\pi/\tau)}) = Q(e^{-2\tau p_0}) = 0 \tag{7-13}$$

The zeros of such a polynomial thus repeat in frequency periods of $2\pi/\tau$, yielding a pole-zero pattern composed of repetitive strips parallel to the σ axis of the p plane (and shown in Figure 7-1). Clearly, one strip suffices to define the network; thus, the transmission line network is, in effect, described by a finite (in one strip) pole-zero pattern.

For the transmission line network of Figure 7-2, with the scattering parameter of Equation (7-4), the denominator is
$$Q(p) = 3 - e^{-4\pi p} \tag{7-14}$$
which, when set equal to zero, yields the equality
$$3 = e^{-4\pi p} = [e^{-4\pi \operatorname{Re} p}][e^{-j4\pi \operatorname{Im} p}] \tag{7-15}$$
characterizing the poles. Here the magnitude is determined by the $\operatorname{Re} p$ via
$$3 = e^{-4\pi \operatorname{Re} p} \tag{7-16}$$
or, equivalently,
$$\operatorname{Re} p = -\ln\frac{3}{4\pi} \tag{7-17}$$
while the phase determines the imaginary part of p via
$$1 = e^{-j4\pi \operatorname{Im} p} \tag{7-18}$$
or
$$\operatorname{Im} p = \frac{n}{4} \qquad n = 0, \pm 1, \pm 2, \cdots \tag{7-19}$$

The poles for this scattering parameter thus appear along the vertical line
$$\sigma = -\ln\frac{3}{4\pi} \tag{7-20}$$
in the complex plane at frequencies
$$j\omega = 0, \pm\tfrac{1}{4}, \pm\tfrac{1}{2}, \pm\tfrac{3}{4}, \pm 1, \pm\tfrac{5}{4}, \cdots \tag{7-21}$$
and are thus, as expected, periodic in frequency. Note, however, that all of the poles are in the LHP, as required of a passive network.

Although we have, by definition, allowed arbitrary interconnection of transmission lines, practical consideration often forces one to consider the special case of cascaded lines possibly with a resistive load, as illustrated in Figure 7-4. The characteristics of such a network may be determined with the aid of the following proposition.

Figure 7-4 Cascaded lines with resistive load.

*Proposition 7-1

Let a 1-port be connected to an arbitrary load (possibly containing transmission lines itself) through a transmission line of length τ and characteristic impedance R (as per Figure 7-5). Then the scattering parameter of the 1-port is

$$s_R(p) = e^{-2\tau p} s_R{}^L(p)$$

where $s_R{}^L(p)$ is the scattering parameter of the load normalized to R. □

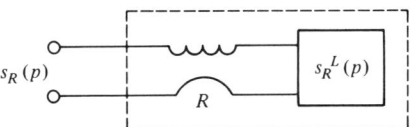

Figure 7-5 Cascade loaded line.

The proposition implies that the scattering parameter seen at the input is the same as that at the load, except for the addition of two delays, one for the incident wave to travel down the line and the other for the reflected wave to return.

For the 1-port of Figure 7-6, the inductive load has the scattering parameter

$$s_e(p) = \frac{p - e}{p + e} \tag{7-22}$$

when normalized to the characteristic impedance of the line e.

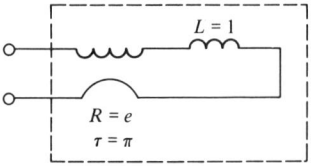

Figure 7-6 Inductor-loaded transmission line.

Applying the proposition now results in the 1-port scattering parameter

$$s_e(p) = \frac{e^{-2\pi p}(p - e)}{(p + e)} = \frac{pe^{2\pi p} - e^{(1-2\pi p)}}{(p + e)} \tag{7-23}$$

The network of Figure 7-6 is not truly a transmission line n-port, since it contains an inductor. Fortunately, much of the transmission line theory applies to such mixed transmission line RLC networks (including Proposition 7-1), which allows one to handle these very commonly encountered mixed devices. A detailed study of this class of networks is given in the next chapter, within the multivariable context. Two useful special cases of the proposition correspond to the use of open and short circuit loads.

*Corollary 7-1

The scattering parameter seen at the input of an open circuited line is

$$s_R(p) = e^{-2\tau p}$$

and that seen at the input of a short circuited line is

$$s_R(p) = -e^{-2\tau p} \qquad \square$$

Here the normalization constant R is taken as the characteristic impedance of the line.

7-3 The Richards Transformation

Since the system function of a network of commensurable transmission lines is rational in $e^{-2\tau p}$, it could readily be reduced to a real rational function via the transformation

$$\lambda \leftrightarrow e^{-2\tau p} \qquad (7\text{-}24)$$

Unfortunately, such a transformation might take a BR scattering matrix to one which was not BR. The seemingly more complex Richards transformation serves the same purpose, while preserving passivity.

Definition 7-3

Let $H(p)$ be the system function for a transmission line n-port, rational in $e^{-2\tau p}$. Then the substitution

$$e^{-2\tau p} \leftrightarrow \frac{\lambda - 1}{\lambda + 1}$$

or, inversely,

$$\lambda \leftrightarrow \coth \tau p$$

is the *Richards transformation*. $\qquad \square$

After carrying out the Richards transformation we use a caret to denote the new matrix. Hence,

$$\hat{H}(\lambda) = H(p)\bigg|_{e^{-2\tau p} = \lambda - 1/\lambda + 1} = H\left(\frac{-1}{2} \ln\left(\frac{\lambda - 1}{\lambda + 1}\right)\right) \qquad (7\text{-}25)$$

Of course, $H(p)$ may represent any of the usual network matrices. The similarity in form between Richards' transformation and Richards' theorem is by no means coincidental, the transformation having been devised as a means of applying the theorem (in its impedance form) to the cascade synthesis of transmission line n-ports [136], [137].

The network of Figure 7-2 has the unit normalized scattering parameter

$$s(p) = \frac{e^{-8\pi p} - 3e^{-12\pi p}}{3 - e^{-4\pi p}} \quad \text{(7-26)}$$

with unit element length

$$\tau = \pi \quad \text{(7-27)}$$

Hence, Richards' transformation results in the substitution

$$e^{-2\pi p} \leftrightarrow \frac{\lambda - 1}{\lambda + 1} \quad \text{(7-28)}$$

and the λ domain scattering parameter

$$\hat{s}(\lambda) = s(p)\bigg|_{e^{-2\pi p} = \frac{\lambda-1}{\lambda+1}}$$

$$= \frac{(\lambda - 1)^4(\lambda + 1)^2 - 3(\lambda - 1)^6}{3(\lambda + 1)^6 - (\lambda + 1)^4(\lambda - 1)^2} \quad \text{(7-29)}$$

In this example, $s(p)$ is rational in $e^{-4\pi p}$, rather than $e^{-2\pi p}$; hence, we may choose a unit element length of 2π, this yielding the scattering parameter

$$\hat{s}(\theta) = s(p)\bigg|_{e^{-4\pi p} = \frac{\theta-1}{\theta+1}}$$

$$= \frac{(\theta - 1)^2(\theta + 1) - 3(\theta - 1)^3}{3(\theta + 1)^3 - (\theta + 1)^2(\theta - 1)} \quad \text{(7-30)}$$

under a modified Richards transformation. Clearly, the advantage of this modified transformation, when applicable, is to halve the degree (in the sense of a rational function) of the rational system function.

Clearly, one has the following proposition.

Proposition 7-2

Let $H(p)$ be the system function representing a transmission line n-port. Then $\hat{H}(\lambda)$ as defined above is a real rational in λ. □

The preservation of passivity is due to the following theorem.

Theorem 7-2

Let $S(p)$ be the scattering matrix of a transmission line n-port. Then $\hat{S}(\lambda)$ is BR (LBR) if $S(p)$ is BR (LBR).

Proof

Since $S(p)$ has the desired properties when p is in the RHP, while $\hat{S}(\lambda)$ is the composite function of $S(p)$ with

$$p = (\tfrac{1}{2}) \ln\left(\frac{\lambda - 1}{\lambda + 1}\right) \tag{7-31}$$

it suffices to show that when λ is in the RHP, so is p, for then

$$\hat{S}(\lambda) = S\left((-\tfrac{1}{2}) \ln\left(\frac{\lambda - 1}{\lambda + 1}\right)\right) \tag{7-32}$$

will have the same properties as $S(p)$ in the RHP. For convenience, we use the inverse form of Equation (7-31),

$$\lambda = \coth \tau p = \frac{\cosh \tau p}{\sinh \tau p} \tag{7-33}$$

Now, when p is on the $j\omega$ axis,

$$\lambda = \frac{\cosh j\omega}{\sinh j\omega} = \frac{\cos \omega}{j \sin \omega} \tag{7-34}$$

is also on the $j\omega$ axis. Since Equation (7-33) is a conformal map [144] and it takes the imaginary axis of the p plane to the imaginary axis of the λ plane, it must take the RHP of the p plane to either the RHP or LHP of the λ plane (since conformal maps preserve boundaries). To find out which is the case, one must test the map at some value in the RHP, say $p = 1$, to determine whether it goes to the right or left half λ plane. We have

$$\coth(1) = \frac{\cosh(1) = e^2 + 1}{\sinh(1) = e^2 - 1} > 0 \tag{7-35}$$

showing that the transformation takes at least one point in the RHP p plane to the RHP λ plane. The above boundary argument thus implies that the mapping takes the entire RHP p plane to the RHP λ plane, which was to be shown. □

It is worthy of note that the theorem is actually necessary and sufficient. That is, if $S(\lambda)$ is BR (LBR), then so is $S(p)$. This is, however, of no relevance to our application, since one is interested in transforming the p plane to the λ plane, but not conversely.

With the aid of Richards' transformation, one may realize a transmission line network simply by realizing its λ domain representation (which is real rational BR or LBR if the transmission line network is BR or LBR, respectively) by RLC techniques, and then substituting the appropriate transmission line devices for the λ domain capacitors and inductors. In the particular case of the cascade structure, Proposition 7-1 translates under the Richards transformation into the equation

$$\hat{s}_R(\lambda) = \frac{\lambda - 1}{\lambda + 1} \hat{s}_R{}^L(\lambda) \tag{7-36}$$

since

$$\frac{\lambda - 1}{\lambda + 1} \leftrightarrow e^{-2\tau p} \tag{7-37}$$

and

$$\hat{s}_R{}^L(\lambda) \leftrightarrow s_R{}^L(p) \tag{7-38}$$

The scattering parameter of a (unit element length) line loaded in a realization of $\hat{s}_R{}^L(\lambda)$ is thus readily calculated in the λ domain. Moreover, Equation (7-36) exhibits the intimate relationship between Richards' theorem and the cascade structure, this being exploited in the following section.

The application of the general formula of Equation (7-36) to the special cases of open and short circuited lines yields

$$\hat{s}_R(\lambda) = \frac{\lambda - 1}{\lambda + 1} \tag{7-39}$$

as the scattering parameter of an open circuited line of unit length and characteristic impedance R, while

$$\hat{s}_R(\lambda) = \frac{1 - \lambda}{1 + \lambda} \tag{7-40}$$

is the scattering parameter of the corresponding short circuited line. It is thus apparent, upon comparison with the usual p domain formulas, that an open circuited line is a λ domain inductor, while a shorted line is a λ domain capacitor. This observation establishes a straightforward means of realizing a transmission line n-port via RLC techniques. That is, one realizes a specified real rational $\hat{S}(\lambda)$ as a λ domain RLC network replacing the λ domain inductors with open circuited lines and the λ domain capacitors with short circuited lines to obtain a realization of the transmission line network in the p domain. Although this result, illustrated in Figure 7-7, is of little practical significance, because of the inherent difficulty of implementing open and short circuited structures with

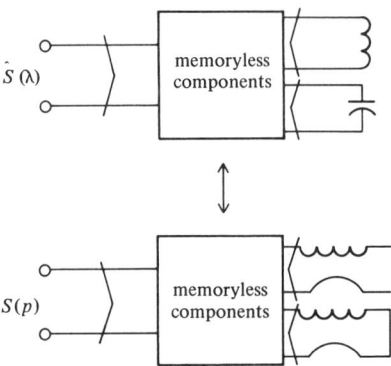

Figure 7-7 Synthesis by Richards' transformation.

many practical lines, it is a useful theoretical tool, since it allows RLC theory to be translated to the transmission line case. Summarizing, we have the following theorem.

Theorem 7-3

A transmission line n-port specified in the λ domain by a matrix $\hat{H}(\lambda)$ with given unit element length can be realized by substituting p domain open and short circuited lines for λ domain reactors in any realization of $\hat{H}(\lambda)$ as a λ domain RLC network. □

In Chapter 6 it was shown that the LBR rational scattering parameter

$$s(p) = \frac{p^2 - 2p + 2}{p^2 + 2p + 2} \tag{7-41}$$

could be realized by the RLC network of Figure 7-8(a); hence, upon implementing the substitution of the theorem, the transmission line n-port represented in the λ domain by

$$\hat{s}(\lambda) = \frac{\lambda^2 - 2\lambda + 2}{\lambda^2 + 2\lambda + 2} \tag{7-42}$$

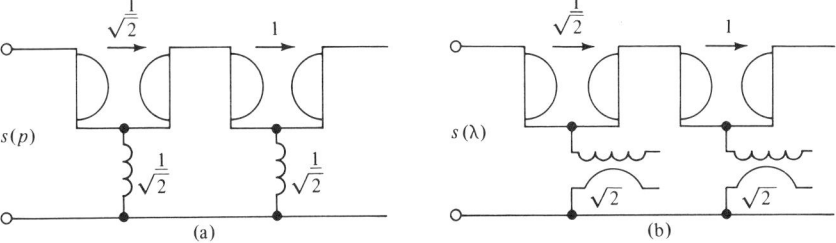

Figure 7-8 (a) Realization of an RLC network. (b) Its equivalent transmission line network.

has the realization of Figure 7-8(b). Here, open circuited lines of unit element length and characteristic impedance $\sqrt{2}$ replace the $1/\sqrt{2}$ λ domain inductors of the RLC realization, as shown in Figure 7-8(a).

7-4 Cascade Synthesis

From a theoretical viewpoint, the synthesis of Theorem 7-3 is complete. Unfortunately, the structures usually employed for RLC synthesis yield physically undesirable, though valid, transmission line realizations. In this section, we consider the specialized, but more practical, structure of cascaded lines, possibly with a resistive load. This structure, illustrated in Figure 7-4, yields realizations which are readily implemented with practical lines.

Initially, we consider the realization of an LBR rational scattering parameter $\hat{s}(\lambda)$ by such a configuration. Here the implication of losslessness is that the load must be either an open or short circuit (the only lossless "resistors"). Starting with $\hat{s}(\lambda)$ and applying Richards' theorem with $k = 1$, one finds that

$$\hat{s}_{z(1)}(\lambda) = \frac{\lambda - 1}{\lambda + 1} \hat{s}_{z(1)}{}^r(\lambda) \tag{7-43}$$

which, upon comparison with Equation (7-36), allows $\hat{s}(\lambda)$ to be realized as a line with characteristic impedance $z(1)$ loaded in a realization of $\hat{s}_{z(1)}{}^r(\lambda)$, as per Figure 7-9. Moreover, since $\hat{s}(\lambda)$ is LBR, Richards' theorem yields degree

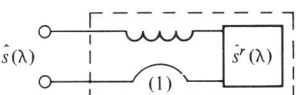

Figure 7-9 Step of cascade synthesis for a transmission line network.

reduction; hence, $\hat{s}^r(\lambda)$ is still LBR and of degree strictly less than that of $s(\lambda)$. Repeating the process until it terminates with a zero degree remainder (that is, an open or short circuit) then yields the final realization, as shown in Figure 7-10. One thus finds, by construction, that every lossless transmission line scattering parameter has a cascade realization. It is significant that the Richards

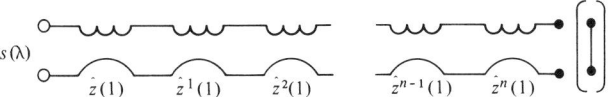

Figure 7-10 Realization of an LBR transmission line network.

theorem approach to synthesis, which required various multiport coupling elements in the RLC case (to realize the product and the required change of normalization), gives a direct cascade realization in the transmission line case without resort to coupling elements of any kind, this because the cascade structure naturally realizes a product, whereas the normalization is built into the characteristic impedance of the lines. The preceding development is summed up by the following theorem.

Theorem 7-4

An LBR scattering parameter $\hat{s}(\lambda)$ can be realized as a cascade of transmission lines loaded in either an open or short circuit by repeated application of Richards' theorem with $k = 1$. □

Corollary 7-2

Every lossless transmission line 1-port has a cascade realization. □

Consider the lossless scattering parameter

$$\hat{s}(\lambda) = \frac{\lambda^2 - 2\lambda + 2}{\lambda^2 + 2\lambda + 2} \tag{7-44}$$

realized by the substitution approach in the preceding section. Here the corresponding impedance is

$$\hat{z}(\lambda) = \frac{\lambda^2 + 2}{2\lambda} \tag{7-45}$$

Hence, with $k = 1$,

$$\hat{z}(1) = \tfrac{3}{2} \tag{7-46}$$

is the normalization required for the first step in the synthesis (and the characteristic impedance of the first of the unit element length lines in the realization). Renormalizing $\hat{z}(\lambda)$ and $\hat{s}(\lambda)$ yields

$$\hat{z}_{(3/2)}(\lambda) = \frac{\lambda^2 + 2}{3\lambda} \tag{7-47}$$

and

$$\hat{s}_{(3/2)}(\lambda) = \left(\frac{\lambda - 1}{\lambda + 1}\right)\left(\frac{\lambda - 2}{\lambda + 2}\right) = \frac{\lambda - 1}{\lambda + 1} \hat{s}_{(3/2)}{}^r(\lambda) \tag{7-48}$$

The first line in the realization thus has characteristic impedance $\tfrac{3}{2}$ and unit length (as do all lines under the commensurable assumption). Now, repeating the process for $\hat{s}^r(\lambda)$ yields, with a normalization of $\tfrac{3}{4}$,

$$\hat{s}_{(3/4)}{}^r(\lambda) = \frac{\lambda - 1}{\lambda + 1}(1) \tag{7-49}$$

showing that the second and last line in the cascade has characteristic impedance $\tfrac{3}{4}$, and that the final memoryless load is an open circuit (characterized by the scattering parameter 1). The cascade realization of the given scattering parameter is thus that which is shown in Figure 7-11. A comparison of this realization

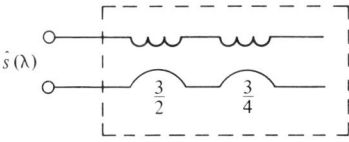

Figure 7-11 Cascade realization of a lossless scattering parameter.

with that obtained by the substitution method of the last section illustrates the clear superiority of the cascade synthesis.

Unlike the lossless case, it is not, in general, possible to realize all lossy scattering parameters in the cascade configuration [48]. The advantages of the cascade structure are, however, such as to merit giving special consideration to that class of lossy scattering parameters which have such a realization. Rather than considering the scattering matrix directly, it is found to be more convenient to study the resistivity operator

$$1 - \hat{s}(\lambda)_*\hat{s}(\lambda) \tag{7-50}$$

upon which a constraint may be obtained by an inductive process of adding cascaded lines, one at a time. To this end, we begin by studying the effect of adding a single line to some arbitrary load, possibly containing more cascaded lines itself. This goal is most easily achieved by a mixed scattering impedance approach, where the load is represented as an impedance, while the resultant 1-port has a scattering representation. Since we consider only 1-ports, this is done without loss of generality.

Applying the general λ domain formula for such a configuration, we obtain, for the scattering parameter seen at the input normalized to R, the characteristic impedance of the line,

$$\hat{s}_R(\lambda) = \frac{\lambda - 1}{\lambda + 1}\hat{s}_R{}^L(\lambda) = \left(\frac{\lambda - 1}{\lambda + 1}\right)\left(\frac{\hat{z}(\lambda) - R}{\hat{z}(\lambda) + R}\right) \tag{7-51}$$

For the purposes of standardization, we now renormalize this back to unity, obtaining

$$\hat{s}(\lambda) = \frac{R(\hat{z}(\lambda) - 1) + (R^2 - \hat{z}(\lambda))}{R(\hat{z}(\lambda) + 1) + (R^2 + \hat{z}(\lambda))} \tag{7-52}$$

Here Equation (7-52) can be obtained from Equation (7-51), either by invoking the change of normalization formula from Chapter 4, or by converting Equation (7-51) to a normalized impedance, scaling, and reconverting back to a scattering parameter. Using Equation (7-52) to calculate the resistivity operator, one obtains

$$1 - \hat{s}(\lambda)_*\hat{s}(\lambda) = \frac{2R^2(\hat{z}(\lambda) + \hat{z}(\lambda)_*)(1 - \lambda^2)}{\text{denominator}} \tag{7-53}$$

Here the denominator has not been given explicitly, since only the zeros of the operator are needed. Of course, one must assume that the numerator and denominator of this operator have no common terms. This is indeed the case when none of the lines have characteristic impedance 1, an assumption we make throughout the remainder of this section. The fundamental characteristic of the resistivity operator of Equation (7-53) is that it has a pair of zeros at $\lambda = \pm 1$ independent of the load and the characteristic impedance of the line (and hence, by exhaustion, dependent on the cascade topology). We have thus proven the first of a series of lemmas needed to obtain the desired result.

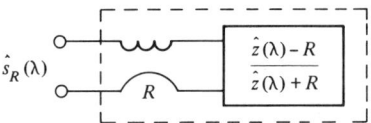

Figure 7-12 One-line cascade load configuration.

Lemma 7-1

Let $\hat{s}(\lambda)$ be the scattering parameter of the configuration of Figure 7-12 (where $R \neq 1$). Then
$$1 - \hat{s}(\lambda)_*\hat{s}(\lambda)$$
has zeros at
$$\lambda = \pm 1$$
independent of the load and the characteristic impedance of the line. □

Since we are using a mixed impedance scattering derivation it is desirable to obtain an impedance formulation of Lemma 7-1.

Lemma 7-2

Let a 1-port have a scattering parameter $\hat{s}(\lambda)$ and impedance $\hat{z}(\lambda)$. Then the zeros of
$$1 - \hat{s}(\lambda)_*\hat{s}(\lambda)$$
and
$$\hat{z}(\lambda) + \hat{z}(\lambda)_*$$
coincide.

Proof

$$\hat{z}(\lambda) = \frac{1 + \hat{s}(\lambda)}{1 - \hat{s}(\lambda)} \tag{7-54}$$

and

$$\hat{z}(\lambda)_* = \frac{1 + \hat{s}(\lambda)_*}{1 - \hat{s}(\lambda)_*} \tag{7-55}$$

Hence,

$$\hat{z}(\lambda) + \hat{z}(\lambda)_* = \frac{1 + \hat{s}(\lambda)}{1 - \hat{s}(\lambda)} + \frac{1 + \hat{s}(\lambda)_*}{1 - \hat{s}(\lambda)_*}$$
$$= \frac{2(1 - \hat{s}(\lambda)_*\hat{s}(\lambda))}{(1 - \hat{s}(\lambda))(1 - \hat{s}(\lambda)_*)} \tag{7-56}$$

Here the poles of the numerator and denominator rational functions coincide, so that the zeros of $\hat{z}(\lambda) + \hat{z}(\lambda)_*$ are exactly those of the resistivity operator, and conversely. □

With these two lemmas established, the desired necessary condition on the scattering parameter of a cascade of lines loaded in a resistor can be obtained.

Proposition 7-3

Let $\hat{s}(\lambda)$ be the scattering parameter of a cascade of n transmission lines (none with characteristic impedance 1) loaded in a resistor. Then
$$1 - \hat{s}(\lambda)_*\hat{s}(\lambda)$$
has n-fold zeros at $\lambda = \pm 1$ (that is, it has a factor of $(1 - \lambda^2)^n$).

Proof

The contention is proven by induction on the number of lines in the cascade. When $n = 1$ it is true, via Lemma 3, with $\hat{z}(\lambda)$ the load resistance. Let us, therefore, assume that it is true for all possible cascades of $n - 1$ lines, and show that it must also hold for the case of n lines in cascade. Employing the configuration of Figure 7-12, let the load represent the first $n - 1$ lines and the resistor, while the external line represents the nth line in the cascade. Now, by the induction hypothesis, the resistivity operator of the load has a factor

$$(1 - \lambda^2)^{n-1} \tag{7-57}$$

Hence, Lemma 7-2 implies that the even part of the impedance

$$\hat{z}(\lambda) + \hat{z}(\lambda)_* \tag{7-58}$$

also has such a factor. Once again, invoking Lemma 7-1, the resistivity operator of the overall 1-port with n lines in cascade has the form [shown in Equation (7-53)]

$$1 - \hat{s}(\lambda)_* \hat{s}(\lambda) = \frac{2R^2(\hat{z}(\lambda) + \hat{z}(\lambda)_*)(1 - \lambda^2)}{\text{denominator}} \tag{7-59}$$

where $(\hat{z}(\lambda) + \hat{z}(\lambda)_*)$ has a factor $(1 - \lambda^2)^{n-1}$. The overall operator thus has a factor $(1 - \lambda^2)^n$, which was to be shown. Since this holds for every n, the proof is complete. $\square$

Upon applying Lemma 7-2 to the result of the proposition, one obtains the corresponding impedance result.

Corollary 7-3

Under the hypothesis of Proposition 7-3, the even part of the impedance of n cascaded lines with a resistive load has n-fold zeros at $\lambda = \pm 1$. $\square$

The network of Figure 7-13 is composed of two unit elements (that is, lines of unit element length) of characteristic impedance $\frac{1}{2}$ loaded in a 1-ohm resistor. Now, the scattering parameter, normalized to $\frac{1}{2}$ for this resistor, is

$$\hat{s}_{(1/2)}(\lambda) = \left(\tfrac{1}{3}\right) \tag{7-60}$$

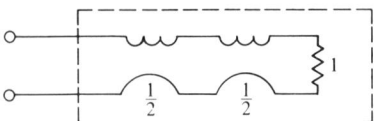

Figure 7-13 Resistor-loaded unit elements.

Hence, according to Equation (7-36), the scattering parameter for the overall 1-port is

$$\hat{s}_{(1/2)}(\lambda) = \left(\frac{\lambda - 1}{\lambda + 1}\right)^2 \left(\frac{1}{3}\right) \tag{7-61}$$

which corresponds to an impedance

$$\hat{z}(\lambda) = \frac{\lambda^2 + \lambda + 1}{\lambda^2 + 4\lambda + 1} \tag{7-62}$$

and the unnormalized scattering parameter

$$\hat{s}(\lambda) = \frac{-3\lambda}{2\lambda^2 + 5\lambda + 2} \tag{7-63}$$

Now, the resistivity operator may be calculated as

$$1 - \hat{s}(\lambda)\hat{s}(\lambda)_* = 1 - \hat{s}(\lambda)\hat{s}(-\lambda)$$
$$= \frac{4\lambda^4 - 8\lambda^2 + 4}{4\lambda^4 - 17\lambda^2 \pm 4} = \frac{4(\lambda^2 - 1)^2}{4\lambda^4 - 17\lambda^2 + 4} \tag{7-64}$$

while the even part of the impedance is found to be

$$\text{Ev } \hat{z}(\lambda) = \tfrac{1}{2}[\hat{z}(\lambda) + \hat{z}(-\lambda)]$$
$$= \frac{\lambda^4 - 2\lambda^2 + 1}{\lambda^4 - 14\lambda^2 + 1} = \frac{(\lambda^2 - 1)^2}{\lambda^4 - 14\lambda^2 + 1} \tag{7-65}$$

both of which do indeed have the required 2-fold zeros at $\lambda = \pm 1$.

Having exhibited a necessary condition on the scattering parameter of the desired configuration, we now proceed to show that it is also sufficient, this being achieved by exhibiting a cascade realization of any such scattering parameter. Since it is not lossless, it is necessary to prove that an application of Richards' theorem, to obtain a cascade realization, also gives degree reduction when a scattering parameter satisfies the conditions of Proposition 7-3. If this is so, as indicated by the following lemma, repeated application of Richards' theorem will yield the required realization, thereby proving the sufficiency of the proposition.

Lemma 7-3

Let $\hat{s}(\lambda)$ be a scattering parameter satisfying the hypotheses of Proposition 7-3, and let $\hat{z}(\lambda)$ be its corresponding impedance. Then

$$\hat{z}(1) = -\hat{z}(1)_* = -\hat{z}(-1)$$

Hence (by Theorem 6-3), Richards' theorem applied to $\hat{s}(\lambda)$ with $k = 1$ yields degree reduction.

Proof

From Corollary 7-3 we know that

$$\hat{z}(\lambda) + \hat{z}(\lambda)_* \tag{7-66}$$

has a zero at $\lambda = 1$. Thus,

$$\hat{z}(1) + \hat{z}(1)_* = \hat{z}(1) + \hat{z}(-1) = 0 \tag{7-67}$$

or, equivalently,
$$\hat{z}(1) = -\hat{z}(-1) \tag{7-68}$$

□

With the completion of the lemma, our main theorem on cascade synthesis may be proven. If one is given a scattering parameter satisfying the conditions of Proposition 7-3, the lemma asserts that it may be realized by repeated application of Richards' theorem, with $k = 1$, until the process terminates with a resistor. Of course, one must assume that the given scattering parameter is BR if the load resistor is to be passive.

Theorem 7-5

The necessary and sufficient conditions for $\hat{s}(\lambda)$ to be the (λ domain) scattering parameter of n transmission lines (none with unity characteristic impedance) in cascade loaded with a positive resistor are that it be BR rational and

$$1 - \hat{s}(\lambda)_* \hat{s}(\lambda)$$

have n-fold zeros at $\lambda = \pm 1$. Furthermore, if $\hat{s}(\lambda)$ satisfies these conditions it can be realized by repeated application of Richards' theorem, with $k = 1$ (which is assured to give degree reduction), until the process terminates with a zero-degree remainder corresponding to a load resistor. □

Consider the bounded, real, rational scattering parameter

$$\hat{s}(\lambda) = \frac{5\lambda^2 - 2\lambda + 5}{7\lambda^2 + 10\lambda + 7} \tag{7-69}$$

Here,

$$1 - \hat{s}(\lambda)_* \hat{s}(\lambda) = \frac{24(1 - \lambda^2)^2}{49\lambda^4 - 2\lambda^2 + 49} \tag{7-70}$$

Hence, this scattering parameter has a cascade realization with two lines. If we change the normalization to

$$\hat{z}(1) = 2 \tag{7-71}$$

the scattering parameter becomes

$$\hat{s}_2(\lambda) = \left(\frac{\lambda - 1}{\lambda + 1}\right)^2 \frac{1}{2} \tag{7-72}$$

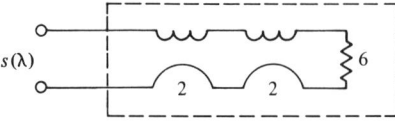

Figure 7-14 Cascade realization of a lossy scattering parameter.

and the first line in the cascade has a characteristic impedance of 2. The remainder is now

$$\hat{s}_2{}^r(\lambda) = \left(\frac{\lambda - 1}{\lambda + 1}\right)\frac{1}{2} \tag{7-73}$$

which clearly corresponds to another line of characteristic impedance 2 loaded in a 6-ohm resistor (having $s_2 = \frac{1}{2}$.) The final realization is thus shown in Figure 7-14.

7-5 Frequency Dependent Lines

In the preceding sections, we have considered only the case of frequency independent uniform transmission lines. The Richards transformation approach to the study of such lines is, however, applicable to a considerably larger class of lines.

*Definition 7-4

A *frequency dependent transmission line* with characteristic impedance R and length τ is a 2-port with scattering matrix

$$S_R(p) = \begin{bmatrix} 0 & e^{-\tau f(p)} \\ e^{-\tau f(p)} & 0 \end{bmatrix}$$

where $f(p)$ is some arbitrary function of p. ☐

Although the lines of Definition 7-4 are certainly not the most general class which might be defined [69], [186], they are sufficient for most practical purposes, including the LC, RC, and lossy LC lines commonly encountered.

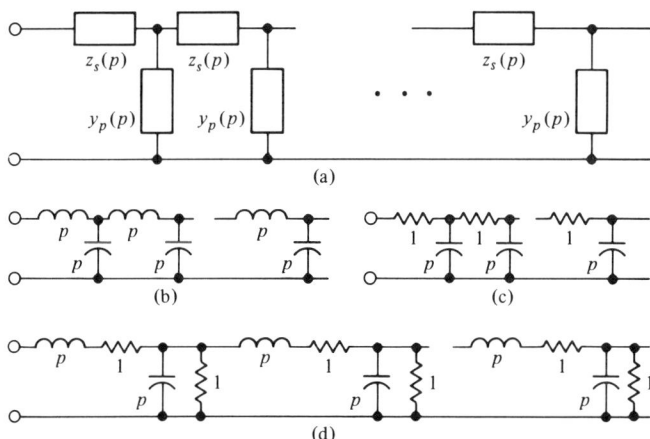

Figure 7-15 Construction of a uniform transmission line as a limit of a sequence of ladder networks. (**a**) General line. (**b**) LC line. (**c**) RC line. (**d**) Lossy LC line.

If one takes the common intuitive approach of a transmission line being the limit of a sequence of ladder networks, such as that shown in Figure 7-15, then the frequency dependence characteristic $f(p)$ may be taken as

$$f(p) = \sqrt{z_s(p)y_p(p)} \tag{7-74}$$

where the square root is interpreted as the "primary" value defined by

$$\sqrt{Me^{j\theta}} = \sqrt{M}\; e^{j\theta/2} \tag{7-75}$$

Now, in the usual frequency independent or LC case, the series ladder components are inductors, whereas the parallel components are capacitors, yielding

$$f_{LC}(p) = \sqrt{p \cdot p} = p \tag{7-76}$$

as we have been using. Similarly, for the uniform RC lines commonly encountered in integrated circuitry [119], the series component is a resistor, while the parallel component is a capacitor; thus,

$$f_{RC}(p) = \sqrt{1 \cdot p} = \sqrt{p} \tag{7-77}$$

Clearly, any such uniform line may be viewed as a frequency dependent line, including lossy lines, inverted CL (series, C, and parallel, L) lines, and so forth.

Of course, any interconnection of frequency dependent lines with a specified $f(p)$ together with resistors and gyrators may be used to define a network of such lines. If, moreover, all the lines of such a network are of commensurable lengths one may let

$$\tau = (\tfrac{1}{2})\; \text{GCD}\; \{\tau_1, \tau_2, \cdots, \tau_m\} \tag{7-78}$$

be the unit element length of the network. As in the uniform case, this choice of τ renders the system function of a network of frequency dependent lines rational in $e^{-2\tau f(p)}$, thereby setting the stage for a Richards-type transformation. Before proceeding with the definition of such a transformation, we consider the restrictions imposed on $f(p)$ by passivity and losslessness.

*Proposition 7-4

A frequency dependent transmission line characterized by

$$S_R(p) = \begin{bmatrix} 0 & e^{-\tau f(p)} \\ e^{-\tau f(p)} & 0 \end{bmatrix}$$

is passive if and only if $f(p)$ is PR, and lossless if and only if $f(p)$ is LPR.   □

Now, proceeding with the definition of the transformation, we have the following.

*Definition 7-5

Let a network be composed of resistors, gyrators, and a finite number of transmission lines with unit element length τ and specified $f(p)$ [$f(p)$ the same for all lines in the network]. Then the Richards transformation for the network is the substitution

$$\frac{\lambda - 1}{\lambda + 1} \leftrightarrow e^{-2\tau f(p)} \qquad \square$$

As in the uniform case, the transformation preserves passivity, though in this case one requires the further condition that the lines themselves be passive.

*Proposition 7-5

Let $S(p)$ be the scattering matrix of a network of frequency dependent lines satisfying the conditions of Definition 7-5, with $f(p)$ PR. Then

$$\hat{S}(\lambda) = S(p) \Big|_{e^{-2\tau f(p)} = \frac{\lambda-1}{\lambda+1}}$$

is BR if $S(p)$ is BR, and it is LBR if $S(p)$ is BR and the network contains no resistors. $\qquad\square$

It should be noted that if $\hat{S}(\lambda)$ is LBR, $S(p)$ may be BR rather than LBR, since the transformation identifies any line, even though it may be lossy, with a "lossless" λ domain device.

These preliminaries now allow synthesis to proceed by exactly the same λ domain techniques employed in the uniform case. In particular, a λ domain inductor corresponds to an open circuited line, while a λ domain capacitor corresponds to a short circuited line; hence, the substitution synthesis holds in the frequency dependent case and follows exactly the same procedure as in the uniform case. In addition, the insertion of a unit element length of line between a port and its load yields

$$\hat{s}_R(\lambda) = \left(\frac{\lambda - 1}{\lambda + 1}\right) \hat{s}_R{}^L(\lambda) \qquad (7\text{-}79)$$

Hence, the Richards theorem cascade synthesis also holds in this case. In this instance, the necessary and sufficient conditions derived in the last section for the λ domain representation of the networks are exactly the same as in the uniform case. Of course, the p domain conditions for the cascade realization change with $f(p)$. We thus make the possibly surprising observation that the λ domain representation is a function only of the transmission line "concept," whereas the apparently more basic p domain is a function of the particular kind of line employed.

Consider the uniform RC transmission line 1-port characterized by the scattering parameter

$$s(p) = \frac{5 \sinh^2 \tau\sqrt{p} - 2 \sinh \tau\sqrt{p} \cosh \tau\sqrt{p} + 5 \cosh^2 \tau\sqrt{p}}{7 \sinh^2 \tau\sqrt{p} + 10 \sinh \tau\sqrt{p} \cosh \tau\sqrt{p} + 7 \cosh^2 \tau\sqrt{p}} \qquad (7\text{-}80)$$

which, upon application of the Richards transformation

$$e^{-2\pi\sqrt{p}} \leftrightarrow \frac{\lambda - 1}{\lambda + 1} \qquad (7\text{-}81)$$

with unit element length 1, translates to the λ domain rational function

$$\hat{s}(\lambda) = \frac{5\lambda^2 - 2\lambda + 5}{7\lambda^2 + 10\lambda + 7} \qquad (7\text{-}82)$$

Now, we have already found a cascade realization of this in the frequency independent LC context, as shown in Figure 7-14. Hence, precisely the same λ domain realization will lead, in the RC case, to the analogous uniform RC realization of Figure 7-16.

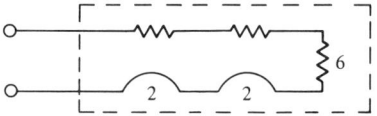

Figure 7-16 Cascade realization of an RC transmission line 1-port.

7-6 Discussion

We began the chapter by considering a class of networks which were apparently quite different from the RLC networks. In particular, the basic frequency dependent component was a 2-port characterized by two independent parameters, rather than the 1-port capacitor which is characterized by a single parameter. However, after carrying out an appropriate transformation, it was found that these transmission line networks could be reduced to the usual RLC rational functions, though of a new variable.

The success of converting the transcendental system functions of the transmission line networks to rational functions indicates that rationality is not merely a property of the RLC components, nor even a property representative of the finite number of poles which characterize any dynamical system. Rather, rationality is indicative of the finite number of components used to construct both the RLC networks and the transmission line networks. This concept is further exploited in Chapter 8, where finite networks of essentially arbitrary devices are studied by means of multivariable rational functions.

In addition to the λ domain approach to the study of transmission line networks taken here, an alternative p domain theory can be derived for the uniform case [88], [89]. This approach, though apparently more direct, does not extend to the frequency dependent case, nor does it permit much of the RLC theory to be applied to the transmission line case.

CHAPTER

8
Multivariable Networks

8-1 Introduction

In the preceding chapters various classes of networks were considered, but they were always composed of components from a single class; that is, we considered RLC and transmission line networks, but not mixed-lumped distributed networks; we considered uniform and frequency dependent lines, but not the mixed case; and so forth. It is these mixed cases of networks, containing components of two or more different types, which are considered in this chapter.

Assuming that we are given a network composed of several types of component, each individually characterized by a frequency domain representation $f_i(p)$, application of the analysis techniques of Chapter 4 leads to the conclusion that the system function of such a network is real rational in the variables $f_i(p)$, $i = 1, 2, \cdots, m$. Now, upon invoking the procedure employed for the transmission line analysis, one makes the transformation

$$\lambda_i = f_i(p) \qquad i = 1, 2, \cdots, m \tag{8-1}$$

after which the transformed system function is rendered real rational in the m variables λ_i, $i = 1, 2, \cdots, m$. Of course, it is not necessary to require that each λ_i be a function of frequency; rather, additional λ_i may be included which are determined entirely by some outside parameter x_j.

$$\lambda_j = f_j(x_j) \qquad j = m+1, m+2, \cdots, n \tag{8-2}$$

Here x_j may be a function of a control setting, temperature, and/or other environmental factors; it is, however, assumed to be independent of time, so as to allow (time-invariant) frequency domain techniques to be employed. In this way, the multivariable networks simultaneously include both mixed and/or variable parameter networks [21], [62], [81], [82], [124], [129], [153].

After formalizing the basic multivariable theory in the next section, we begin a study of synthesis by variable decomposition in the next three sections [114], [200]. In essence, this amounts to the realization of an m-variable network as an $(m-1)$-variable network loaded in a 1-variable network, as shown in Figure 8-1. Here the $(m-1)$ variables, denoted by λ, are separated from the

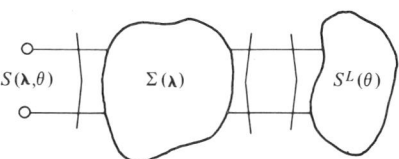

Figure 8-1 Multivariable decomposition scheme.

load variable θ by the decomposition. This thus yields yet another example of synthesis by cascade loading. The decomposition is a multistep process. First, one constructs "any" decomposition, this then being modified to achieve a minimal (with respect to the number of load ports) decomposition. Finally, this minimal decomposition is used as a starting point from which a "good" equivalent decomposition and the final realization are obtained. With the basic synthesis formulated in the general case, in the last two sections we consider its application to some useful special cases. These include variable parameter RLC networks, where a "single control" realization for a large class of networks is obtained, and a study of mixed-lumped distributed networks, which is the practical application which has motivated much of the multivariable theory. As in the pure transmission line case, the cascade structure is of special significance; hence, the synthesis of a cascade of lines with a lumped load is given special attention.

8-2 Multivariable Functions

Unlike the transmission line theory, where we begin with a specific frequency domain formulation and convert it to the λ domain, in the multivariable case we begin with a function of λ, relating it back to the p domain only after its λ domain properties have been examined. This is necessitated by the fact that we do not wish to assume any specific properties about the p domain components; rather, they are simply represented by

$$\lambda_i = f_i(p) \qquad i = 1, 2, \cdots, m \tag{8-3}$$

where $f_i(p)$ is left unspecified. For brevity, Equation (8-3) is usually replaced by the vector notation

$$\boldsymbol{\lambda} = \boldsymbol{f}(p) \tag{8-4}$$

Definition 8-1

A multivariable n-port is one whose system function can be represented, after an appropriate transformation, by a matrix of real rational functions in the several variables $\boldsymbol{\lambda}$. That is,

$$H(\lambda_1, \lambda_2, \cdots, \lambda_m) = H(\boldsymbol{\lambda})$$

where

$$\boldsymbol{\lambda} = \begin{bmatrix} \lambda_1 \\ \lambda_2 \\ \cdot \\ \cdot \\ \cdot \\ \lambda_m \end{bmatrix}$$

□

Possibly the most common example of a multivariable network is the mixed RLC-transmission line n-port. In that case, one lets

$$\lambda_1 = p \tag{8-5}$$

be the usual frequency variable representing the RLC devices, and

$$\lambda_2 = \cosh \tau p \tag{8-6}$$

be the Richards transformation variable corresponding to the lines. Clearly, the two variables are not independent, both being functions of p. This, however, causes no difficulty in practice, where we usually assume independence (or, more accurately, we do not assume dependence) [153]. Sometimes it is even convenient to transform the p variable rather than using it directly. In particular, it is often convenient [200] to let

$$\lambda_1 = \frac{p+1}{p-1} \tag{8-7}$$

rather than as assumed in Equation (8-5).

For the network of Figure 8-2, one may represent the RLC load by the usual p variable, the LC transmission line of unit length by the Richards transformation variable

$$\lambda_1 = \cosh p \tag{8-8}$$

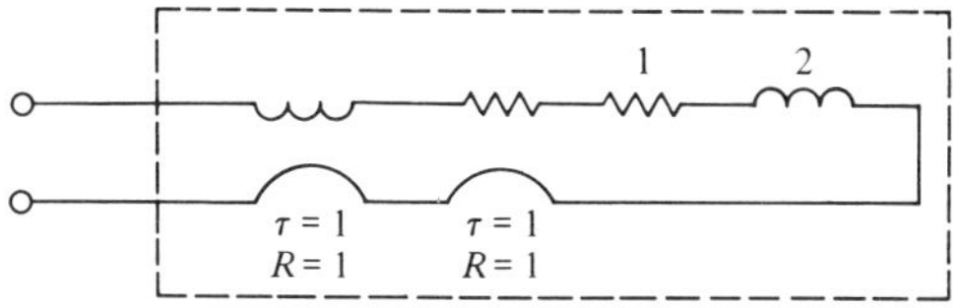

Figure 8-2 Mixed transmission line-lumped network.

and the RC line of length $\sqrt{2}$ by the variable

$$\lambda_2 = \coth \sqrt{2}p \tag{8-9}$$

Now, applying the analysis formula for cascade transmission lines of the preceding chapter, the scattering parameter of this 1-port is found to be

$$\hat{s}(\lambda) = \frac{(\lambda_1 - 1)(\lambda_2 - 1)p}{(\lambda_1 + 1)(\lambda_2 + 1)(p + 1)} \tag{8-10}$$

Note here that even though the lines of this network have different frequency dependence characteristics, a rational representation is still possible, so long as a different variable is used for each type of line. Similarly, two LC or RC lines with the same frequency dependence characteristic, but noncommensurable lengths, are representable by rational multivariable functions [165].

As with the basic definitions (and for the same reasons), it is desirable to define bounded reality independently of the p domain (and thus of passivity), with the relationship between these λ domain concepts and the usual passivity conditions being considered later. For the purpose of notational brevity in the definition, we say that the λ is in the RHP if all of its entries are in the RHP.

Definition 8-2

A multivariable matrix $H(\lambda)$ is *bounded real* (BR) if it satisfies the following conditions:

1. $H(\lambda)$ is analytic for λ in the RHP
2. $1 - H(\lambda)^* H(\lambda) \geq 0$ for λ in the RHP □

Note that real rationality is built into the definition of a multivariable matrix and need not be included as a separate condition for bounded reality. In the definition, the adjoint is, as usual, taken as the complex conjugate transpose

$$H(\lambda)^* = \overline{H(\lambda)'} = H(\bar{\lambda})' \tag{8-11}$$

Of course, one could define a multivariable Hurwitz conjugate by

$$H(\lambda)_* = H(-\lambda)' \tag{8-12}$$

As in the RLC case, this satisfies

$$H(\mathbf{j}\omega)^* = H(\mathbf{j}\omega)_* \tag{8-13}$$

Hence, the multivariable Hurwitz conjugate may be used to replace the adjoint (thereby preserving multivariable rationality). Clearly, the usual RLC arguments regarding the Hurwitz conjugate apply to the multivariable case [153].

*Proposition 8-1

A multivariable matrix, $H(\lambda)$ is BR if and only if it satisfies the following conditions:

1. $H(\lambda)$ has no poles in the (closed) RHP
2. $1 - H(\mathbf{j\omega})_* H(\mathbf{j\omega}) \geq 0$ for all imaginary axis $\mathbf{j\omega}$ (that is, each entry in the vector $\mathbf{j\omega}$ is imaginary).

Continuing the RLC analogy, the multivariable LBR conditions may be defined as follows.

Definition 8-3

A multivariable matrix $H(\lambda)$ is LBR if it is BR and

$$1 - H(\mathbf{j\omega})^* H(\mathbf{j\omega}) = 1 - H(\mathbf{j\omega})_* H(\mathbf{j\omega}) = 0$$

for all imaginary axis $\mathbf{j\omega}$. □

As in the RLC case, the real rationality of $H(\lambda)$ implies that the condition of Definition 8-3 is equivalent to the requirement that

$$H(\lambda)_* H(\lambda) = 1 \qquad (8\text{-}14)$$

for all RHP λ. Of course, analogous multivariable PR and LPR conditions can be obtained [153]; these are, however, not required for our present purposes, and will not be given.

For the network of Figure 8-2, having the scattering parameter shown in Equation (8-10), the poles occur whenever λ_1, λ_2, or p is equal to -1 independently of the other variables. Hence, for any pole, at least one of the three variables has a negative real part and the scattering matrix is, therefore, analytic in the closed RHP (that is, where all three variables have nonnegative real parts). For the positivity condition we have

$$1 - s(\lambda)^* s(\lambda) = 1 - \frac{(\lambda_1 + 1)(\lambda_2 + 1)p}{(\lambda_1 - 1)(\lambda_2 - 1)(p - 1)} \frac{(\lambda_1 - 1)(\lambda_2 - 1)p}{(\lambda_1 + 1)(\lambda_2 + 1)(p + 1)}$$

$$= 1 - \frac{p^2}{p^2 - 1} = \frac{-1}{p^2 - 1} \qquad (8\text{-}15)$$

Evaluating this on the $\mathbf{j\omega}$ axis yields

$$1 - s(\mathbf{j\omega})_* s(\mathbf{j\omega}) = \frac{-1}{-\omega^2 - 1} = \frac{1}{\omega^2 + 1} \qquad (8\text{-}16)$$

which is clearly nonnegative for all ω; hence, the scattering parameter of Equation (8-10) is bounded real, as may be expected from the fact that it represents a passive 1-port.

The scattering parameter of the preceding example had poles wherever one of the variables was equal to -1 independently of the value of the other variables. In particular, if one variable, say λ_1, is equal to -1, the function has a pole even when the other variables, λ_2 and p, are in the RHP. This is, however, not contradictory to the RHP analyticity requirement of Proposition 8-1, because the RHP for the variable vector λ is that region where all three variables are individually in the RHP, whereas, in the example, at least one of the variables is in the LHP at each pole. Now, consider the function

$$h(\lambda) = \frac{1}{\lambda_1 + \lambda_2 - 1} \tag{8-17}$$

for which the poles occur when

$$\lambda_1 + \lambda_2 = 1 \tag{8-18}$$

Unlike the previous example, the poles of this function are determined simultaneously by both variables and may be in the RHP, for instance, if

$$\lambda_1 = \lambda_2 = \tfrac{1}{2} \tag{8-19}$$

in which case both variables are individually in the RHP. On the other hand, the function

$$h(\lambda) = \frac{1}{\lambda_1 + \lambda_2 + 1} \tag{8-20}$$

is analytic in the RHP, since the poles occur when

$$\lambda_1 + \lambda_2 = -1 \tag{8-21}$$

which requires that at least one of the variables has a negative real part; hence, the vector λ is not in the RHP at any pole.

There are certainly many forms of multivariable network which may give rise to BR matrices that are unrelated to the p domain and passivity. If, however, one is interested in a multivariable network formed from a p domain matrix by the substitutions

$$\lambda_i = f_i(p) \qquad i = 1, 2, \cdots, m \tag{8-22}$$

then, with some assumptions, the resultant λ matrix will be BR if the original p domain network was passive.

*Proposition 8-2

Let $\hat{S}(\lambda)$ be a multivariable scattering matrix formed from a p domain scattering matrix $S(p)$ by the substitutions

$$\lambda_i = f_i(p) \qquad i = 1, 2, \cdots, m$$

where for every RHP λ_0, such that $\lambda_0 = f(p_0)$, p_0 is also in the RHP (that is, the inverse of f takes the RHP λ space into the RHP p plane). Then $\hat{S}(\lambda)$ is BR. □

A proof of this proposition can be obtained by the same argument used for the corresponding Richards transformation result for transmission lines (Theorem 7-2), and will not be given here. Applying the proposition to a mixed RLC-transmission line network, one finds that its 2-variable representation

$$S(\lambda) = S(p, \lambda) \tag{8-23}$$

will be BR if the network is passive, since Richards' transformation preserves the RHP.

A corresponding result holds in the lossless case if one makes the additional assumption that the inverses of the transformations

$$\lambda_i = f_i(p) \qquad i = 1, 2, \cdots, m \tag{8-24}$$

take the RHP λ space onto (rather than into) the RHP p plane. In fact, such an assumption also yields a necessity condition for the passive case.

*Corollary 8-1

Let $\hat{S}(\lambda)$ be the multivariable scattering matrix formed from a scattering matrix $S(p)$ by the substitutions

$$\lambda_i = f_i(p) \qquad i = 1, 2, \cdots, m$$

where for every RHP λ_0 there exists at least one p_0 in the RHP, such that $\lambda_0 = f(p_0)$, and moreover, every such p_0 (if more than one exists) is in the RHP. Then $S(p)$ is BR (LBR) if and only if $\hat{S}(\lambda)$ is BR (LBR). □

Possibly the most important application of Proposition 8-2 and its corollary is to the case of a network composed of several different types of transmission line with different frequency dependence characteristics or unit element lengths [165]. In this case, the f_i correspond to the Richards transformations

$$\lambda_i = \cosh \tau_i h_i(p) \tag{8-25}$$

Now, if the transmission lines are to be passive, the $h_i(p)$ functions are PR, and the condition of the proposition holds, hence, the λ domain representation of such a network is BR if its p domain representation is BR.

Finally, the possibility of defining a multivariable-degree concept must be considered. This is, however, complicated by the fact that one variable may "kill" another, so that the poles of one variable are functions of the others. Since it is desired that the λ_i degree be related to the number of λ_i domain reactors needed to realize the given function, one must take a "worst case" viewpoint. That is, the degree of the λ_i variable should correspond to the number of λ_i domain reactors required with the worst possible choice of the other variables.

*Definition 8-4

Let $H(\lambda)$ be a multivariable matrix. Then the λ_i degree of $H(\lambda)$ is

$$\delta_{\lambda_i} H(\lambda) = \max_{\lambda_j : j \neq i} \delta H(\lambda_i)$$

(that is, the maximum possible 1-variable degree with the other variables arbitrarily fixed). □

*Definition 8-5

Let $H(\lambda)$ be a multivariable matrix. Then the degree of $H(\lambda)$ is

$$\delta H(\lambda) = \sum_{i=1}^{m} \delta_{\lambda_i} H(\lambda_i)$$

□

The scattering parameter

$$s(\lambda) = \frac{(\lambda_1 - 1)(\lambda_2 - 1)p}{(\lambda_1 + 1)(\lambda_2 + 1)(p + 1)} \qquad (8\text{-}26)$$

is degree 1 in each of its variables and, hence, has an overall degree of three. Note that, for this function, if

$$\lambda_2 = \lambda_1 = 1 \qquad (8\text{-}27)$$

the degree in p of the resultant function is zero, not 1. The correct p degree is, however, 1, since by Definition 8-4, the p degree is defined to be the maximum possible degree of the 1-variable function resulting from fixing the other 2 variables arbitrarily. In this case, with λ_1 and λ_2 as in Equation (8-27), the p degree is zero; however, for different values of λ_1 and λ_2, say

$$\lambda_1 = \lambda_2 = 0 \qquad (8\text{-}28)$$

the p degree is 1, which is correct.

8-3 Multivariable Decomposition

In this section we begin the study of multivariable synthesis by decomposition. For this purpose, it is assumed that one is given an m-variable network made up of devices from m component classes, each represented by a different variable

$$\lambda_i \qquad i = 1, 2, \cdots, m \qquad (8\text{-}29)$$

The goal is to realize the m-variable network as an $(m - 1)$-variable coupling network loaded in a 1-variable network. We denote the $m - 1$ variables of the coupling network by λ and the load variable by θ. The desired decomposition

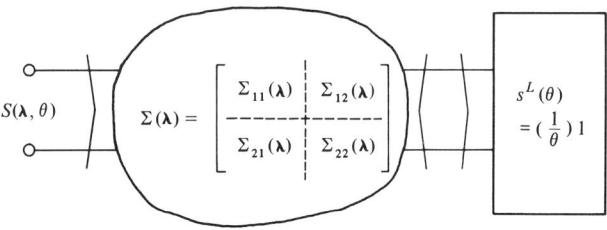

Figure 8-3 Decomposition of a multivariable network.

is then that illustrated in Figure 8-3. Rather than accepting any 1-variable load, we require that it have the special form with scattering matrix

$$S^L(\theta) = \left(\frac{1}{\theta}\right)1 \tag{8-30}$$

so as to assure that it may be readily realized by decoupled "$1/\theta$" devices.

Application of the cascade loading formula from Chapter 4 to the configuration of Figure 8-3 yields the representation

$$S(\lambda,\theta) = \Sigma_{11}(\lambda) + \Sigma_{12}(\lambda)(1 - S^L(\theta)\Sigma_{22}(\lambda))^{-1}S^L(\theta)\Sigma_{21}(\lambda) \tag{8-31}$$

for the multivariable network realized by the configuration. Moreover, the requirement that

$$S^L(\theta) = \left(\frac{1}{\theta}\right)1 \tag{8-32}$$

allows this representation to be further simplified to

$$\begin{aligned}S(\lambda,\theta) &= \Sigma_{11}(\lambda) + \left(\frac{1}{\theta}\right)\Sigma_{12}(\lambda)\left(1 - \left(\frac{1}{\theta}\right)\Sigma_{22}(\lambda)\right)^{-1}\Sigma_{21}(\lambda) \\ &= \Sigma_{11}(\lambda) + \Sigma_{12}(\lambda)(\theta 1 - \Sigma_{22}(\lambda))^{-1}\Sigma_{21}(\lambda)\end{aligned} \tag{8-33}$$

The decomposition problem may thus be viewed as that of writing a given multivariable function in the form of Equation (8-33), from which it can then by readily realized in the desired form by inspection. Of course, one would like the coupling network

$$\Sigma(\lambda) = \begin{bmatrix} \Sigma_{11}(\lambda) & \Sigma_{12}(\lambda) \\ \hline \Sigma_{21}(\lambda) & \Sigma_{22}(\lambda) \end{bmatrix} \tag{8-34}$$

to be passive or lossless if $S(\lambda,\theta)$ is passive or lossless. This aspect of the decomposition problem will, however, be deferred to the following sections, with the decomposition alone being considered here. We will, however, attempt to obtain a $\Sigma(\lambda)$ which is analytic in the RHP in preparation for our later consideration of passivity and losslessness.

Clearly, it is not possible to write all multivariable functions in the form of Equation (8-33).

Consider, for example,

$$s(\lambda,\theta) = \lambda + \theta \tag{8-35}$$

which at

$$\theta = \infty \tag{8-36}$$

satisfies

$$s(\lambda, \infty) = \infty \tag{8-37}$$

independent of λ. Now, if $S(\lambda,\theta)$ has a decomposition as per Equation (8-33), then

$$s(\lambda, \infty) = \Sigma_{11}(\lambda) \tag{8-38}$$

which is not infinite independent of λ (though $\Sigma_{11}(\lambda)$ may have poles at certain values of λ). Since the existence of a decomposition leads to a contradiction, one must conclude that the function $S(\lambda,\theta) = \lambda + \theta$ has no decomposition. Extrapolating the argument leads to the conclusion that any multivariable matrix with a decomposition must satisfy the condition

$$S(\lambda, \infty) = \Sigma_{11}(\lambda) \tag{8-39}$$

and, hence, does not have a pole at infinity independent of λ.

Proposition 8-3

A necessary condition for a multivariable matrix $S(\lambda,\theta)$ to be decomposable as

$$S(\lambda,\theta) = \Sigma_{11}(\lambda) + \Sigma_{12}(\lambda)(\theta 1 - \Sigma_{22}(\lambda))^{-1}\Sigma_{21}(\lambda)$$

is that it does not have a pole at $\theta = \infty$ which is independent of λ. □

Fortunately, the condition of the proposition is the only restriction on $S(\lambda, \infty)$ needed to assure the existence of a decomposition [114]. This is, however, easier said than done. In fact, the remainder of this section is devoted to a constructive proof of this fact, together with its corollaries. This is of more than academic interest, since the construction of the decomposition is a necessary preliminary to the multivariable synthesis of the following sections. It is worthy of note that the restriction of the proposition is quite minimal, this because any BR or LBR matrix is of necessity finite at infinity, whereas a general multivariable matrix can be converted to a BR matrix by the active synthesis procedure of Chapter 5.

As a first step in obtaining the decomposition, we consider the $\Sigma_{11}(\lambda)$ term which was shown to be given by

$$\Sigma_{11}(\lambda) = S(\lambda, \infty) \tag{8-40}$$

in the preceding development.

Proposition 8-4

For every decomposition of $S(\lambda,\theta)$ as

$$S(\lambda,\theta) = \Sigma_{11}(\lambda) + \Sigma_{12}(\lambda)(\theta 1 - \Sigma_{22}(\lambda))^{-1}\Sigma_{21}(\lambda)\Sigma_{11}(\lambda)$$

$$S(\lambda, \infty) = \Sigma_{11}(\lambda)$$

where $\Sigma_{11}(\lambda)$ is analytic in the RHP if $S(\lambda,\theta)$ is analytic in the RHP. □

Consistent with the proposition, one may restrict consideration to the (apparently) simpler decomposition

$$\ddot{S}(\lambda,\theta) = \Sigma_{12}(\lambda)(\theta 1 - \Sigma_{22}(\lambda))^{-1}\Sigma_{21}(\lambda) \tag{8-41}$$

where

$$\ddot{S}(\lambda,\theta) = S(\lambda,\theta) - S(\lambda, \infty) \tag{8-42}$$

The particular advantage of this step [besides determining $\Sigma_{11}(\lambda)$] is that

$$\ddot{S}(\lambda, \infty) = S(\lambda, \infty) - S(\lambda, \infty) = 0 \tag{8-43}$$

thereby allowing one to expand $S(\lambda,\theta)$ in a Laurant series about infinity [144]. One may thus write

$$\ddot{S}(\lambda,\theta) = \frac{A_0(\lambda)}{\theta} + \frac{A_1(\lambda)}{\theta^2} + \cdots = \sum_{k=0}^{\infty} \frac{A_k(\lambda)}{\theta^{k+1}} \tag{8-44}$$

If, moreover, it is assumed that a decomposition exists and one expands the right side of Equation (8-41) (with the usual binomial expansion for an inverse), we obtain

$$\ddot{S}(\lambda,\theta) = \sum_{k=0}^{\infty} \Sigma_{12}(\lambda) \left(\frac{(\Sigma_{22}(\lambda))^k}{\theta^{k+1}} \right) \Sigma_{21}(\lambda) \tag{8-45}$$

The decomposition problem is thus reduced to that of comparing the coefficients of the two series. In particular, one desires to find three matrices $\Sigma_{12}(\lambda)$, $\Sigma_{22}(\lambda)$, and $\Sigma_{21}(\lambda)$ such that

$$A_k(\lambda) = \Sigma_{12}(\lambda)(\Sigma_{22}(\lambda))^k \Sigma_{21}(\lambda) \tag{8-46}$$

where the $A_k(\lambda)$ is a specified infinite series of matrices. Moreover, if $S(\lambda,\theta)$ is analytic in the RHP [hence also $\ddot{S}(\lambda,\theta)$], we require that the Σ's be analytic in the RHP.

Clearly, the problem has no general solution since the $\Sigma(\lambda)$ matrices have only a finite number of parameters, whereas the series may be characterized by infinitely many independent parameters. In the present case, however, the series of $A_k(\lambda)$ is derived from a rational matrix of finite order, and hence has only a finite number of independent parameters. In fact, if $\ddot{S}(\lambda,\theta)$ is an rth order function of θ, one may write all of the $A_k(\lambda)$ from the first r terms of the series.

Lemma 8-1

Let

$$\ddot{S}(\lambda,\theta) = \frac{B_0(\lambda)\theta^r + B_1(\lambda)\theta^{r-1} + \cdots + B_r(\lambda)}{a_0(\lambda)\theta^r + a_1(\lambda)\theta^{r-1} + \cdots + a_r(\lambda)}$$

where

$$a_0(\lambda)\theta^r + a_1(\lambda)\theta^{r-1} + \cdots + a_r(\lambda)$$

is a common denominator of $\ddot{S}(\lambda,\theta)$ which also has the Laurant expansion

$$\ddot{S}(\lambda,\theta) = \sum_{k=0}^{\infty} \frac{A_k(\lambda)}{\theta^{k+1}}$$

Then

$$a_0(\lambda) A_k(\lambda) = -\sum_{i=0}^{r} a_i(\lambda) A_{k-i}(\lambda)$$

for k greater than or equal to r.

Proof

Equating the two representations of $\ddot{S}(\lambda,\theta)$ yields

$$\frac{B_0(\lambda)\theta^r + B_1(\lambda)\theta^{r-1} + \cdots + B_r(\lambda)}{a_0(\lambda)\theta^r + a_1(\lambda)\theta^{r-1} + \cdots + a_r(\lambda)} = \frac{A_0(\lambda)}{\theta} + \frac{A_1(\lambda)}{\theta} + \cdots = \sum_{k=0}^{\infty} \frac{A_k(\lambda)}{\theta^{k+1}} \tag{8-47}$$

Now, multiplying through by the denominator polynomial and combining the coefficients of like powers of θ results in

$$B_0(\lambda)\theta^r + B_1(\lambda)\theta^{r-1} + \cdots + B_r(\lambda) = \sum_{k=0}^{\infty}\left(\sum_{i=0}^{r} a_i(\lambda)A_{k-i}(\lambda)\right)\theta^{r-k-1} \quad \text{(8-48)}$$

Since the left side of Equation (8-48) has no negative powers of θ, the coefficients of the negative powers on the right side must be zero. Hence,

$$\sum_{i=0}^{r} a_i(\lambda)A_{k-i}(\lambda) = 0 \qquad k \geq r \quad \text{(8-49)}$$

or, equivalently,

$$a_0(\lambda)A_k(\lambda) = -\sum_{i=1}^{r} a_i(\lambda)A_{k-i}(\lambda)$$

which was to be shown. □

Consistent with the lemma, any sequence of r consecutive $A_k(\lambda)$ suffice to describe the network. For our purposes, it is found that the most convenient way to represent such a sequence is via the $nr \times nr$ Hankel matrix [where n is the dimension of $\breve{S}(\lambda,\theta)$] [114], [194], [200]

$$G_r(\lambda) = \begin{bmatrix} A_0(\lambda) & A_1(\lambda) & \cdots & A_{r-1}(\lambda) \\ A_1(\lambda) & A_2(\lambda) & \cdots & A_r(\lambda) \\ A_2(\lambda) & & & \\ \vdots & & & \\ A_{r-1}(\lambda) & A_r(\lambda) & \cdots & A_{2r-2}(\lambda) \end{bmatrix} \quad \text{(8-50)}$$

More generally, since the sequence of $A_k(\lambda)$ need not start with $A_0(\lambda)$, one may deal with

$$G_{r,i}(\lambda) = \begin{bmatrix} A_i(\lambda) & A_{1+i}(\lambda) & \cdots & A_{r-1+i}(\lambda) \\ A_{1+i}(\lambda) & A_{2+i}(\lambda) & \cdots & A_{r+1}(\lambda) \\ A_{2+1}(\lambda) & & & \\ \vdots & & & \\ A_{r-1+i}(\lambda) & A_{r+i}(\lambda) & \cdots & A_{2r-2+i}(\lambda) \end{bmatrix} \quad \text{(8-51)}$$

Clearly,
$$G_r(\lambda) = G_{r,0}(\lambda) \tag{8-52}$$

Now, the lemma implies that the various $G_{r,i}(\lambda)$ are not independent. In fact, if one lets

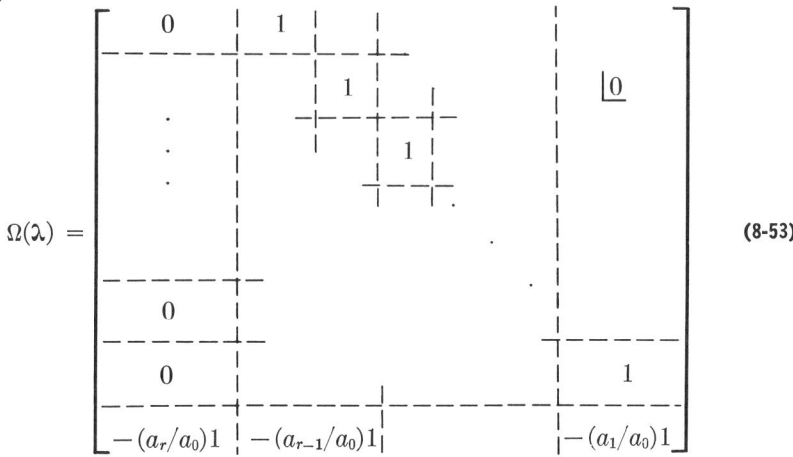

$$\Omega(\lambda) = \tag{8-53}$$

be the $nr \times nr$ companion matrix (where 1 and 0 represent $n \times n$ identity and zero matrices, respectively) associated with the denominator polynomial

$$a_0(\lambda)\theta^r + a_1(\lambda)\theta^{r-1} + \cdots + a_r(\lambda) \tag{8-54}$$

then the relationship between them is given by the following lemma.

Lemma 8-2
$$\Omega(\lambda)G_{r,i}(\lambda) = G_{r,i+1}(\lambda) \qquad \square$$

A proof of the lemma can be obtained by direct calculation, with the aid of the identity of Lemma 8-1, and will not be given in detail. Repeated application of Lemma 8-2 yields the more general relation

$$\Omega(\lambda)^k G_{r,i}(\lambda) = G_{r,i+k}(\lambda) \tag{8-55}$$

Here, if $S(\lambda,\theta)$ is analytic in the RHP, so are both the A's and the a's; hence, so are both $G_{r,i}(\lambda)$ and $\Omega(\lambda)$. With Lemma 8-2 as a tool, the desired decomposition can be obtained by letting

$$\Sigma_{12}(\lambda) = [1 \quad 0 \quad \cdots \quad 0] \tag{8-56}$$

be the $nr \times nr$ matrix with an $n \times n$ identity matrix in the leading position,

$$\Sigma_{22}(\lambda) = \Omega(\lambda) \tag{8-57}$$

and

$$\Sigma_{21}(\lambda) = G_r(\lambda)\Sigma_{12}(\lambda)' \tag{8-58}$$

Now, all three of these matrices are clearly analytic in the RHP if $S(\lambda,\theta)$ is analytic there, and, moreover,

$$\Sigma_{12}(\lambda)\Sigma_{22}(\lambda)^i\Sigma_{21}(\lambda) = [1 \mid 0 \mid \cdots \mid 0]\Omega(\lambda)^i G_r(\lambda) \begin{bmatrix} 1 \\ 0 \\ \cdot \\ \cdot \\ \cdot \\ 0 \end{bmatrix}$$

(8-59)

$$= [1 \mid 0 \mid \cdots \mid 0]G_{r,i}(\lambda) \begin{bmatrix} 1 \\ 0 \\ \cdot \\ \cdot \\ \cdot \\ 0 \end{bmatrix} = A_i(\lambda)$$

which is, by Equation (8-46), exactly the condition needed to assure that

$$\ddot{S}(\lambda,\theta) = \Sigma_{12}(\lambda)(\theta 1 - \Sigma_{22}(\lambda))^{-1}\Sigma_{21}(\lambda) \qquad (8\text{-}60)$$

In Equation (8-59), the last equality is due to the fact that the post- and premultiplication of $G_{r,i}(\lambda)$ by

$$[1 \mid 0 \mid \cdots \mid 0] \qquad (8\text{-}61)$$

and its transpose picks out the upper left entry, which is $A_i(\lambda)$ [104], [105]. Combining this development with our previous choice of $\Sigma_{11}(\lambda)$ now leads to the required decomposition theorem.

Theorem 8-1

A necessary and sufficient condition for a multivariable matrix $S(\lambda,\theta)$ to be decomposable as

$$S(\lambda,\theta) = \Sigma_{11}(\lambda) + \Sigma_{12}(\lambda)(\theta 1 - \Sigma_{22}(\lambda))^{-1}\Sigma_{21}(\lambda)$$

is that $S(\lambda,\theta)$ does not have a pole at $\theta = \infty$ independent of λ. If this condition is satisfied, one decomposition is given by

$$\Sigma_{11}(\lambda) = S(\lambda, \infty)$$
$$\Sigma_{12}(\lambda) = [1 \mid 0 \mid \cdots \mid 0]$$
$$\Sigma_{21}(\lambda) = G_r(\lambda)\Sigma_{12}(\lambda)'$$

and
$$\Sigma_{22}(\lambda) = \Omega(\lambda)$$

where $G_r(\lambda)$ and $\Omega(\lambda)$ are the Hankel and companion matrices defined above. Moreover, these matrices are analytic in the RHP if $S(\lambda,\theta)$ is analytic in the RHP. □

Although a rather long and tedious derivation was employed to obtain Theorem 8-1, it is important to note that the application of the theorem is quite straightforward. In fact, the decomposition may be written from $S(\lambda,\theta)$ and its first r Laurant coefficients by inspection. Consider, for example, the LBR 2-variable scattering parameter

$$s(\lambda,\theta) = \left(\frac{\lambda-1}{\lambda+1}\right)^2 \left(\frac{\theta-1}{\theta+1}\right) \tag{8-62}$$

Here,
$$\Sigma_{11}(\lambda) = s(\lambda, \infty) = \left(\frac{\lambda-1}{\lambda+1}\right)^2 \tag{8-63}$$

and
$$\ddot{s}(\lambda,\theta) = \left(\frac{\lambda-1}{\lambda+1}\right)^2 \left(\frac{-2}{\theta+1}\right)$$
$$= -2\left(\frac{\lambda-1}{\lambda+1}\right)^2 \left(\frac{1}{\theta} - \frac{1}{\theta^2} + \frac{1}{\theta^3} - \cdots\right) \tag{8-64}$$

Thus,
$$G_r(\lambda) = A_0(\lambda) = -2\left(\frac{\lambda-1}{\lambda+1}\right)^2 \tag{8-65}$$

while
$$\Omega(\lambda) = -1 \tag{8-66}$$

Hence,
$$\Sigma_{11}(\lambda) = \left(\frac{\lambda-1}{\lambda+1}\right)^2 \tag{8-67}$$

$$\Sigma_{12}(\lambda) = 1 \tag{8-68}$$

$$\Sigma_{21}(\lambda) = -2\left(\frac{\lambda-1}{\lambda+1}\right)^2 \tag{8-69}$$

and
$$\Sigma_{22}(\lambda) = -1 \tag{8-70}$$

is the required decomposition. Physically, $S(\lambda,\theta)$ may thus be realized as shown in Figure 8-4, with the λ domain coupling network determined by the decomposition, and a $(1/\theta)$ load.

It is significant to note that the decomposition does not merely split off the θ domain portion of the network, but, in fact, splits off one of the simplest possible loads, this being realizable by inspection. One difficulty with the decomposition of Theorem 8-1 is that it does not minimize the number of load ports (and hence the number of θ domain components) needed in the realization.

It is, however, possible to obtain a minimal decomposition from the nonminimal one of the theorem [87]. In that case, one lets $B(\lambda)$ and $C(\lambda)$ be matrices, analytic in the RHP, which diagonalizes $G_r(\lambda)$ into the form

$$B(\lambda)G_r(\lambda)C(\lambda) = 1 \oplus 0 = \left[\begin{array}{c|c} 1 & 0 \\ \hline 0 & 0 \end{array}\right] \tag{8-71}$$

and then uses these to derive Corollary 8-2 [87], [114].

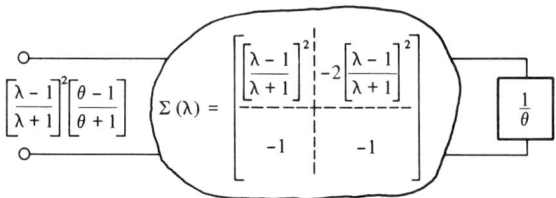

Figure 8-4 Realization of a 2-variable scattering parameter by decomposition.

*Corollary 8-2

Under the hypotheses of Theorem 8-1, a minimal realization is given by

$$\Sigma_{11}(\lambda) = S(\lambda, \infty)$$

$$\Sigma_{12}(\lambda) = [1 \mid 0 \mid \cdots \mid 0]G_r(\lambda)C(\lambda) \begin{bmatrix} 1 \\ \hline 0 \\ \hline \cdot \\ \cdot \\ \cdot \\ \hline 0 \end{bmatrix}$$

$$\Sigma_{21}(\lambda) = [1 \mid 0 \mid \cdots \mid 0]B(\lambda)G_r(\lambda) \begin{bmatrix} 1 \\ \hline 0 \\ \hline \cdot \\ \cdot \\ \cdot \\ \hline 0 \end{bmatrix}$$

and

$$\Sigma_{22}(\lambda) = [1 \mid 0 \mid \cdots \mid 0] B(\lambda)\Omega(\lambda)G_r(\lambda)C(\lambda) \begin{bmatrix} 1 \\ \hline 0 \\ \hline \cdot \\ \cdot \\ \cdot \\ \hline 0 \end{bmatrix}$$

where $B(\lambda)$ and $C(\lambda)$ diagonalize $G_r(\lambda)$ as per Equation (8-71). □

Although the decompositions thus far constructed realize the desired m-variable network as an $(m-1)$-variable coupling network with a minimal decoupled load, there is no assurance that the properties of the original network will be preserved by the coupling network. In particular, if the original network is BR (LBR), then is the coupling network BR (LBR)? This is certainly not the case with the decomposition of the example in which an active coupling network was obtained for an LBR scattering parameter. If, however, one begins with a decomposition constructed via Theorem 8-1 or its corollary, and uses it as a starting point from which to construct an alternate decomposition, it might be possible to obtain a "good" equivalent decomposition. We, therefore, consider the problem of constructing all equivalent decompositions (good and bad) from a given one. For this purpose it is most convenient to represent a decomposition symbolically as the $(m-1)$-variable coupling network

$$\Sigma(\lambda) = \begin{bmatrix} \Sigma_{11}(\lambda) & \mid & \Sigma_{12}(\lambda) \\ \hline \Sigma_{21}(\lambda) & \mid & \Sigma_{22}(\lambda) \end{bmatrix} \quad (8\text{-}72)$$

which it induces. Now, if two coupling networks (or decompositions) are to be equivalent, they must realize the same (given) $S(\lambda,\theta)$ when loaded by

$$S^L(\theta) = \left(\frac{1}{\theta}\right)1 \quad (8\text{-}73)$$

For this it is required that

$$\begin{aligned} S(\lambda,\theta) &= \Sigma_{11}(\lambda) + \Sigma_{12}(\lambda)(\theta 1 - \Sigma_{22}(\lambda))^{-1}\Sigma_{21}(\lambda) \\ &= \hat{\Sigma}_{11}(\lambda) + \hat{\Sigma}_{12}(\lambda)(\theta 1 - \hat{\Sigma}_{22}(\lambda))^{-1}\hat{\Sigma}_{21}(\lambda) \end{aligned} \quad (8\text{-}74)$$

Clearly, a necessary condition for equivalence is that

$$\Sigma_{11}(\lambda) = \hat{\Sigma}_{11}(\lambda) = S(\lambda, \infty) \quad (8\text{-}75)$$

For our purposes, one is primarily interested in generating equivalent decompositions of the same size as the given decomposition, and, in particular, in generating minimal RHP analytic decompositions from a given such decomposition (since we do not want to increase the number of load ports in the process). This class of decomposition equivalences is characterized by the following theorem [87], [114], [194], [200].

*Theorem 8-2

If

$$\Sigma(\lambda) = \left[\begin{array}{c|c} \Sigma_{11}(\lambda) & \Sigma_{12}(\lambda) \\ \hline \Sigma_{21}(\lambda) & \Sigma_{22}(\lambda) \end{array}\right]$$

is a decomposition of $S(\lambda,\theta)$, and $T(\lambda)$ is an invertible multivariable matrix of appropriate dimension,

$$\hat{\Sigma}(\lambda) = \left[\begin{array}{c|c} \Sigma_{11}(\lambda) & \Sigma_{12}(\lambda)T(\lambda)^{-1} \\ \hline T(\lambda)\Sigma_{21}(\lambda) & T(\lambda)\Sigma_{22}(\lambda)T(\lambda)^{-1} \end{array}\right]$$

$$= \left[\begin{array}{c|c} 1 & 0 \\ \hline 0 & T(\lambda) \end{array}\right] \Sigma(\lambda) \left[\begin{array}{c|c} 1 & 0 \\ \hline 0 & T(\lambda)^{-1} \end{array}\right]$$

is also a decomposition. Furthermore, if $\Sigma(\lambda)$ is a minimal decomposition, then all other minimal decompositions are obtained by such a transformation, and if $\Sigma(\lambda)$, $T(\lambda)$, and $T(\lambda)^{-1}$ are analytic in the RHP, then so is $\hat{\Sigma}(\lambda)$. ◻

The first part of the theorem can be proven by substitution into Equation (8-74) and observing that $T(\lambda)$ cancels. The more complex necessary condition for a minimal decomposition, however, requires certain controllability arguments [87] which we do not wish to consider here. A proof is, however, given by Newcomb [114].

For the scattering parameter

$$s(\lambda,\theta) = \left(\frac{\lambda-1}{\lambda+1}\right)^2 \left(\frac{\theta-1}{\theta+1}\right) \qquad (8\text{-}76)$$

the decomposition corresponding to the coupling network

$$\Sigma(\lambda) = \left[\begin{array}{c|c} \left(\frac{\lambda-1}{\lambda+1}\right)^2 & -2\left(\frac{\lambda-1}{\lambda+1}\right)^2 \\ \hline -1 & -1 \end{array}\right] \qquad (8\text{-}77)$$

was found previously. Now, if one chooses for $T(\lambda)$ the function

$$T(\lambda) = \sqrt{2}\left(\frac{\lambda-1}{\lambda+1}\right) \qquad (8\text{-}78)$$

the equivalent decomposition characterized by the coupling network

$$\hat{\Sigma}(\lambda) = \left[\begin{array}{c|c} \left(\dfrac{\lambda-1}{\lambda+1}\right)^2 & -\sqrt{2}\left(\dfrac{\lambda-1}{\lambda+1}\right) \\ \hline -\sqrt{2}\left(\dfrac{\lambda-1}{\lambda+1}\right) & -1 \end{array}\right] \tag{8-79}$$

is obtained. Although equivalent, the new coupling network is reciprocal, whereas the original was not; thus, an application of the equivalence theory may yield improved decompositions. Of course, our primary interest is to apply the equivalent decomposition theory in such a way as to generate passive and lossless decompositions when they exist. This subject will be treated in more detail in Section 8-5.

Although one is primarily interested in minimal decompositions, the preceding theorem can be modified to allow the generation of all possible decompositions of $S(\lambda,\theta)$ from a given minimal decomposition. In this case we have the following result [12].

*Theorem 8-3

Let

$$\Sigma(\lambda) = \left[\begin{array}{c|c} \Sigma_{11}(\lambda) & \Sigma_{12}(\lambda) \\ \hline \Sigma_{21}(\lambda) & \Sigma_{22}(\lambda) \end{array}\right]$$

be a minimal decomposition of $S(\lambda,\theta)$. Then all other (minimal or not) decompositions of $S(\lambda,\theta)$ are given by

$$\hat{\Sigma}(\lambda) = \left[\begin{array}{c|c} \hat{\Sigma}_{11}(\lambda) & \hat{\Sigma}_{12}(\lambda) \\ \hline \hat{\Sigma}_{21}(\lambda) & \hat{\Sigma}_{22}(\lambda) \end{array}\right]$$

where

$$\hat{\Sigma}_{11}(\lambda) = \Sigma_{11}(\lambda)$$

$$\hat{\Sigma}_{12}(\lambda) = [0 \mid \Sigma_{12}(\lambda) \mid A(\lambda)]T(\lambda)^{-1}$$

$$\hat{\Sigma}_{21}(\lambda) = T(\lambda)\left[\begin{array}{c} B(\lambda) \\ \hline \Sigma_{21}(\lambda) \\ \hline 0 \end{array}\right]$$

and

$$\hat{\Sigma}_{22}(\lambda) = T(\lambda)\left[\begin{array}{c|c|c} C(\lambda) & D(\lambda) & E(\lambda) \\ \hline 0 & \Sigma_{22}(\lambda) & F(\lambda) \\ \hline 0 & 0 & G(\lambda) \end{array}\right]T(\lambda)^{-1}$$

Here, $A(\lambda)$, $B(\lambda)$, $C(\lambda)$, $D(\lambda)$, $E(\lambda)$, $F(\lambda)$, and $G(\lambda)$ are arbitrary matrices of conformable dimension, and $T(\lambda)$ is an arbitrary nonsingular matrix of conformable dimension. Moreover, if $\Sigma(\lambda)$, the A, B, $\cdots$, G matrices, and $T(\lambda)$ and its inverse are all analytic in the RHP, so is $\hat{\Sigma}(\lambda)$. □

8-4 Passive Multivariable Synthesis

As in our previous special cases, the techniques of Chapter 5 suffice for the synthesis of multivariable π and/or active scattering matrices; hence, we may begin our discussion of multivariable synthesis with a BR scattering matrix $S(\lambda)$. As in the RLC case, one can invoke the techniques of Chapter 5 to realize $S(\lambda)$ as a lossless multivariable coupling network loaded in 1-ohm resistors, as illustrated in Figure 8-5. Also, as in that case, if one replaces the adjoint

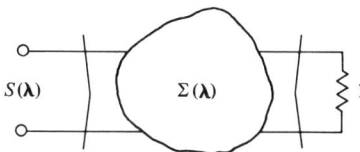

Figure 8-5 Cascade realization of a BR $S(\lambda)$.

of the general theory by the Hurwitz conjugate, so as to preserve multivariable rationality, then $\Sigma(\lambda)$ must satisfy

$$\Sigma_{11}(\lambda) = S(\lambda) \tag{8-80}$$

$$\Sigma_{12}(\lambda)\Sigma_{12}(\lambda)_* = 1 - S(\lambda)S(\lambda)_* \tag{8-81}$$

$$\Sigma_{21}(\lambda)_*\Sigma_{21}(\lambda) = 1 - S(\lambda)_*S(\lambda) \tag{8-82}$$

and

$$\Sigma_{22}(\lambda) = -\Sigma_{21}(\lambda)S(\lambda)_*\Sigma_{12}(\lambda)_*^{-R} \tag{8-83}$$

if its LBR character is to be assured. Unlike the RLC case, no general factorization for multivariable matrices of the type required by Equations (8-81) and (8-82) is known, nor is the existence of such even assured; hence, the validity of this approach to multivariable passive synthesis is not assured. If, however, a factorization can be found which renders $\Sigma(\lambda)$ analytic in the RHP, then we have the following proposition.

Proposition 8-5

Let $S(\lambda)$ be a multivariable BR scattering matrix and $\Sigma(\lambda)$ be a multivariable scattering matrix which is analytic in the RHP and satisfies Equations (8-80) through (8-83). Then $\Sigma(\lambda)$ is LBR, and a realization of it loaded in 1-ohm resistors realizes $S(\lambda)$. □

Consider the 3-variable scattering parameter for the network of Figure 8-2,

$$s(\lambda) = \frac{(\lambda_1 - 1)(\lambda_2 - 1)p}{(\lambda_1 + 1)(\lambda_2 + 1)(p + 1)} \tag{8-84}$$

with the resistivity operators

$$1 - s(\lambda)_* s(\lambda) = 1 - s(\lambda)s(\lambda)_* = \frac{-1}{p^2 - 1} = \frac{1}{1 - p^2} \tag{8-85}$$

Clearly, one choice for the factorization of this function is

$$\Sigma_{12}(\lambda) = \Sigma_{21}(\lambda) = \frac{1}{1 + p} \tag{8-86}$$

which has no RHP poles. Unfortunately, such a factorization leads to

$$\Sigma_{22}(\lambda) = -\Sigma_{21}(\lambda)S(\lambda)_*\Sigma_{12}(\lambda)_*{}^{-R}$$
$$= \frac{(\lambda_1 + 1)(\lambda_2 + 1)p}{(\lambda_1 - 1)(\lambda_2 - 1)(p + 1)} \tag{8-87}$$

which has an RHP pole at

$$\lambda_1 = 1 \tag{8-88}$$

$$\lambda_2 = 2 \tag{8-89}$$

$$p = 1 \tag{8-90}$$

Hence, the coupling network obtained via such a factorization is not LBR. If, rather, we choose the more complex factorizations

$$\Sigma_{12}(\lambda) = \frac{1}{p + 1} \tag{8-91}$$

and

$$\Sigma_{21}(\lambda) = \frac{(\lambda_1 - 1)(\lambda_2 - 1)}{(\lambda_1 + 1)(\lambda_2 + 1)(p + 1)} \tag{8-92}$$

which are also analytic in the RHP, we obtain

$$\Sigma_{22}(\lambda) = \frac{p}{p + 1} \tag{8-93}$$

and the coupling network

$$\Sigma(\lambda) = \begin{bmatrix} \dfrac{(\lambda_1 - 1)(\lambda_2 - 1)p}{(\lambda_1 + 1)(\lambda_2 + 1)(p + 1)} & \dfrac{1}{p + 1} \\ \dfrac{(\lambda_1 - 1)(\lambda_2 - 1)}{(\lambda_1 + 1)(\lambda_2 + 1)(p + 1)} & \dfrac{p}{p + 1} \end{bmatrix} \tag{8-94}$$

which is LBR and realizes the given BR 3-variable function under resistive loading.

In essence, the synthesis procedure of Proposition 8-5 can be viewed as a trade-off wherein a BR scattering matrix is converted to an LBR matrix at the cost of increasing the number of ports. In the multivariable case, an alternative approach might be to convert a BR matrix to an LBR matrix by increasing the number of variables. This is indeed a possibility, as is exhibited by the following proposition.

*Proposition 8-6

Let
$$S(\lambda) = \frac{M(\lambda)}{n(\lambda)}$$
be a BR m-variable matrix. Then
$$\hat{S}(\lambda,\theta) = \left(M(\lambda) + n(\lambda)_*\frac{\theta-1}{\theta+1}1\right)\left(M(\lambda)_*\frac{\theta-1}{\theta+1} + n(\lambda)1\right)^{-1}$$
exists and is an LBR $(m+1)$-variable matrix. Moreover, if the θ domain reactors in a realization of $\hat{S}(\lambda,\theta)$ are replaced by 1-ohm resistors, $S(\lambda)$ is realized.

□

The proposition thus yields an alternative approach to multivariable passive synthesis wherein the multivariable factorization is eliminated. Of course, the resultant LBR matrix has one more variable than did the original BR matrix.

Operationally, the passive synthesis procedure via Proposition 8-6 is extremely simple, as is illustrated by the following examples. Consider first the 2-variable function
$$s(\lambda) = \frac{\lambda_1\lambda_2}{1 + \lambda_1\lambda_2} \tag{8-95}$$

Here,
$$M(\lambda) = M(\lambda)_* = \lambda_1\lambda_2 \tag{8-96}$$
and
$$n(\lambda) = n(\lambda)_* = 1 + \lambda_1\lambda_2 \tag{8-97}$$

Hence,
$$\hat{s}(\lambda,\theta) = \frac{(\lambda_1\lambda_2) + (1+\lambda_1\lambda_2)\dfrac{\theta-1}{\theta+1}}{(\lambda_1\lambda_2)\dfrac{\theta-1}{\theta+1} + (1+\lambda_1\lambda_2)} \tag{8-98}$$
$$= \frac{2\lambda_1\lambda_2\theta + \theta - 1}{2\lambda_1\lambda_2\theta + \theta + 1}$$

is the required 3-variable LBR scattering parameter.

One application of Proposition 8-6 might be to convert a 1-variable BR RLC scattering parameter to a 2-variable LBR matrix, whence it can be realized by purely lossless techniques. In this way, one hopefully might be able to by-pass the 1-variable factorization required by the RLC synthesis. For example, consider the BR RLC scattering parameter

$$s(p) = \frac{p}{p+1} \tag{8-99}$$

which is converted by the procedure of Proposition 8-6 into the 2-variable LBR scattering parameter

$$s(p,\theta) = \frac{2p + \theta - 1}{2p + \theta + 1} \tag{8-100}$$

with

$$M(p) = -M(p)_* = p \tag{8-101}$$

$$n(p) = p + 1 \tag{8-102}$$

and

$$n(p)_* = -p + 1 \tag{8-103}$$

8-5 Lossless Multivariable Synthesis

With the synthesis of a BR matrix reduced to that of an LBR matrix, we may now proceed to complete the realization via the decomposition process, thus reducing an LBR $S(\lambda)$ m-variable matrix to a memoryless (zero-variable) coupling network in m steps of decomposition, the latter being readily realized. Unfortunately, the decomposition cannot be applied directly, since the realization it induces requires use of an active load

$$S^L(\theta) = \left(\frac{1}{\theta}\right)\mathbf{1} \tag{8-104}$$

Therefore, it is not, in general, possible to find a lossless coupling network which transforms this active load into a given LBR m-variable matrix. Rather than using the load of Equation (8-104) if a completely lossless realization of $S(\lambda,\theta)$ is to be obtained, one would prefer to extract a load of θ domain inductors

$$S^L(\theta) = \left(\frac{\theta - 1}{\theta + 1}\right)\mathbf{1} \tag{8-105}$$

which is itself LBR. This may, in fact, be achieved with the multivariable decomposition of Section 8-3 by the simple artifice of making the substitution

$$\tilde{\theta} = \frac{\theta + 1}{\theta - 1} \tag{8-106}$$

which converts $S(\lambda,\theta)$ to an equivalent

$$\tilde{S}(\lambda,\tilde{\theta}) = S(\lambda,\theta) \Big|_{\frac{\theta+1}{\theta-1} = \tilde{\theta}} \tag{8-107}$$

If one were now to decompose $\tilde{S}(\lambda,\tilde{\theta})$, a realization with load

$$S^L(\tilde{\theta}) = \left(\frac{1}{\tilde{\theta}}\right)\mathbf{1} \tag{8-108}$$

would be obtained by the process of Section 8-3. Now, if this were to be converted back to the θ domain, it would result in a realization of the original matrix $S(\lambda,\theta)$ with load

$$S^L(\theta) = \left(\frac{\theta-1}{\theta+1}\right)1 = \left(\frac{1}{\tilde{\theta}}\right)1 \qquad (8\text{-}109)$$

as shown in Figure 8-3, this being exactly the desired load of θ domain inductors. The LBR synthesis problem is then to find a decomposition $\Sigma(\lambda)$ of $\tilde{S}(\lambda,\tilde{\theta})$ which is LBR if $S(\lambda,\theta)$ is LBR. This is somewhat complicated by the fact that an LBR $S(\lambda,\theta)$ does not lead to an LBR $\tilde{S}(\lambda,\tilde{\theta})$ (since the transformation from θ to $\tilde{\theta}$ takes the RHP θ plane to the region of the $\tilde{\theta}$ plane, which is outside the unit circle). Fortunately, however, the relative simplicity of the transformation allows one to translate the LBR properties of $S(\lambda,\theta)$ to corresponding (but non-LBR) properties of $\tilde{S}(\lambda,\tilde{\theta})$. In particular, the condition

$$S(\lambda,\theta)_* S(\lambda,\theta) = 1 \qquad (8\text{-}110)$$

is equivalent to

$$\tilde{S}\left(-\lambda,\frac{1}{\tilde{\theta}}\right)' \tilde{S}(\lambda,\tilde{\theta}) = 1 \qquad (8\text{-}111)$$

since

$$\frac{-\theta+1}{-\theta-1} = \frac{\theta-1}{\theta+1} = \left(\frac{1}{\tilde{\theta}}\right) \qquad (8\text{-}112)$$

while the analyticity and boundedness conditions on $S(\lambda,\theta)$ in the RHP are translated to similar constraints for λ in the RHP and $\tilde{\theta}$ outside the unit circle.

Returning to Equation (8-111), we would like to discern the implications of this constraint on a decomposition of $\tilde{S}(\lambda,\tilde{\theta})$. This may be achieved by observing that Equation (8-111) is equivalent to

$$\tilde{S}(\lambda,\tilde{\theta})^{-1} = \tilde{S}\left(-\lambda,\frac{1}{\tilde{\theta}}\right)' \qquad (8\text{-}113)$$

Hence, if we are given any decomposition of $\tilde{S}(\lambda,\tilde{\theta})$, we may calculate $\tilde{S}(\lambda,\tilde{\theta})^{-1}$ and $\tilde{S}(-\lambda,1/\tilde{\theta})'$ in terms of the decomposition and set them equal, as per Equation (8-113). This then results in a set of constraints on the $\Sigma(\lambda)$ matrices of the decomposition of an LBR $S(\lambda,\theta)$.

For convenience, we assume that we are given a minimal decomposition of $\tilde{S}(\lambda,\tilde{\theta})$ as

$$\tilde{S}(\lambda,\tilde{\theta}) = \Sigma_{11}(\lambda) + \Sigma_{12}(\lambda)(\tilde{\theta}1 - \Sigma_{22}(\lambda))^{-1}\Sigma_{21}(\lambda) \qquad (8\text{-}114)$$

which, upon application of matrix algebra, yields

$$\begin{aligned}\tilde{S}\left(-\lambda,\frac{1}{\tilde{\theta}}\right)' &= \Sigma_{11}(\lambda)_* + \Sigma_{21}(\lambda)_*\left(\left(\frac{1}{\tilde{\theta}}\right)1 - \Sigma_{22}(\lambda)_*\right)^{-1}\Sigma_{12}(\lambda)_* \\ &= \Sigma_{11}(\lambda)_* - \Sigma_{21}(\lambda)_*\Sigma_{22}(\lambda)_*^{-1}\Sigma_{12}(\lambda)_* \\ &\quad - \Sigma_{21}(\lambda)_*\Sigma_{22}(\lambda)_*^{-1}(\tilde{\theta}1 - \Sigma_{22}(\lambda)_*)^{-1}\Sigma_{22}(\lambda)_*^{-1}\Sigma_{12}(\lambda)_*\end{aligned} \qquad (8\text{-}115)$$

and

$$\tilde{S}\left(-\lambda,\frac{1}{\tilde{\theta}}\right)' = \tilde{S}(\lambda,\tilde{\theta})^{-1} = \Sigma_{11}(\lambda)^{-1} \tag{8-116}$$
$$+ \Sigma_{11}(\lambda)^{-1}\Sigma_{12}(\lambda)\left(\tilde{\theta}1 - \Sigma_{22}(\lambda) + \Sigma_{21}(\lambda)\Sigma_{22}(\lambda)^{-1}\Sigma_{12}(\lambda)\right)\Sigma_{21}(\lambda)\Sigma_{11}(\lambda)^{-1}$$

Here $\Sigma_{11}(\lambda)^{-1}$ is assured to exist by the LBR character of $S(\lambda,\theta)$ [200], while the requirement that $\Sigma_{22}(\lambda)_*^{-1}$ exists is an assumption which will be dropped later. In essence, Equations (8-115) and (8-116) are two distinct decompositions of the matrix $\tilde{S}(\lambda,\theta)^{-1} = \tilde{S}(-\lambda,1/\tilde{\theta})'$, given by

$$\sigma_1(\lambda) = \left[\begin{array}{c|c} \Sigma_{11}(\lambda)_* - \Sigma_{21}(\lambda)_*\Sigma_{22}(\lambda)_*^{-1}\Sigma_{12}(\lambda)_* & -\Sigma_{21}(\lambda)_*\Sigma_{22}(\lambda)_*^{-1} \\ \hline \Sigma_{22}(\lambda)_*^{-1}\Sigma_{12}(\lambda)_* & \Sigma_{22}(\lambda)_*^{-1} \end{array}\right] \tag{8-117}$$

and

$$\sigma_2(\lambda) = \left[\begin{array}{c|c} \Sigma_{11}(\lambda)^{-1} & \Sigma_{11}(\lambda)^{-1}\Sigma_{12}(\lambda) \\ \hline \Sigma_{21}(\lambda)\Sigma_{11}(\lambda)^{-1} & \Sigma_{22}(\lambda) - \Sigma_{21}(\lambda)\Sigma_{11}(\lambda)^{-1}\Sigma_{12}(\lambda) \end{array}\right] \tag{8-118}$$

Upon invoking the equivalent decomposition theory of Theorem 8-2, we are assured of the existence of a nonsingular matrix $T(\lambda)$ (since the decomposition used in the above development is minimal), such that

$$\sigma_2(\lambda) = \left[\begin{array}{c|c} 1 & 0 \\ \hline 0 & T(\lambda) \end{array}\right] \sigma_1(\lambda) \left[\begin{array}{c|c} 1 & 0 \\ \hline 0 & T(\lambda)^{-1} \end{array}\right] \tag{8-119}$$

or, equivalently,

$$\Sigma_{11}(\lambda)_* - \Sigma_{21}(\lambda)_*\Sigma_{22}(\lambda)_*^{-1}\Sigma_{12}(\lambda) = \Sigma_{11}(\lambda)^{-1} \tag{8-120}$$

$$\Sigma_{21}(\lambda)_*\Sigma_{22}(\lambda)_*^{-1} = \Sigma_{11}(\lambda)^{-1}\Sigma_{12}(\lambda)T(\lambda) \tag{8-121}$$

$$\Sigma_{22}(\lambda)_*^{-1}\Sigma_{12}(\lambda)_* = T(\lambda)^{-1}\Sigma_{21}(\lambda)\Sigma_{11}(\lambda)^{-1} \tag{8-122}$$

and

$$\Sigma_{22}(\lambda)_*^{-1} = T(\lambda)^{-1}\left(\Sigma_{22}(\lambda) - \Sigma_{21}(\lambda)\Sigma_{11}(\lambda)^{-1}\Sigma_{12}(\lambda)\right)T(\lambda) \tag{8-123}$$

These four equations thus sum up the implications of the LBR property that

$$\tilde{S}\left(-\lambda,\frac{1}{\tilde{\theta}}\right)'\tilde{S}(\lambda,\tilde{\theta}) = 1 \tag{8-124}$$

on its decompositions. If they are satisfied by any decomposition, the corresponding matrix must satisfy the LBR adjoint condition, and conversely.

Of course, one is not really interested in decomposing $\tilde{S}(-\lambda,1/\tilde{\theta})$; rather, we want to decompose $\tilde{S}(\lambda,\tilde{\theta})$ itself. However, with the aid of the four equations, (8-120) through (8-123), this can be achieved. The following lemma, due to Youla [200], is fundamental to that goal.

Lemma 8-3

Let $S(\lambda,\theta)$ be an LBR multivariable matrix, and let $\Sigma(\lambda)$ be any minimal decomposition of

$$\tilde{S}(\lambda,\tilde{\theta}) = S(\lambda,\theta)\Big|_{\frac{\theta+1}{\theta-1}=\tilde{\theta}}$$

Then $\Sigma(\lambda)_*^{-1}$ is also a minimal decomposition of $\tilde{S}(\lambda,\tilde{\theta})$; that is, there exists a nonsingular matrix $K(\lambda)$ such that

$$\Sigma(\lambda)_*^{-1} = \begin{bmatrix} 1 & 0 \\ \hline 0 & K(\lambda) \end{bmatrix} \Sigma(\lambda) \begin{bmatrix} 1 & 0 \\ \hline 0 & K(\lambda)^{-1} \end{bmatrix}$$

Proof

With the aid of the formula for the inverse of a 2×2 partitioned matrix [36], $\Sigma(\lambda)_*^{-1}$ may be calculated, and is found to be

$$\Sigma(\lambda)^{*-1}$$

$$= \begin{bmatrix} \Sigma_{11}(\lambda)_* & \Sigma_{21}(\lambda)_* \\ \hline \Sigma_{12}(\lambda)_* & \Sigma_{22}(\lambda)_* \end{bmatrix}^{-1}$$

$$= \begin{bmatrix} (\Sigma(\lambda)_*^{-1})_{11} & (\Sigma(\lambda)_*^{-1})_{12} \\ \hline (\Sigma(\lambda)_*^{-1})_{21} & (\Sigma(\lambda)_*^{-1})_{22} \end{bmatrix} \quad \text{(8-125)}$$

$$= \begin{bmatrix} (\Sigma_{11*} - \Sigma_{21*}\Sigma_{22*}^{-1}\Sigma_{12*})^{-1} & -\Sigma_{11*} - (\Sigma_{22*}\Sigma_{22*}^{-1}\Sigma_{12*})^{-1}\Sigma_{21*}\Sigma_{22*}^{-1} \\ \hline \Sigma_{22*}^{-1}\Sigma_{12*}(\Sigma_{11*} - \Sigma_{21*}\Sigma_{22*}^{-1}\Sigma_{12*}) & \Sigma_{22*}^{-1} + \Sigma_{22*}^{-1}\Sigma_{12*} \\ & (\Sigma_{11*} - \Sigma_{21*}\Sigma_{22*}^{-1}\Sigma_{12*})^{-1}\Sigma_{21*}\Sigma_{22*}^{-1} \end{bmatrix}$$

where the explicit λ dependence has been deleted. Now, since $S(\lambda,\theta)$ is LBR, Equations (8-120) through (8-123) hold and may be invoked to simplify Equation (8-125). In particular,

$$(\Sigma(\lambda)_*^{-1})_{11} = (\Sigma_{11*} - \Sigma_{21*}\Sigma_{22*}^{-1}\Sigma_{12*})^{-1} = \Sigma_{11}(\lambda)^{-1} \quad \text{(8-126)}$$

$$(\Sigma(\lambda)_*^{-1})_{12} = (\Sigma_{11*} - \Sigma_{21*}\Sigma_{22*}^{-1}\Sigma_{12*})^{-1}\Sigma_{21*}\Sigma_{22*}^{-1}$$
$$= -\Sigma_{11}\Sigma_{11}^{-1}\Sigma_{12}T = -\Sigma_{12}(\lambda)T(\lambda) \quad \text{(8-127)}$$

$$(\Sigma(\lambda)_*^{-1})_{21} = -\Sigma_{22*}^{-1}\Sigma_{12*}(\Sigma_{11*} - \Sigma_{21*}\Sigma_{22*}^{-1}\Sigma_{12*})^{-1}$$
$$= -T^{-1}\Sigma_{21}\Sigma_{11}^{-1}\Sigma_{11} = T(\lambda)^{-1}\Sigma_{21}(\lambda) \quad \text{(8-128)}$$

and

$$(\Sigma(\lambda)_*^{-1})_{22} = \Sigma_{22*}^{-1}\Sigma_{12*}(\Sigma_{11*} - \Sigma_{21*}\Sigma_{22*}^{-1}\Sigma_{12*})^{-1}\Sigma_{21*}\Sigma_{22*}^{-1} + \Sigma_{22*}^{-1}$$
$$= T^{-1}\Sigma_{22} - \Sigma_{21}\Sigma_{11}^{-1}\Sigma_{12}T + T^{-1}\Sigma_{21}\Sigma_{11}^{-1}\Sigma_{11}\Sigma_{11}^{-1}\Sigma_{12}T$$
$$= T^{-1}\Sigma_{22}T - T^{-1}\Sigma_{21}\Sigma_{11}^{-1}\Sigma_{12}T + T^{-1}\Sigma_{21}\Sigma_{11}^{-1}\Sigma_{12}T \quad \text{(8-129)}$$
$$= T(\lambda)^{-1}\Sigma_{22}(\lambda)T(\lambda)$$

Substituting Equations (8-126) through (8-129) into Equation (8-125) now yields

$$\Sigma(\lambda)_*^{-1} = \left[\begin{array}{c|c} \Sigma_{11}(\lambda) & -\Sigma_{12}(\lambda)T(\lambda) \\ \hline -T(\lambda)^{-1}\Sigma_{21}(\lambda) & T(\lambda)^{-1}\Sigma_{22}(\lambda)T(\lambda) \end{array}\right] \quad (8\text{-}130)$$

If one now lets

$$K(\lambda) = -T(\lambda)^{-1} \quad (8\text{-}131)$$

this becomes

$$\Sigma(\lambda)_*^{-1} = \left[\begin{array}{c|c} 1 & 0 \\ \hline 0 & K(\lambda) \end{array}\right] \Sigma(\lambda) \left[\begin{array}{c|c} 1 & 0 \\ \hline 0 & K(\lambda)^{-1} \end{array}\right] \quad (8\text{-}132)$$

which, by Theorem 8-3, assures that $\Sigma(\lambda)_*^{-1}$ is a decomposition, as was to be shown. Finally, we observe that although the above argument only applies when $\Sigma_{22}(\lambda)_*^{-1}$ exists, the general result may be obtained from this special case upon invoking a limiting argument on the frequency variable. The details of this argument appear in reference [200], and will not be given here. □

Although the proof of the lemma is rather long and tedious, it should not come as a surprise since, ultimately, we desire to find a $\Sigma(\lambda)$ such that

$$\Sigma(\lambda)_*^{-1} = \Sigma(\lambda) \quad (8\text{-}133)$$

in which case the lemma would be obvious.

Consider, for example, the 2-variable LBR scattering parameter

$$s(p,\theta) = \frac{2p + \theta - 1}{2p + \theta - 1} \quad (8\text{-}134)$$

which was obtained via the passive synthesis procedure of the last section from the RLC BR scattering parameter

$$s(p) = \frac{p}{p+1} \quad (8\text{-}135)$$

Here p represents the usual frequency domain variable and θ is the artificial variable corresponding to the resistance. Letting

$$\tilde{\theta} = \frac{\theta + 1}{\theta - 1} \quad (8\text{-}136)$$

converts this to

$$\tilde{s}(p,\tilde{\theta}) = \frac{p(\tilde{\theta} - 1) + 1}{p(\tilde{\theta} - 1) - \tilde{\theta}} \quad (8\text{-}137)$$

Now, an application of the decomposition of Section 8-3 yields

$$\Sigma_{11}(p) = \tilde{s}(p, \infty) = \frac{p}{p+1} \quad (8\text{-}138)$$

$$\Sigma_{22}(p) = \Omega(p) = \frac{p}{p+1} \quad (8\text{-}139)$$

$$G_r(p) = A_0(p) = \frac{p}{(p+1)^2} \quad (8\text{-}140)$$

$$\Sigma_{12}(p) = 1 \quad (8\text{-}141)$$

and
$$\Sigma_{21}(p) = G_r(p)\Sigma_{12}(p)' = \frac{1}{(p+1)^2} \tag{8-142}$$

Hence, a decomposition is given by

$$\Sigma(p) = \left[\begin{array}{c|c} \dfrac{p}{p+1} & 1 \\ \hline \dfrac{1}{(p+1)^2} & \dfrac{p}{(p+1)^2} \end{array}\right] \tag{8-143}$$

To verify the lemma we calculate $\Sigma(p)_*^{-1}$, for which purpose

$$\Sigma(p)_* = \left[\begin{array}{c|c} \dfrac{p}{p+1} & \dfrac{1}{(p-1)^2} \\ \hline 1 & \dfrac{p}{p-1} \end{array}\right] \tag{8-144}$$

and

$$\Sigma(p)_*^{-1} = \left[\begin{array}{c|c} \dfrac{p}{p+1} & \dfrac{-1}{(p+1)(p-1)} \\ \hline -\dfrac{p-1}{p+1} & \dfrac{p}{p+1} \end{array}\right] \tag{8-145}$$

Now, if one lets

$$K(p) = 1 - p^2 = (1-p)(1+p) \tag{8-146}$$

then

$$\Sigma(p)_*^{-1} = \left[\begin{array}{c|c} 1 & 0 \\ \hline 0 & (1-p^2) \end{array}\right] \Sigma(p) \left[\begin{array}{c|c} 1 & 0 \\ \hline 0 & \dfrac{1}{(1-p^2)} \end{array}\right] \tag{8-147}$$

which is as is implied by the lemma.

With the formulation of this fundamental lemma we are now prepared to complete the LBR synthesis. This is achieved by starting with the minimal decomposition of Section 8-3, which is analytic in the RHP, and then applying the equivalent decomposition theorem to find an LBR decomposition. In effect, one must find a nonsingular $L(\lambda)$ such that

$$\Sigma^L(\lambda) = \left[\begin{array}{c|c} 1 & 0 \\ \hline 0 & L(\lambda) \end{array}\right] \Sigma(\lambda) \left[\begin{array}{c|c} 1 & 0 \\ \hline 0 & L(\lambda)^{-1} \end{array}\right] \tag{8-148}$$

where $L(\lambda)$, together with its inverse, is analytic in the RHP. Now, if $\Sigma^L(\lambda)$ is LBR, then

$$\Sigma^L(\lambda) = \Sigma^L(\lambda)_*^{-1} = \left[\begin{array}{c|c} 1 & 0 \\ \hline 0 & L(\lambda)_*^{-1} \end{array}\right] \Sigma(\lambda)_*^{-1} \left[\begin{array}{c|c} 1 & 0 \\ \hline 0 & L(\lambda)_* \end{array}\right] \tag{8-149}$$

Furthermore, if $S(\lambda,\theta)$ is LBR, the lemma assures the existence of a nonsingular $K(\lambda)$ such that

$$\Sigma(\lambda)_*^{-1} = \begin{bmatrix} 1 & 0 \\ \hline 0 & K(\lambda) \end{bmatrix} \Sigma(\lambda) \begin{bmatrix} 1 & 0 \\ \hline 0 & K(\lambda)^{-1} \end{bmatrix} \quad (8\text{-}150)$$

which, when substituted into Equation (8-149), yields

$$\Sigma^L(\lambda) = \begin{bmatrix} 1 & 0 \\ \hline 0 & L(\lambda)_*^{-1} K(\lambda) \end{bmatrix} \Sigma(\lambda) \begin{bmatrix} 1 & 0 \\ \hline 0 & K(\lambda)^{-1} L(\lambda)_* \end{bmatrix} \quad (8\text{-}151)$$

Comparison of Equations (8-148) and (8-151) now leads to the conclusion that

$$L(\lambda) = L(\lambda)_*^{-1} K(\lambda) \quad (8\text{-}152)$$

or, equivalently, that

$$K(\lambda) = L(\lambda)_* L(\lambda) \quad (8\text{-}153)$$

The $K(\lambda)$ matrix of the lemma thus serves as a basis for choosing $L(\lambda)$, which in turn yields the LBR synthesis. Since the $\Sigma(\lambda)$ obtained by the method of Section 8-3 is assured to be analytic in the RHP, the $\Sigma^L(\lambda)$ obtained by Equation (8-148) will be LBR if and only if one can find a factorization $L(\lambda)$ of $K(\lambda)$ which, together with its inverse, is analytic in the RHP. We have thus proven the following theorem.

Theorem 8-4

Let $S(\lambda,\theta)$ be LBR and $\Sigma(\lambda)$ be a minimal decomposition of $\tilde{S}(\lambda,\tilde{\theta})$ which is analytic in the RHP (as per Section 8-3), such that

$$\Sigma(\lambda)_*^{-1} = \begin{bmatrix} 1 & 0 \\ \hline 0 & K(\lambda) \end{bmatrix} \Sigma(\lambda) \begin{bmatrix} 1 & 0 \\ \hline 0 & K(\lambda)^{-1} \end{bmatrix}$$

Then if $L(\lambda)$ and its inverse are both analytic in the RHP and satisfy

$$K(\lambda) = L(\lambda)_* L(\lambda)$$

$$\Sigma^L(\lambda) = \begin{bmatrix} 1 & 0 \\ \hline 0 & L(\lambda) \end{bmatrix} \Sigma(\lambda) \begin{bmatrix} 1 & 0 \\ \hline 0 & L(\lambda)^{-1} \end{bmatrix}$$

is an LBR minimal decomposition of $\tilde{S}(\lambda,\tilde{\theta})$. □

Although the derivation of the theorem is complex, it leads to a straightforward synthesis procedure. After obtaining $\Sigma(\lambda)$ by the method of Section 8-3, one may calculate $\Sigma(\lambda)_*^{-1}$ and then find $K(\lambda)$ by solving

$$\Sigma(\lambda)_*^{-1} = \begin{bmatrix} 1 & 0 \\ \hline 0 & K(\lambda) \end{bmatrix} \Sigma(\lambda) \begin{bmatrix} 1 & 0 \\ \hline 0 & K(\lambda)^{-1} \end{bmatrix} \quad (8\text{-}154)$$

This is then factored to find the $L(\lambda)$ which is required to calculate $\Sigma^L(\lambda)$ via

$$\Sigma^L(\lambda) = \begin{bmatrix} 1 & 0 \\ \hline 0 & L(\lambda) \end{bmatrix} \Sigma(\lambda) \begin{bmatrix} 1 & 0 \\ \hline 0 & L(\lambda)^{-1} \end{bmatrix} \quad (8\text{-}155)$$

Of course, unlike the RLC case, no general multivariable factorization is known, so the procedure is not assured to work (except in the 2-variable case, where the usual 1-variable RLC factorization of Theorem 6-9 may be applied).

Applying the theorem to our previous example, with

$$\Sigma(p) = \begin{bmatrix} \dfrac{p}{p+1} & 1 \\ \hline \dfrac{1}{(p+1)^2} & \dfrac{p}{p+1} \end{bmatrix} \quad (8\text{-}156)$$

and

$$K(p) = (1 - p^2) = (1-p)(1+p) = L(p)_*L(p) \quad (8\text{-}157)$$

we find that

$$\Sigma^L(p) = \begin{bmatrix} 1 & 0 \\ \hline 0 & (p+1) \end{bmatrix} \begin{bmatrix} \dfrac{p}{p+1} & 1 \\ \hline \dfrac{1}{(p+1)^2} & \dfrac{p}{p+1} \end{bmatrix} \begin{bmatrix} 1 & 0 \\ \hline 0 & \dfrac{1}{p+1} \end{bmatrix}$$

$$= \begin{bmatrix} \dfrac{p}{p+1} & \dfrac{1}{p+1} \\ \hline \dfrac{1}{p+1} & \dfrac{p}{p+1} \end{bmatrix} \quad (8\text{-}158)$$

which is indeed LBR. Realizing $\Sigma^L(p)$ as a series inductor between two ports now yields the required realization of the given LBR scattering parameter,

$$s(p,\theta) = \dfrac{2p + \theta - 1}{2p + \theta + 1} \quad (8\text{-}159)$$

when one port is loaded with a θ domain inductor, as shown in Figure 8-6(a).

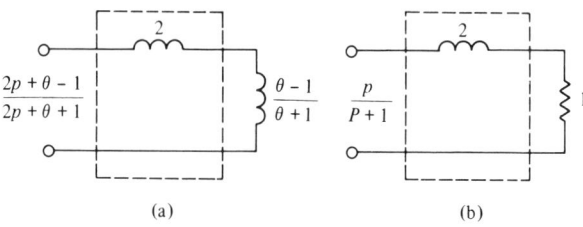

Figure 8-6 (a) Realization of a 2-variable scattering parameter. (b) The corresponding 1-variable RLC scattering parameter.

If, moreover, the θ domain inductor is replaced with a 1-ohm resistor, then the original 1-variable scattering parameter

$$s(p) = \frac{p}{p+1} \quad \text{(8-160)}$$

from which $S(p,\theta)$ was derived (via the passive synthesis procedure of the last section) is realized, this being shown in Figure 8-6(b).

8-6 Mixed RLC Transmission Line Networks

Possibly the most commonly encountered class of multivariable matrices are those representing networks containing both lumped and distributed devices. In this case, the multivariable system function has the form

$$S(p,\lambda) \quad \text{(8-161)}$$

where p is the usual frequency domain variable and

$$\lambda = \cosh \tau p \quad \text{(8-162)}$$

is the Richards transformation variable representing the transmission lines which are assumed to be commensurable in length with unit element length τ. Of course, in the case of networks containing lines with several different frequency dependence characteristics and/or noncommensurable lengths, a multivariable representation

$$S(p,\boldsymbol{\lambda}) \equiv S(p,\lambda_1,\lambda_2,\cdots,\lambda_m) \quad \text{(8-163)}$$

may be employed where

$$\lambda_i = \cosh \tau_i f_i(p) \qquad i = 1, 2, \cdots, m \quad \text{(8-164)}$$

In any event, with these definitions such a network can be readily realized by the synthesis techniques of the preceding sections. That is, we employ the decomposition to extract the various λ_i domain inductors, these being realized by open circuited lines of length τ_i and frequency dependence characteristic $f_i(p)$. The final p domain remainder may be realized by the usual RLC techniques or, alternatively, by an additional step of decomposition wherein the p domain inductors are extracted, leaving a zero variable memoryless coupling network, which is readily realized.

If, for example, one were to interpret the LBR scattering parameter

$$s(p,\lambda) = \frac{2p + \lambda - 1}{2p + \lambda + 1} \quad \text{(8-165)}$$

as an RLC transmission line function, the decomposition of the previous section would yield the realization of Figure 8-7. Here, the decomposition extracts a λ domain inductor (open circuited line) leaving a 2-port RLC coupling network.

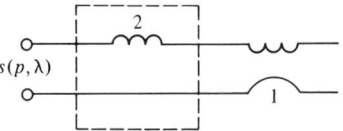

Figure 8-7 Realization of a mixed RLC-transmission line network.

Although these techniques are completely sufficient to realize a mixed RLC transmission line network with arbitrarily many types of line, they are, like the pure transmission line networks, impractical. Rather, the practical considerations of implementing a transmission line network demand that one restrict consideration to a cascade configuration, possibly with lumped discontinuities and/or load, as shown in Figure 8-8. Here λ_i represents the type

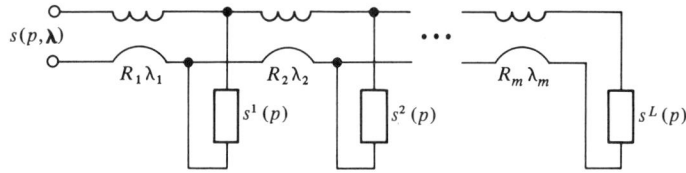

Figure 8-8 Cascade of lines with lumped discontinuities and load.

of line and R_i its characteristic impedance. Of course, there may be more than one line of the same type.

At the time of this writing, the necessary and sufficient conditions on a scattering parameter for it to be realizable by such a configuration are as yet unknown. If, however, we restrict ourselves to the simpler configuration of Figure 8-9, where lumped elements appear only in the load, a reasonably com-

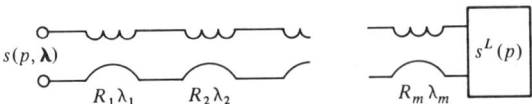

Figure 8-9 Cascade of lines with lumped load.

plete theory is available. Since the derivation of this theory is a natural generalization of the corresponding case where the cascade lines have only a resistive load, we will merely sketch the results, leaving it for the reader to fill in the details.

Following the pattern established in the resistive load case of Chapter 7, we consider the effect on the resistivity operator of adding a single λ_i domain line to an arbitrary multivariable load (possibly containing more lines in cascade). This is illustrated by the configuration of Figure 8-10. As in the special case, a mixed scattering impedance analysis leads to a resistivity operator

$$1 - s(p,\lambda)_* s(p,\lambda) = \frac{2R^2(z^L(p,\lambda) + z^L(p,\lambda)_*)(1 - \lambda_i^2)}{\text{denominator}} \tag{8-166}$$

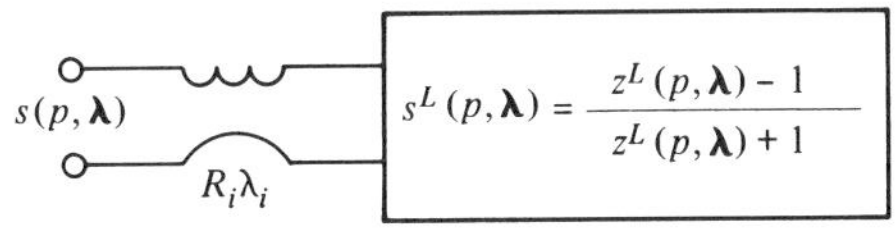

[diagram: $s(p,\lambda)$ with $R_i \lambda_i$ feeding $s^L(p,\lambda) = \dfrac{z^L(p,\lambda) - 1}{z^L(p,\lambda) + 1}$]

Figure 8-10 One-line cascade configuration.

The effect of the cascade topology is thus the addition of a $(1 - \lambda_i^2)$ to the resistivity operator, independent of the load and the characteristic impedance of the line.

Now, since the zeros of

$$z^L(p,\lambda) + z^L(p,\lambda)_* \tag{8-167}$$

are the same as those of

$$1 - s^L(p,\lambda)_* s^L(p,\lambda) \tag{8-168}$$

(see Lemma 7-2), an inductive process of adding lines (of any λ_i and characteristic impedance, except one) leads to the following proposition.

*Proposition 8-7

Let $s(p,\lambda)$ be the scattering parameter of a cascade of lines (as per Figure 8-9) having n_i lines for each variable λ_i (of any characteristic impedance, except one) arranged in any order and terminated in a lumped load. Then

$$1 - s(p,\lambda)_* s(p,\lambda)$$

has n_i-fold zeros at $\lambda_i = \pm 1$ independent of the other variables; that is, it has a factor

$$\prod_{i=1}^{m} (1 - \lambda_i^2)^{n_i}$$

Of course, for a network with only one type of line, this reduces to the usual n-fold zeros at $\lambda = \pm 1$. □

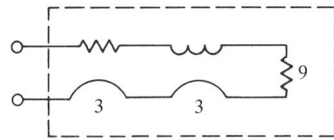

Figure 8-11 Cascaded LC and RC transmission lines loaded in a resistor.

For the 1-port of Figure 8-11, if one represents the LC line by the variable λ_1 and the RC line by λ_2, one obtains the scattering parameter

$$s(\lambda) = \frac{8\lambda_1\lambda_2 + 8}{10\lambda_1\lambda_2 + 6\lambda_1 + 6\lambda_2 + 10} \tag{8-169}$$

and the resistivity operator

$$1 - s(\lambda)s(\lambda)_* = \frac{36\lambda_1^2\lambda_2^2 - 36\lambda_1^2 - 36\lambda_2^2 + 36}{100\lambda_1^2\lambda_2^2 + 128\lambda_1\lambda_2 - 36\lambda_1^2 - 36\lambda_2^2 + 100}$$

$$= \frac{36(\lambda_1^2 - 1)(\lambda_2^2 - 1)}{100\lambda_1^2\lambda_2^2 + 128\lambda_1\lambda_2 - 36\lambda_1^2 - 36\lambda + 100} \tag{8-170}$$

Here the resistivity operator does indeed have the required 1-fold zeros at

$$\lambda_1 = \pm 1 \tag{8-171}$$

and

$$\lambda_2 = \pm 1 \tag{8-172}$$

As before, the proposition is also sufficient, this following from a cascade realization of any scattering parameter which satisfies the hypothesis of the proposition. To this end, we have the following multivariable form of the powerful Richards theorem [153], [165]. We denote

$$z(\lambda)\bigg|_{\lambda_i=k} = z(\lambda_1,\lambda_2,\cdots,\lambda_{i-1},k,\lambda_{i-1},\cdots,\lambda_m) \tag{8-173}$$

by $z(k)_i$.

*Proposition 8-8

Let a multivariable 1-port be characterized by a scattering parameter $s(\lambda)$ and impedance parameter $z(\lambda)$. Then if for some strictly positive real k $z(k)_i$ is a constant independent of the other variables, the scattering parameter

$$s_{z(k)_i}{}^r(\lambda) = \frac{\lambda_i + k}{\lambda_i - k} s_{z(k)_i}(\lambda)$$

is BR (LBR) if $s(\lambda)$ is BR (LBR) and is degree no greater than $s(\lambda)$. Furthermore, if

$$z(k)_i = -z(-k)_i$$

the degree of $s^r(\lambda)$ is strictly less than that of $s(\lambda)$. □

With Richards' theorem established by an argument similar to that used in the 1-variable case, we may proceed to formulate the cascade synthesis for a scattering parameter satisfying the hypothesis of Proposition 8-7. Now, if

$$1 - s(p,\lambda)_* s(p,\lambda) \tag{8-174}$$

has a factor $(1 - \lambda_i^2)$, it can be verified [165] that Richards' theorem, with $k = 1$, yields a factorization

$$s_{z(1)_i}(p,\pmb{\lambda}) = \left(\frac{\lambda_i - 1}{\lambda_i + 1}\right) s_{z(1)_i}{}^r(p,\pmb{\lambda}) \tag{8-175}$$

with degree reduction; hence the realization of Figure 8-12.

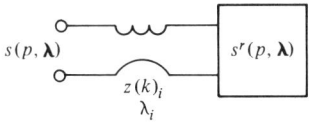

Figure 8-12 One step of multivariable cascade synthesis.

Repeated application of the process for each factor

$$(1 - \lambda_i^2) \tag{8-176}$$

in the resistivity operator then yields the required multivariable cascade realization. We note that the cascade lines in the network may have arbitrary lengths and frequency dependence characteristic and, moreover, may be extracted in arbitrary order. In this case, however, the characteristic impedances of the lines, $z(1)_i$, may be a function of the order in which they are extracted.

*Theorem 8-5

The necessary and sufficient conditions for $s(p,\pmb{\lambda})$ to be the scattering parameter of

$$\sum_{i=1}^{m} n_i$$

transmission lines (n_i of which have a λ_i domain representation) in cascade terminated in a lumped RLC load are that it be a BR multivariable function for which

$$1 - s(p,\pmb{\lambda})_* s(p,\pmb{\lambda})$$

has n_i-fold zeros at $\lambda_i = \pm 1$, $i = 1, 2, \cdots, m$. Furthermore, if $s(p,\pmb{\lambda})$ satisfies these conditions, it can be realized by repeated application of Richards' theorem, with $k = 1$, until the process terminates with a p domain remainder which may be realized as an RLC load. □

Consider the 3-variable scattering parameter

$$s(p,\lambda_1,\lambda_2) = \frac{3(\lambda_1 - 1)(\lambda_2 - 1)^2 p + (\lambda_1 + 1)(\lambda_2 + 1)^2(p + 1)}{(\lambda_1 - 1)(\lambda_2 - 1)^2 p + 3(\lambda_1 + 1)(\lambda_2 + 1)^2(p + 1)}$$

for which the numerator of the corresponding resistivity operator has the factor

$$(1 - \lambda_1^2)(1 - \lambda_2^2)^2 \tag{8-177}$$

and

$$z(p,1,\lambda_2) = z(1)_1 = 2 \tag{8-178}$$

If the normalization of $s(p,\lambda_1,\lambda_2)$ is now changed to $z(1)_1 = 2$, we obtain

$$s_2(p,\lambda_1,\lambda_2) = \frac{(\lambda_1 - 1)(\lambda_2 - 1)^2 p}{(\lambda_1 - 1)(\lambda_2 - 1)^2(p + 1)} \tag{8-179}$$

which may be realized as a λ_1 domain line cascaded (in any order) with two λ_2 domain lines, all of characteristic impedance 2, this loaded in a realization of

$$s_2(p) = \frac{p}{p + 1} \tag{8-180}$$

The final realization is then that shown in Figure 8-13. Here, the RL realization of the lumped load characterized by

$$s_2(p) = \frac{p}{p + 1}$$

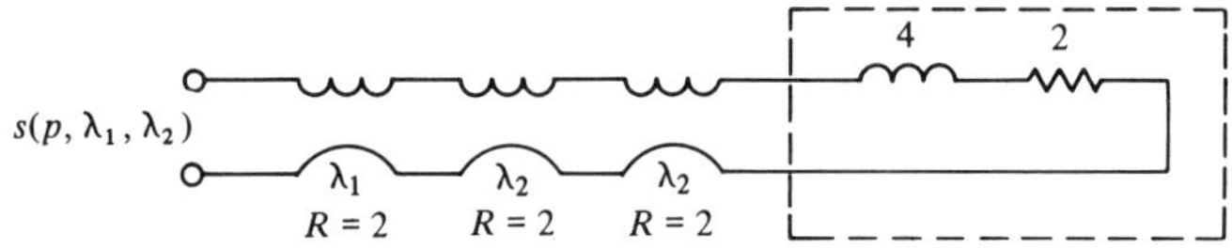

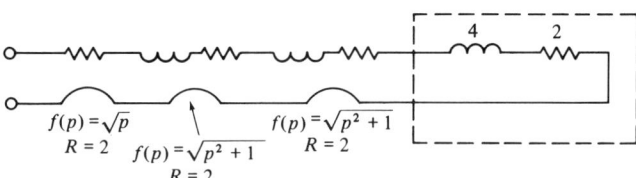

Figure 8-13 Cascade realization of an RLC-transmission line network.

is that obtained in the example of the last section via the multivariable decomposition with an appropriate renormalization. Assuming that the physical interpretation of the λ_1 variable is that of an RC line, while the λ_2 variable corresponds to a lossy LC line, the p domain realization of Figure 8-14 is obtained.

Figure 8-14 Realization of a 3-variable 1-port with an RC line and two lossy LC lines terminated in a lumped load.

8-7 Variable Parameter Networks

In addition to the RLC transmission line networks, another application of the multivariable concept is to variable parameter networks, that is, time-invariant RLC networks, the characteristics of which depend on the value of a (time-independent) parameter x. Such a network might then be represented by a rational scattering parameter, the coefficients of which are functions of x,

$$s(p,x) = \frac{\sum_{i=0}^{k} f_i(x) p^i}{\sum_{i=k+1}^{m} f_i(x) p^{i-k-1}} \tag{8-181}$$

This may readily be converted to a multivariable function by letting

$$f_i(x) = \frac{\lambda_i - 1}{\lambda_i + 1} \tag{8-182}$$

and forming the scattering parameter

$$s(p,\pmb{\lambda}) = \frac{\sum_{i=0}^{k} \frac{\lambda_i - 1}{\lambda_i + 1} p^i}{\sum_{i=k+1}^{m} \frac{\lambda_i - 1}{\lambda_i + 1} p^{i-k-1}} \tag{8-183}$$

Now, if this network is realized by any of the multivariable techniques, and the λ_i domain reactors are then replaced with variable resistors tapered so that their scattering parameters vary as

$$f_i(x) \tag{8-184}$$

a realization of the original $s(p,x)$ is obtained.

Although the preceding formulation is quite general, it requires the use of a great many variables. A more practical scheme suggested by Delansky [61], [62] makes use of the fact that the parameter coefficients are not, in general, independent. Consider, for example, the case when all of the parameters are rational in x [or, more generally, some monotonic function $f(x)$]. If one lets

$$\frac{\lambda - 1}{\lambda + 1} = x \tag{8-185}$$

then the substitution

$$S(p,\lambda) = S(p,x)\bigg|_{\frac{\lambda-1}{\lambda+1}=x} \tag{8-186}$$

yields a real, rational 2-variable matrix. 2-variable synthesis techniques may thus be applied to realize this rather large class of variable parameter functions, this being achieved by realizing the 2-variable matrix $S(p,\lambda)$ and replacing the λ domain reactors with potentiometers which vary linearly in x (or if $\lambda - 1/\lambda + 1 = f(x)$, a potentiometer with taper f may be employed). The realization of a variable parameter network by such an approach is illustrated in Figure 8-15. Here we note that the realization generally requires many variable re-

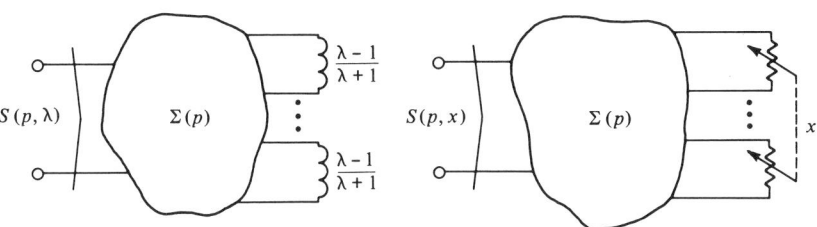

Figure 8-15 Realization of a variable parameter network with (a) λ domain inductors and (b) variable resistors.

sistors; they all, however, vary identically with the parameter x and, hence, may be "gang tuned" by a single control. Although a degree of approximation may be required to place a given variable parameter network into a form to which Delansky's synthesis is applicable, the resultant realization will have the very desirable single control characteristic. Moreover, it may be implemented with 2-variable rather than m-variable techniques.

Of course, the technique can readily be extended to networks which vary with several parameters by letting

$$x_i = \frac{\lambda_i - 1}{\lambda_i + 1} \qquad i = 1, 2, \cdots, m \tag{8-187}$$

As before, replacement of the λ_i domain reactors with potentiometers which vary linearly with x_i yields the desired realization. In this case, the various parameters are independent, but all of the x_i variations for a given i are identical; hence, "gang tuning" allows a realization with but one control per parameter.

Upon making the substitutions

$$x_1 = \frac{\lambda_1 - 1}{\lambda_1 + 1} \tag{8-188}$$

and

$$x_2 = \frac{\lambda_2 - 1}{\lambda_2 + 1} \tag{8-189}$$

the variable scattering parameter

$$s(p,x_1,x_2) = \frac{(2x_1x_2^2 + 1)p + 1}{(x_1x_2^2 + 3)p + 1} \tag{8-190}$$

becomes

$$s(p,\lambda_1,\lambda_2) = \frac{3(\lambda_1 - 1)(\lambda_2 - 1)^2 p + (\lambda_1 + 1)(\lambda_2 + 1)^2(p + 1)}{(\lambda_1 - 1)(\lambda_2 - 1)^2 p + 3(\lambda_1 + 1)(\lambda_2 + 1)^2(p + 1)} \tag{8-191}$$

which was realized in the last section. An alternative realization for the variable parameter application is that shown in Figure 8-16.

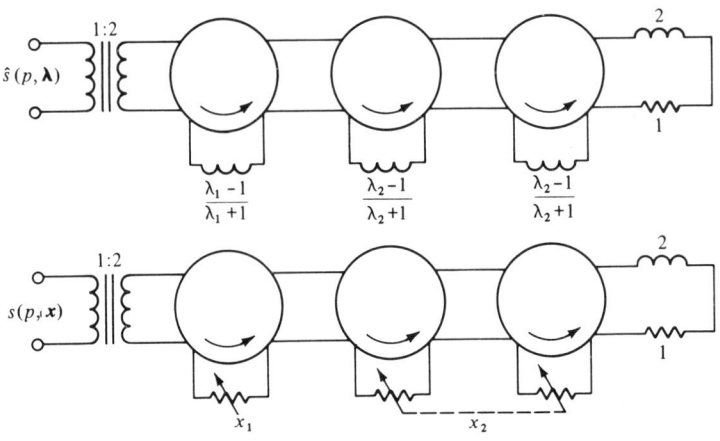

Figure 8-16 Realization of a variable parameter network with λ_i domain inductors and variable resistors.

8-8 Discussion

Although the chapter began with an apparently restricted class of rational functions, though of several variables, it was found that an appropriate application of substitution techniques allowed a large number of generalized networks to be included. In fact, the restrictions were really only two-fold. First, so as to assure the validity of the frequency domain techniques employed, we considered only time-invariant networks (in particular, the parameters x_i in the variable parameter case must be independent of time). Secondly, the assumption of multivariable rationality implies that the network has only a finite number of components, though the components themselves may or may not be finite. The multivariable theory, although originally developed with the mixed RLC transmission line case in mind, thus yields a basis for the study of essentially arbitrary time-invariant networks. In particular, the various commonly encountered classes of mixed networks are readily included.

One rather powerful application of the multivariable theory which we have not considered at all is the synthesis of RLC networks and/or time-invariant dynamical systems. In that case, the 1-variable special case of the multivariable theory is employed. In particular, the decomposition synthesis for an RLC network leads to a realization similar to that illustrated in Figure 8-17, where

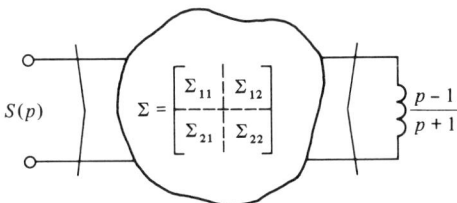

Figure 8-17 Realization of an RLC network by reactance extraction.

the desired $S(p)$ is coupled to an inductive load with a memoryless coupling network. Moreover, there is a one-to-one correspondence between the decompositions of such an $S(p)$ and the dynamical systems which realize it as an input-output system function. The theory of multivariable networks is thus a natural extension of the commonly encountered state-space approach to dynamical system theory. This approach is considered in further detail in Chapter 9, in the context of the reactance extraction synthesis.

CHAPTER

9
Finite Time-Variable Networks; State-Space Techniques

9-1 Introduction

In Chapter 6 we considered what might be termed traditional network theory, though by a nontraditional approach. A natural generalization of the RLC networks, allowing their parameters to vary with time, is the subject of the present chapter. The resultant finite networks (that is, networks composed of a finite number of time-variable RLC devices) may be successfully studied by a number of techniques. Possibly the most natural approach is to replace the rational system functions of Chapter 6 by their time-variable generalization [55], [68], [207], yields, for this case, system functions which are rational in p but have time-variable (real) coefficients. Unfortunately, the pathological character of the time-variable system function renders this approach of little value.

Alternatively, one may choose to represent a finite network by the two variable impulse response kernels of Chapter 2 [6], [56], [168] or the differential operator of the network [145], [172]. In fact, it is most convenient to take a mixed approach, where both the impulse response and its associated differential operator are employed, the latter in the form of a matrix first-order dynamical system [19], [117],

$$\dot{X} = F(t)X + G(t)a \qquad (9\text{-}1)$$

$$b = H(t)'X + J(t)a \qquad (9\text{-}2)$$

This particular form for the differential operator, although formally less like the RLC rational functions than the ordinary input-output operator, is more compatible with the impulse response representation and also serves as a natural intermediary in the synthesis process.

In the first section, the underlying definitions for a finite network are made and various equivalences between classes of finite RLC networks are exhibited. In particular, it is shown that every such network can be realized with constant resistors and capacitors, together with time-variable gyrators or time-variable ideal transformers [168].

In the second and third sections, the representation of finite networks by differential operators, dynamical systems, or 2-variable impulse responses is considered, along with a study of the various interrelationships between these representations. In particular, it is shown that the impulse response of a finite n-port always has the separable form [207]

$$H(t,q) = \sum_{i=0}^{n} J_i(t)\delta^{(i)}(t-q) + \phi(t)\theta(q)'u(t-q) \qquad (9\text{-}3)$$

and, in most cases, has the even more restrictive proper form

$$H(t,q) = J(t)\delta(t-q) + \phi(t)\theta(q)'U(t-q) \qquad (9\text{-}4)$$

The following section deals with the process of synthesis by reactance extraction [19], [117]. This method, which is in some respect reminiscent of the techniques used to simulate a differential equation on an analog computer, yields active syntheses for arbitrary finite n-ports, and passive syntheses for "most" passive n-ports. Unlike most of our previous results, the dynamical system which is used as an intermediary in this procedure is more easily employed in the immittance than the scattering case; hence, the main results of this section are stated in terms of immittance matrices.

In the last section, the application of the reactance extraction syntheses, developed for time-variable synthesis, to the time-invariant special case is considered, these yielding the so-called state-space syntheses [17], [182] for RLC networks. Although this approach is no more general than that of Chapter 6, it is computationally straightforward and gives an insight into the properties of reciprocal n-ports [182], [199].

Unlike the preceding chapters, where an essentially complete theory is available, the time-variable theory of the present chapter is relatively new and still incomplete. In particular, many of the results are stated only as sufficient conditions with various "pathological" cases being excluded from the result. In fact, this is the case with the passive and lossless syntheses, where the synthesis is carried out in terms of the solution of specified nonlinear matrix equations. The existence of such a solution is, however, not assured for all time-variable passive n-ports; hence, the techniques are not assured to yield realizations for all such passive n-ports. They are, however, quite general, the main exceptions being those n-ports whose "degree" is a function of time. In fact, virtually nothing is known about this class of network which is exempted from most of the development [145]. Similarly, the scattering theory for time-variable networks is less fully developed than the corresponding immittance theory; hence, the latter takes precedence in much of the formulation.

Although, as presented, the results are incomplete, the techniques employed are powerful in themselves and are indicative of those procedures which may be expected to ultimately yield a complete theory of finite RLC time-variable n-ports.

9-2 Time-Variable RLC Networks

For the RLC networks it was found that resistors, capacitors, and gyrators were sufficient to realize all members of the class; hence, one might expect that the time-variable RLC networks could be defined as those networks which can be constructed using time-variable resistors, capacitors, and gyrators. While such an approach is valid, it is not necessary, being completely sufficient to consider networks composed of constant resistors and capacitors, together with time-variable gyrators or time-variable ideal transformers. This is due to the fact that the time-variable resistor and capacitor can be realized, themselves, with fixed resistors and capacitors, in addition to time-variable transformers [4], [7], [8], [168]. Equivalently, one can construct a variable transformer from two variable gyrators [92] and, in turn, use these to realize the variable resistor and capacitor.

Consider, for example, a passive resistor with resistance

$$r(t) \geq 0 \tag{9-5}$$

Clearly, one may define a 2-port ideal transformer with turns ratio $n(t)$ such that

$$n^2(t) = r(t) \tag{9-6}$$

(in fact, the turns ratio may be taken as the square root of the resistance). Now, if such an ideal transformer is loaded in a 1-ohm resistor, as shown in Figure 9-1(a), then the 1-port so realized is characterized by the impedance

$$\begin{aligned} z(t,q) &= n(t)\delta(t - q)n(q) = n^2(t)\delta(t - q) \\ &= r(t)\delta(t - q) \end{aligned} \tag{9-7}$$

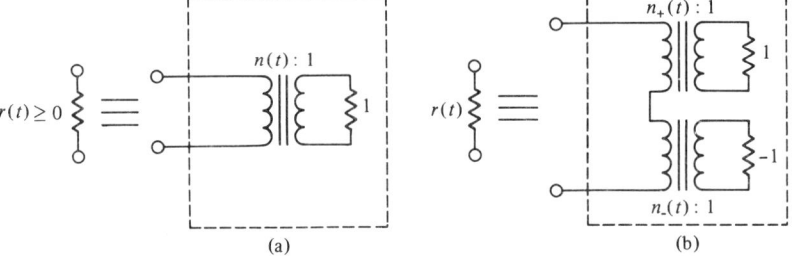

Figure 9-1 Equivalent circuits for (a) passive and (b) general time-variable resistors.

which is precisely the impedance of the given time-variable resistor. Figure 9-1(a) thus represents an equivalent circuit for a passive time-variable resistor, wherein the time variations appear only in the transformers. Of course, if one replaces the transformer of this equivalent with the gyrator circuit of Figure 9-2, an alternate equivalent circuit, is obtained, wherein all of the time variations appear in the gyrators [93].

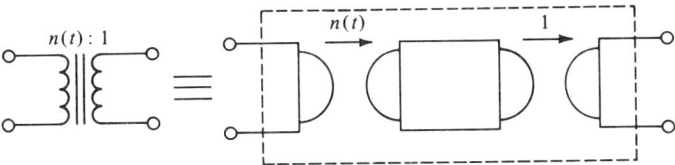

Figure 9-2 Gyrator equivalent of time-variable ideal transformer.

Since both components in the equivalent circuit of Figure 9-1(a) are passive, one would not expect such a configuration to be capable of realizing a general resistance. If, however, one decomposes the resistance of an arbitrary resistor as

$$f(t) = r_+(t) + r_-(t) \tag{9-8}$$

where $r_+(t)$ is positive and $r_-(t)$ is negative, then one may define two ideal transformers with turns ratios defined by

$$n_+^2(t) = r_+(t) \tag{9-9}$$

and

$$n_-^2(t) = -r_-(t) \tag{9-10}$$

Now, the series configuration of Figure 9-1(b), the n_+ transformer loaded in a 1-ohm resistor and the n_- transformer loaded with a -1-ohm resistor, has the impedance

$$\begin{aligned} z(t,q) &= n_+(t)\delta(t - q)n_+(q) + n_-(t)[-\delta(t - q)]n_-(q) \\ &= r_+(t)\delta(t - q) - r_-(t)\delta(t - q) = r(t)\delta(t - q) \end{aligned} \tag{9-11}$$

and, hence, is an equivalent circuit for the general time-variable resistor. Of course, as before, the time-variable transformers may be replaced by the gyrator configuration of Figure 9-2. Thus it is possible to realize the resistor using equivalent circuits wherein all time variations are restricted to either the ideal transformers or the gyrators. It is noteworthy that these equivalences could have been obtained directly from the synthesis procedures of Chapter 5 which yield complete (causal) realizations in the memoryless case. The realizations of Figure 9-1 are, however, simpler and more direct than those which would have been obtained from the syntheses of Chapter 5.

For the inductor, a similar set of equivalents can be found. In the passive case,

$$l(t) \geq 0 \tag{9-12}$$

and

$$\dot{l}(t) \geq 0 \tag{9-13}$$

Hence, one may define two ideal transformers by

$$n_1^2(t) = \dot{l}(t) \tag{9-14}$$

and

$$n_2^2(t) = \frac{\dot{l}(t)}{2} \tag{9-15}$$

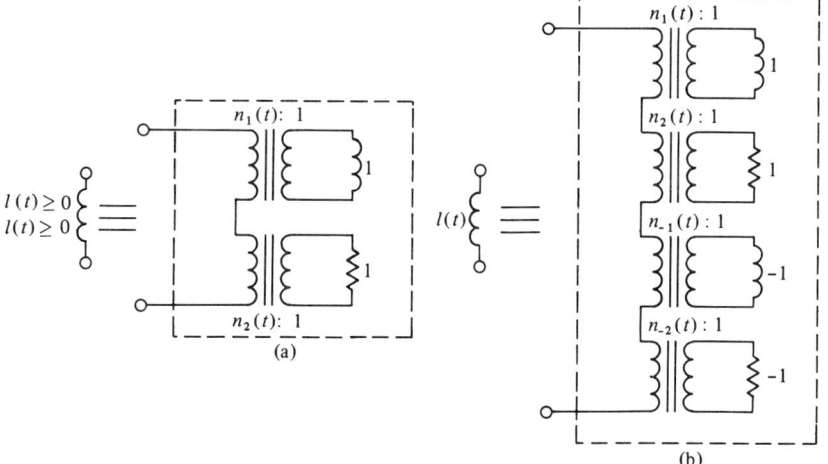

Figure 9-3 Equivalent circuits for (a) passive and (b) general time-variable inductors.

which, when loaded by a unit inductor and a unit resistor, respectively, as shown in Figure 9-3(a), yields the impedance

$$\begin{aligned} z(t,q) &= n_1(t)\dot{\delta}(t-q)n_1(q) + n_2(t)\delta(t-q)n_2(q) \\ &= \dot{\delta}(t-q)n_1^2(q) - \frac{n_1^2(t)\delta(t-q)}{2} + n_2^2(t)\delta(t-q) \\ &= \dot{\delta}(t-q)l(q) - \frac{\dot{l}(t)}{2}\delta(t-q) + \frac{\dot{l}(t)}{2}\delta(t-q) \\ &= \dot{\delta}(t-q)l(q) \end{aligned} \tag{9-16}$$

and thus realizes the passive inductor. This technique may be generalized to realize a general (not necessarily passive) inductor by a procedure similar to that used for the resistor. That is, the inductance and its derivative are split into positive and negative parts with the overall realization being obtained as the series configuration of ideal transformers of Figure 9-3(b), loaded in a 1-henry inductor, a −1-henry inductor, a 1-ohm resistor, and a −1-ohm resistor, respectively. The verification that this configuration does indeed yield the required equivalent circuit is left to the reader as an exercise. Of course, the time-variable ideal transformers may be replaced by gyrators, as in the resistive case. Alternatively, the entire configuration may be cascaded with a gyrator, yielding a realization for a time-variable capacitor.

Proposition 9-1

Time-variable inductors, capacitors, and resistors can be realized with constant inductors (or capacitors) and resistors, together with time-variable gyrators (or time-variable ideal transformers and constant gyrators). ☐

The use of time-variable gyrators or ideal transformers as the only variable element in a finite network is not only sufficient, but, if one desires to realize a lossless network with lossless components, is also necessary. This can be discerned from the results of Chapter 1, where it was shown that a resistor is never lossless (except in the trivial case of an open or short circuit), whereas the inductor and capacitor are lossless if and only if they are constant. This, then, leaves time-variable transformers and/or gyrators as the only remaining lossless variable components. Consistent with these arguments, we thus define a finite time-variable RLC network as follows.

Definition 9-1

A *finite time-variable* RLC *n-port* is one which can be realized (as per Chapter 4) by a finite number of constant resistors and capacitors, together with (possibly) time-variable gyrators. ☐

As in the RLC case, inductors, transformers, circulators, nullators, and so forth, are included in the finite n-ports since they can be constructed from the three basic elements of the definition [44], [110]. We will, therefore, employ these elements freely, allowing the reader to substitute the indicated equivalent circuits if desired. The time-variable n-ports include the time-invariant as well as time-variable networks; we are, however, primarily interested in the time-variable case (since finite time-invariant networks fall within the class of RLC networks already considered), and thus refer to the entire class as "time-variable."

The addition of time variability to the gyrators of an RLC network, while apparently a minor modification, in fact has a number of far-reaching effects. Some of these, such as the pathological character of the time-variable system function [55], [68] and the possible failure of the scattering matrix to exist [9], have already been encountered. An additional and possibly surprising property of the time-variable RLC n-ports is that the interconnection of lossless n-ports may not be lossless [1], [5]. This is illustrated by the network of Figure 9-4. Here the ideal transformer has a unity turns ratio before zero and a zero turns ratio after zero. Now, if an L_2 incident wave is applied before time zero,

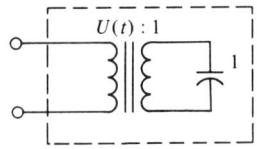

Figure 9-4 Network of lossless components which is not lossless.

a charge will accumulate on the capacitor and be trapped by the change of turns ratio (since after time zero the transformer looks like an open circuit to the capacitor). It is thus possible for an L_2 incident wave to deliver energy to this network which is not eventually returned; hence, it is not lossless. Though this network is not lossless, it does not dissipate energy; rather, it is stored permanently. We thus say that such a network is nondissipative [1].

Another similar pathology of the finite networks created by the introduction of time variability is that the interconnection of reciprocal time-variable networks may not be reciprocal. In fact, Anderson [4] has exhibited a network of capacitors and time-variable ideal transformers which realize a (time-invariant)

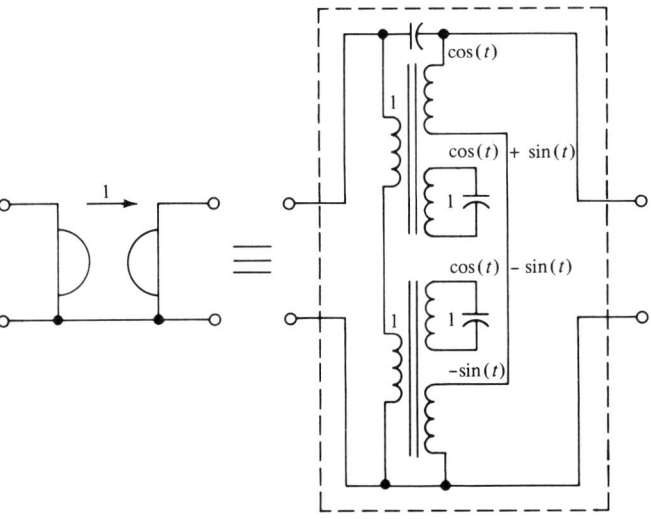

Figure 9-5 Transformer-capacitor gyrator realization.

unit gyrator. The reader may verify that this network, illustrated in Figure 9-5, does, in fact, realize a unit gyrator at its 2 ports. This being the case, Proposition 9-1 may be strengthened to read as follows.

Proposition 9-2

Time-variable capacitors, inductors, resistors, and gyrators can be realized with constant resistors, capacitors, and time-variable ideal transformers. □

Note that this condition is stronger than the second condition of Proposition 9-1, since that condition required both time-variable ideal transformers and constant gyrators.

9-3 Differential Operators and Dynamical Systems

Possibly the most significant characteristic of the time-variable RLC networks is that their input-output relation may be characterized by an ordinary differential equation. In fact, these networks are the most general class of network for which this is true [168], and they could have been defined by this property. If one applies essentially the same argument used in Proposition 6-1 to exhibit the rationality of the RLC system function, the more general time-variable result of Proposition 9-3 is obtained.

Proposition 9-3

Let a finite n-port be characterized by a network matrix $H(t,q)$ such that

$$y(t) = \int_{-\infty}^{\infty} H(t,q)u(q)dq$$

where y and u are some specified network variables. Then the relation between y and u is characterized by an ordinary differential operator

$$H(D,t) = \left(\sum_{i=0}^{n} A_i(t)D^i\right)^{-1}\left(\sum_{i=0}^{m} B_i(t)D^i\right)$$

such that

$$\left(\sum_{i=0}^{n} A_i(t)D^i\right)y = \left(\sum_{i=0}^{m} B_i(t)D^i\right)u$$

Here the $A_i(t)$ and $B_i(t)$ are $n \times n$ matrices of time functions. □

As in Chapter 1, D is the derivative operator and D^{-1} is the integral operator, both on the space D_+. Restriction of these operators to D_+ assures that they are well defined, even without explicit consideration of their initial conditions, since all D_+ signals are zero prior to some time.

The differential operators which characterize the various elementary RLC devices were considered in Chapter 1, where they are tabulated (Table 1-1). According to the proposition, however, all interconnections of such also have differential operator characterizations. Consider, for example, the time-variable RLC 1-port of Figure 9-6. Here the series RL portion of the network is characterized by the impedance

$$Z(D,t) = e^{2t}D + 4e^{2t} \tag{9-17}$$

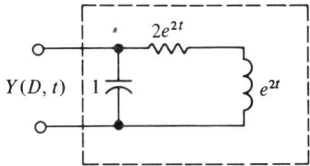

Figure 9-6 Finite 1-port.

which, in combination with the parallel capacitor, yields the admittance operator

$$Y(D,t) = (e^{2t}D + 4e^{2t})^{-1}(e^{2t}D^2 + 4e^{2t}D + 1) \tag{9-18}$$

One difficulty in dealing with differential operators is that they may be unbounded (that is, have an infinite norm). In fact, the derivative itself is unbounded [207]. Although there are techniques for dealing with such operators, they are more complex than those for the bounded case; hence, one would like a means of distinguishing between the bounded and unbounded operators. This can, in fact, be done in terms of the relative orders m and n of the numerator and denominator of the differential operator [201]. For these to be well defined, however, it is necessary that the order of the denominator not be a function of time; hence, we make the following definition.

Definition 9-2

A differential operator

$$\left(\sum_{i=0}^{n} A(t)D^i\right)^{-1}\left(\sum_{i=0}^{m} B_i(t)D^i\right)$$

is *regular* if the leading denominator coefficient matrix $A_n(t)$ is nonsingular for all time. Otherwise, it is *irregular*. □

Although there are a number of real n-ports which have irregular operator characterizations (for instance, the admittance $Y(D,t)$ of an inductor for which the inductance goes to zero at certain times), such operators are rather pathological mathematically, and will be disregarded. Unless otherwise stated, all differential operators will be presumed regular. With this restriction to regular operators, we may now define the following.

Definition 9-2

A regular differential operator

$$\left(\sum_{i=0}^{n} A_i(t)D^i\right)^{-1}\left(\sum_{i=0}^{m} B_i(t)D^i\right)$$

is *proper* if $n \geq m$, *strictly proper* if $n > m$, and *improper* if $n < m$. □

The admittance of Equation (9-18) is a regular operator since

$$A_1 = e^{2t} \tag{9-19}$$

is nonsingular for all t, but improper since the numerator order exceeds that of the denominator. With these definitions, we may now proceed to the study of the boundedness characteristics of the differential operators. This is aided by the following lemma.

Lemma 9-1

Every proper differential operator can be decomposed into the sum of a strictly proper operator and a memoryless (algebraic) operator.

Proof

If $m \neq n$, the lemma is obvious. In the case where $m = n$, consider an operator

$$\left(\sum_{i=0}^{n} A_i(t)D^i\right)^{-1}\left(\sum_{i=0}^{n} B_i(t)D^i\right) \tag{9-20}$$

and decompose it as

$$\left\{\left(\sum_{i=0}^{n} A_i(t)D^i\right)^{-1}\left(\sum_{i=0}^{n} B_i(t)D^i\right) - \left(A_n(t)^{-1}B_n(t)\right)\right\} + A_n(t)^{-1}B_n(t) \tag{9-21}$$

Now, some algebra will reveal that this is the required decomposition, since the subtraction of $A_n(t)^{-1}B_n(t)$ from the given proper operator cancels the leading numerator term, leaving the required strictly proper operator [207]. □

A similar argument will reveal the following lemma.

Lemma 9-2

Every regular differential operator can be decomposed into the sum of a strictly proper operator and $m - n + 1$ derivative (including the zeroth derivative) terms [207]. □

Applying Lemma 9-2 to the improper admittance operator of Equation (9-18) yields the decomposition

$$Y(D,t) = D + (D + 4)^{-1}(e^{-2t}) \tag{9-22}$$

Note that, in this case, the zeroth-order (that is, memoryless) term does not appear, though, in general, all of the derivatives from the $(m - n)$th through the zeroth appear. For convenience, the notation

$$H(D,t) = J_{m-n}(t)D^{m-n} \cdots J_1(t)D + J(t) + R(D,t) \tag{9-23}$$

is adopted for such a decomposition.

Proposition 9-4

Every improper differential operator is unbounded (that is, has infinite norm).

Proof

Let an arbitrary operator $H(D,t)$ be decomposed as

$$H(d,t) = \sum_{i=0}^{m-n} J_i(t)D^i + R(D,t) \tag{9-24}$$

where the remainder $R(D,t)$ is strictly proper. Now, apply an input which "looks" like the function

$$\sin(\omega t) \tag{9-25}$$

over some interval (outside of which it is zero). Now, the last term in Equation (9-24) yields a response

$$\omega^{(m-n)} \sin^{(m-n)}(\omega t) \tag{9-26}$$

to this input which is clearly unbounded and may be made to dominate the solution of the equation; hence, the operator is unbounded. □

Proposition 9-4 is not necessary and sufficient, since a proper operator with ill-behaved coefficient matrices may be unbounded [158], [160]. It does, however, allow one to neglect the improper operators in many circumstances where physical arguments assure boundedness. In particular, we have [150] the following corollary.

*Corollary 9-1

The scattering operator of a nonenergetic time-variable RLC n-port is always proper. □

In many circumstances one would prefer to deal with first-order matrix differential equations, rather than the higher order operators thus far considered. For this reason, we define a dynamical system as follows.

Definition 9-3

A *dynamical system* is a pair of equations

$$\dot{X} = F(t)X + G(t)u$$

$$y = H(t)'X + J(t)u + J_1(t)\dot{u} + \cdots + J_\gamma(t)u^{(\gamma)}$$

where X is the state vector, u the input, and y the output. □

The dynamical system is said to decompose the input-output relation determined by the equation between u and y. Any dynamical system which realizes a given input-output relation is said to be a decomposition of that relation, and is denoted by the $(\gamma + 4)$-tuple of matrices

$$R(t) = \{F(t) \cdot G(t), H'(t), J(t), J_1(t), \cdots, J_\gamma(t)\} \tag{9-27}$$

A decomposition is often termed a realization [19], [117], [158] of the system. This, however, tends to be confused with the realization of a network in the synthesis context; hence, we adopt the term decomposition. Since the decomposition of a time-invariant input-output relation is intimately related to the (1-variable special case of the) multivariable decomposition considered in the previous chapter, the similarity in terminology in this respect causes no difficulty.

Although the dynamical system is always first order (in X), one can always convert higher order regular differential operators into an equivalent dynamical system by increasing the size of the matrices considered [134]. For instance, the regular "one-sided" operator

$$\left(\sum_{i=0}^{n} A_i(t) D^i\right)^{-1} B_0 \tag{9-28}$$

can always be converted to the so-called companion form of a dynamical system

[207], where the state vector is taken as a column vector composed of the output y and its first $n-1$ derivatives,

$$\begin{bmatrix} y_0 \\ y_1 \\ y_2 \\ \cdot \\ \cdot \\ \cdot \\ y_{n-1} \end{bmatrix} = \begin{bmatrix} 0 & 1 & & & & 0 \\ \hline & \cdot & & & & \\ & & \cdot & & & \\ \hline 0 & & & & & 1 \\ \hline -A_n^{-1}A_0 & A_n^{-1}A_1 & \cdot & \cdot & A_n^{-1}A_{n-1} \end{bmatrix} \begin{bmatrix} y_0 \\ y_1 \\ y_2 \\ \cdot \\ \cdot \\ \cdot \\ y_{n-1} \end{bmatrix} + \begin{bmatrix} 0 \\ \hline \cdot \\ \cdot \\ 0 \\ \hline A_n^{-1}B_0 \end{bmatrix} u \quad \textbf{(9-29)}$$

$$y = [1 \mid 0 \mid 0 \mid \cdot \quad \cdot \quad \cdot \mid 0 \mid 0] \begin{bmatrix} y_0 \\ y_1 \\ y_2 \\ \cdot \\ \cdot \\ y_{n-1} \end{bmatrix} + [0]u \quad \textbf{(9-30)}$$

In general, it is always possible to convert a regular differential operator into an equivalent dynamical system which realizes the same input-output relation as the given operator [134]. In the case of an operator more general than that of Equation (9-28), the technical details required to carry out the transformation tend to overwhelm the concept and will, therefore, not be given. The reader may, however, verify the following.

*Theorem 9-1

The input-output relation characterized by any regular differential operator (hence, that of a time-variable RLC n-port) may be realized by a dynamical system. Moreover, if the operator is strictly proper, all of the $J(t)$ matrices are zero, and if the operator is proper, all but the first (zeroth order) $J(t)$ terms are zero. □

Many different proofs of this result are known, as, in fact, many different decompositions exist for a given operator. Porter [134] gives a constructive proof which allows the dynamical system to be written by inspection of the coefficient matrices of the adjoint of the given operator. Consistent with the theorem, a dynamical system of the form

$$\dot{X} = F(t)X + G(t)u \quad \textbf{(9-31)}$$

$$y = H(t)'X \quad \textbf{(9-32)}$$

is termed *strictly proper* since this is the form of dynamical system obtained from a strictly proper operator, and a system of the form

$$\dot{X} = F(t)X + G(t)u \quad \textbf{(9-33)}$$

$$y = H(t)'X + J(t)u \quad \textbf{(9-34)}$$

is said to be *proper*. Similarly, the general system with more than one $J(t)$ matrix is termed *improper*.

The admittance of Equation (9-18) is split [in Equation (9-22)] into the sum of a strictly proper operator and a derivative term. Now, the strictly proper operator is one-sided and, thus, the formula of Equations (9-29) and (9-30) yields the decomposition

$$\dot{i}_0 = [-4]i_0 + [e^{-2t}]v \tag{9-35}$$

$$i = [1]i_0 \tag{9-36}$$

for this operator. Representing the derivative as a $J_1(t)$ term, the resultant improper dynamical system for the given admittance is

$$\dot{i}_0 = [-4]i_0 + [e^{-2t}]v \tag{9-37}$$

$$i = [1]i_0 + [0]v + [1]\dot{v} \tag{9-38}$$

Although there are a number of techniques for obtaining a dynamical system from a specified differential operator, it would be desirable to obtain the dynamical system directly from the network it represents rather than going through the intermediary of an input-output differential operator. In many instances this can, indeed, be achieved [117] and, in fact, the resultant technique is operationally superior to the indirect process.

Let us begin by considering the case where one desires to obtain a dynamical system which decomposes the impedance matrix $Z(t,q)$ of an RLC network. Given an RLC realization of such an impedance, the equivalent circuits of the preceding section allow one to construct an equivalent realization wherein all of the reactors have been replaced by unit inductors, together with time-variable transformers and gyrators. If one now extracts these unit inductors out of the

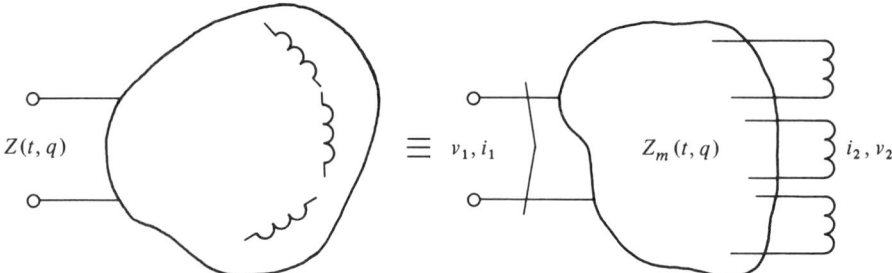

Figure 9-7 Reactance extraction for the analysis of an RLC network.

realization, as shown in Figure 9-7, the network is then in the form of a memoryless coupling network loaded in unit inductors. Assuming that this memoryless coupling network has an impedance matrix

$$Z(t,q) = \begin{bmatrix} Z_{11}(t) & | & Z_{12}(t) \\ -\,-\,-\, & | & -\,-\,-\, \\ Z_{21}(t) & | & Z_{22}(t) \end{bmatrix} \delta(t-q) \tag{9-39}$$

the network variables are constrained by the equality

$$v_1 = Z_{11}(t)i_1 + Z_{12}(t)i_2 \tag{9-40}$$

$$v_2 = Z_{21}(t)i_1 + Z_{22}(t)i_2 \tag{9-41}$$

Substituting the constraint

$$v_2 = -\dot{i}_2 \tag{9-42}$$

imposed on the network variables by the load into Equation (9-41), and rearranging the terms of the resultant equation, yields

$$\dot{i}_2 = [-Z_{22}(t)]i_2 + [-Z_{21}(t)]i_1 \tag{9-43}$$

$$v_1 = [Z_{12}(t)]\dot{i}_2 + [Z_{22}(t)]i_1 \tag{9-44}$$

which is the required decomposition of the impedance matrix. Here,

$$X = \dot{i}_2 \tag{9-45}$$

and

$$R(t) = \{-Z_{22}(t), -Z_{21}(t), Z_{12}(t), Z_{11}(t)\} \tag{9-46}$$

Of course, the entire technique is based on the assumption that the memoryless coupling network has an impedance matrix, which is not always the case. In fact, even if the given n-port has an impedance matrix, the memoryless coupling network is not assured to be so characterized, since the extraction process may leave a degenerate network [92], [104], [105]. Given that such an impedance matrix exists, the advantage of the approach is that one need only analyze the memoryless portion of the network.

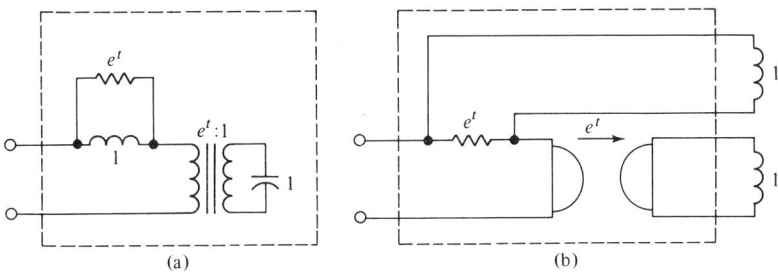

Figure 9-8 (a) Finite network. (b) Its equivalent after reactance extraction.

Consider the network of Figure 9-8(a) which, after an appropriate transformation, takes the form of Figure 9-8(b). Now, upon analyzing the resultant memoryless coupling network, the impedance matrix

$$Z(t,q) = \begin{bmatrix} e^t & e^t & e^t \\ \hline e^t & e^t & 0 \\ e^t & e^t & 0 \end{bmatrix} \delta(t-q) \tag{9-47}$$

318   FINITE TIME-VARIABLE NETWORKS; STATE-SPACE TECHNIQUES

is found. Upon letting

$$X = \begin{bmatrix} i_2 \\ i_3 \end{bmatrix} \tag{9-48}$$

the dynamical system

$$\begin{bmatrix} \dot{i}_2 \\ \dot{i}_3 \end{bmatrix} = \begin{bmatrix} -e^t & 0 \\ -e^t & 0 \end{bmatrix} \begin{bmatrix} i_2 \\ i_3 \end{bmatrix} + \begin{bmatrix} -e^t \\ -e^t \end{bmatrix} i_1 \tag{9-49}$$

$$v_1 = \begin{bmatrix} e^t & e^t \end{bmatrix} \begin{bmatrix} \dot{i}_2 \\ i_3 \end{bmatrix} + [e^t] i_1 \tag{9-50}$$

is obtained. One approach to surmounting the problem of whether or not the memoryless coupling network has an impedance matrix might be to work with the scattering variables rather than the impedance variables. Unfortunately, the differential form of the dynamical system requires that one represent the inductors via their immittance variables (so as to assure that one load variable be the derivative of the other). One can, however, extract capacitors rather than inductors, in which case the memoryless coupling network is required to have an admittance matrix rather than an impedance matrix. Alternatively, a hybrid representation can be employed, where one set of variables is employed at the input ports and another at the load ports. For instance, if one would like to find a dynamical system relating the scattering variables of a finite n-port, the inductors may be extracted as per Figure 9-7, but rather than calculating the impedance matrix of the coupling network, one calculates a hybrid matrix $H(t,q)$ such that

$$b_1 = H_{11}(t)a_1 + H_{12}(t)i_2 \tag{9-51}$$

and

$$v_2 = H_{21}(t)a_1 + H_{22}(t)i_2 \tag{9-52}$$

Given the existence of this matrix, the dynamical system

$$R(t) = \{-H_{22}(t), -H_{21}(t), H_{12}(t), H_{11}(t)\} \tag{9-53}$$

is, as before, a decomposition of the n-port, this time with the scattering variables as the input-output variables.

Consider, for example, the unit capacitor of Figure 9-9(a). Now, the memory-

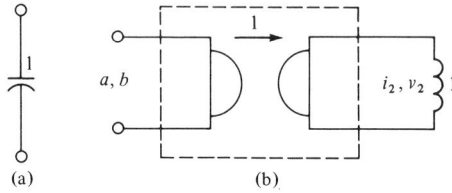

Figure 9-9 (a) Unit capacitor. (b) Its equivalent.

less coupling network shown in Figure 9-9(b) is just a gyrator and, hence, has the hybrid matrix

$$H(t,q) = \left[\begin{array}{c|c} -1 & -2 \\ \hline 1 & 1 \end{array}\right] \delta(t-q) \quad (9\text{-}54)$$

and the dynamical system

$$\dot{i}_2 = [-1]i_2 + [-1]a_1 \quad (9\text{-}55)$$

$$b_1 = [-2]i_2 + [-1]a_1 \quad (9\text{-}56)$$

Clearly, the variety of hybrid matrices which can be employed in such an analysis is great. However, one must always represent the load ports by an impedance if inductors are extracted or, alternatively, by an admittance if capacitors are extracted. Consider, for example, the 1-port of Figure 9-6, for which we have already calculated a dynamical system of equations, (9-37) and (9-38), from its differential operator. Clearly, the parallel capacitor simply corresponds to the first derivative term $J_1(t)$, which may be included after the analysis is complete. Neglecting this component, therefore, and drawing the remainder of the network in the form of a memoryless network with a capacitive load, as per Figure 9-10, the admittance matrix

$$Y(t,q) = \left[\begin{array}{c|c} 0 & -e^{-t} \\ \hline e^{-t} & 3 \end{array}\right] \delta(t-q) \quad (9\text{-}57)$$

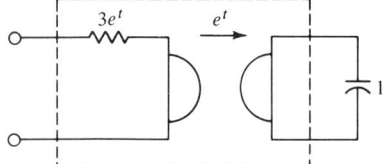

Figure 9-10 Decomposition of the network of Figure 9-6.

is found for the coupling network. Now, following the usual formulation, the corresponding dynamical system

$$\dot{v}_2 = [-3]v_2 + [-e^{-t}]v_1 \quad (9\text{-}58)$$

$$i_1 = [-e^{-t}]v_2 + [0]v_1 \quad (9\text{-}59)$$

for the RL portion of the network is found. Adding the derivative term associated with the parallel capacitor now results in the improper system

$$\dot{v}_2 = [-3]v_2 + [-e^{-t}]v_1 \quad (9\text{-}60)$$

$$i_1 = [-e^{-t}]v_2 + [0]v_1 + [1]\dot{v}_1 \quad (9\text{-}61)$$

representing the overall admittance.

Summarizing the preceding development, we have the following proposition.

Proposition 9-5

Let a finite network be split into a memoryless coupling network loaded in unit inductors (capacitors), where the coupling network is characterized by a hybrid matrix $H(t,q)$, with the input ports characterized by some arbitrary variables u and y, and the load ports by the impedance (admittance) variables. Then

$$R(t) = \{-H_{22}(t), -H_{21}(t), H_{12}(t), H_{11}(t)\}$$

is a dynamical system decomposing the input-output relation imposed by the network on the u and y variables, with the inductor currents (capacitor voltages) taken as the state vectors. □

On the surface, the dynamical system of Equations (9-60) and (9-61) appears to be quite different from the system of Equations (9-37) and (9-38) constructed for the same admittance. Both, however, are correct and, in fact, differ only by a transformation of the state variable. Since the state vector is essentially a mathematical artifice to which no physical significance [87] need be attributed (though it is often convenient to take them as capacitor voltages or inductor currents), one may transform the state vector while fixing the input-output variables to obtain alternative dynamical systems which decompose the same input-output relation. We say that two such systems are equivalent [193].

Definition 9-4

Two dynamical systems

$$R(t) = \{F(t), G(t), H(t)', J(t), \cdots, J_\gamma(t)\}$$

and

$$\boldsymbol{R}(t) = \{\boldsymbol{F}(t), \boldsymbol{G}(t), \boldsymbol{H}(t)', \boldsymbol{J}(t), \cdots, \boldsymbol{J}_\gamma(t)\}$$

with state vectors X and $\boldsymbol{X}$, respectively, are *equivalent* if the relationship they impose on the input and output vectors, u and y, coincide. □

Although a given input-output relation has many decompositions, one can make certain generalizations concerning all such decompositions. The reader may verify [193], [207] the following proposition.

*Proposition 9-6

If $R(t)$ and $\boldsymbol{R}(t)$ are equivalent, $\gamma = \boldsymbol{\gamma}$ and

$$J_i(t) = \boldsymbol{J}_i(t) \quad i = 0, 1, 2, \cdots, \gamma$$
□

In general, very little can be said about the $F(t)$, $G(t)$, and $H(t)'$ matrices of equivalent decompositions. In fact, the corresponding state vectors need not even have the same dimension. For our purposes, we are primarily interested in equivalent dynamical systems of the same dimension and, moreover, in the special case where the two dynamical systems differ only by a nonsingular transformation of the state vector such as

$$\boldsymbol{X} = T(t)X \tag{9-62}$$

where $T(t)$ is a square matrix of differentiable functions which has a continuous inverse for all time. Application of such a transformation yields the following theorem.

Theorem 9-2

Let $T(t)$ be a square matrix of differentiable time functions which has a continuous inverse for all t. Then the dynamical systems

$$\{F(t),G(t),H(t)',J(t),\cdots,J_\gamma(t)\}$$

and

$$\{T(t)F(t)T^{-1}(t) + T(\dot{t})T^{-1}(t),T(t)G(t),H(t)'T^{-1}(t),J(t),\cdots,J_\gamma(t)\}$$

are equivalent.

Proof

Let us apply the transformation

$$\mathbf{X} = T(t)X \tag{9-63}$$

to the state vector of the dynamical system

$$\dot{X} = F(t)X + G(t)u \tag{9-64}$$

$$y = H(t)'X + J(t)u + \cdots + J_\gamma(t)u^{(\gamma)} \tag{9-65}$$

Now,

$$\dot{\mathbf{X}} = T(\dot{t})X + T(t)\dot{X} \tag{9-66}$$

and multiplying Equation (9-64) by $T(t)$ and substituting Equation (9-66) yields

$$\dot{\mathbf{X}} = T(t)F(t)X + T(\dot{t})X + T(t)G(t)u \tag{9-67}$$

Substituting Equation (9-63) into this now yields

$$\dot{\mathbf{X}} = [T(t)F(t)T^{-1}(t) + T(\dot{t})T^{-1}(t)]\mathbf{X} + [T(t)G(t)]u \tag{9-68}$$

while substituting Equation (9-63) into Equation (9-64) yields

$$y = [H(t)'T^{-1}(t)]\mathbf{X} + J(t)u + \cdots + J_\gamma(t)U^{(\gamma)} \tag{9-69}$$

□

Equations (9-68) and (9-69) are clearly a dynamical system with input u and output y equivalent to the given one (since the transformation of state variable carried out does not effect the input and output vectors), as required by the theorem.

If one lets

$$T(t) = -e^{-t} \tag{9-70}$$

and applies the theorem to the dynamical system

$$\{-3,-e^{-t},-e^{-t},0,1\} \tag{9-71}$$

of Equations (9-60) and (9-61) for the 1-port of Figure 9-6, the equivalent system with

$$F(t) = (-e^{-t})(-3)(-e^t) + (e^{-t}) = -4 \tag{9-72}$$

$$G(t) = (-e^{-t})(-e^{-t}) = e^{-2t} \tag{9-73}$$

$$H(t)' = (-e^{-t})(-e^t) = 1 \tag{9-74}$$

$$J(t) = 0 \tag{9-75}$$

$$J_1(t) = 1 \tag{9-76}$$

is obtained. For brevity, we say

$$\{-3, -e^{-t}, -e^{-t}, 0, 1\} = \{-4, e^{-2t}, 1, 0, 1\} \tag{9-77}$$

Now, this latter realization is precisely that obtained in Equations (9-37) and (9-38); hence, the two realizations for this network are indeed equivalent.

It is noteworthy that the equivalence theory for the dynamical system decompositions of an input-output relation is similar to the equivalence theory for the multivariable decompositions derived in the preceding chapter [12], [193]. In fact, in the time-invariant case, $T(t)$ is constant; hence, the derivative term in Theorem 9-2 does not appear, in which case the two equivalence theories are identical, upon making the identifications

$$F \leftrightarrow \Sigma_{22} \tag{9-78}$$

$$G \leftrightarrow \Sigma_{21} \tag{9-79}$$

$$H' \leftrightarrow \Sigma_{12} \tag{9-80}$$

and

$$J \leftrightarrow \Sigma_{12} \tag{9-81}$$

This analogy is more than coincidental and, in fact, the 1-variable case of the multivariable decomposition may be used to find a dynamical system decomposition of a time-invariant input-output relation under the identification of Equations (9-78) through (9-81) [114], [117], [199]. To see this, consider an arbitrary system function $H(p)$ for a time-invariant n-port characterized by a time-invariant dynamical system

$$\dot{X} = FX + Gu \tag{9-82}$$

$$y = H'X + Ju \tag{9-83}$$

Since the dynamical system is time invariant, we may identify the differentiation of Equation (9-82) with its corresponding system function p, in which case the equation becomes

$$pX = FX + Gu \tag{9-84}$$

or, equivalently,

$$X = (p1 - F)^{-1}Gu \tag{9-85}$$

which, upon substitution into Equation (9-83), yields the given system function $H(p)$ via

$$H(p) = H'(p1 - F)^{-1}G + J \tag{9-86}$$

Now, Equation (9-86) yields a formula for calculating the system from its dynamical system decomposition or, alternatively, the decomposition may be obtained from the system function by writing it in the form of Equation (9-86). Once again, invoking the analogy between the dynamical system decomposition and the multivariable decomposition of Equations (9-78) through (9-81), Equation (9-86) is seen to be just the 1-variable special case of the multivariable decomposition. We have thus proven the following proposition.

Proposition 9-7

Let $H(p)$ be a system function decomposed via Theorem 8-1 as

$$H(p) = \Sigma_{11} + \Sigma_{12}(p1 - \Sigma_{22})^{-1}\Sigma_{21}$$

Then a time-invariant dynamical system decomposition for the input-output relation defined by $H(p)$ is

$$\dot{X} = \Sigma_{22}X + \Sigma_{21}u$$

$$y = \Sigma_{12}X + \Sigma_{11}u \qquad \square$$

In this special case we thus have still another means for constructing a dynamical system which decomposes a given input-output relation—via the multivariable decomposition.

9-4 Separable Impulse Responses

The purpose of this section is to determine the special characteristics which separate the impulse response (i.e., time-domain representation) of a time-variable RLC n-port from the general class of 2-variable impulse responses studied in Chapter 2. Our starting point is the dynamical system decomposition of a finite network,

$$\dot{X} = F(t)X + G(t)u \qquad (9\text{-}87)$$

$$y = H(t)'X + J(t)u + \cdots + J_\gamma(t)u^{(\gamma)} \qquad (9\text{-}88)$$

Clearly, the effect of the $J(t)$ matrices on the impulse response is to induce additive terms of the form

$$J(t)\delta(t-q) + J_1(t)\dot{\delta}(t-q) + \cdots + J_\gamma(t)\delta^{(\gamma)}(t-q) \qquad (9\text{-}89)$$

where the higher order delta functions correspond to the input differentiation associated with the $J(t)$ terms [207]. Our main problem is, therefore, that of calculating the impulse response associated with the strictly proper system

$$\dot{X} = F(t)X + G(t)u \qquad (9\text{-}90)$$

$$y = H(t)'X \qquad (9\text{-}91)$$

Since the output y from such a system may be calculated from the input by "direct" integration, once the impulse response kernel is known, it might be expected that the differential equation, (9-91), must be solved in calculating the impulse response $H(t,q)$ for such a system. This is indeed the case, for which reason one defines the fundamental solution matrix associated with a dynamical system [207] as follows.

Definition 9-5

Let
$$\dot{X} = F(t)X + G(t)u$$
$$y = H(t)'X$$

be a strictly proper dynamical system. Then an $m \times m$ matrix, where m is the dimension of $F(t)$, $\Psi(t)$, which satisfies the equation

$$\dot{\Psi}(t) = F(t)\Psi(t)$$

with

$$\Psi(0) = 1$$

is a set of *fundamental solutions* for the system. □

For the present purposes, one is more interested in the form and properties of the fundamental solutions than their actual values. $\Psi(t)$, however, always exists (if $F(t)$ is continuous), is invertible, and may be written as an infinite series in $F(t)$ [207], while if $F(t_1)$ commutes with $F(t_2)$ for all t_1 and t_2,

$$\Psi(t) = e^{\int_0^t F(\gamma)d\gamma} \qquad (9\text{-}92)$$

where γ is a dummy variable of integration [207].

Although the formula of Equation (9-92) is more symbolic than practical, it does yield a convenient means for calculating the fundamental solutions in the scaler case. Consider the dynamical system with

$$F(t) = -e^{-t} \qquad (9\text{-}93)$$

Now, according to Equation (9-92),

$$\Psi(t) = e^{\int_0^t -e^{-\gamma}d\gamma} = e^{e^{-\gamma}|_0^t}$$
$$= e^{(e^{-t}-1)} \qquad (9\text{-}94)$$

Once the fundamental solutions for a dynamical system have been found, they may be used as a multiplying factor in solving Equation (9-90) for X. For this purpose, one employs the following lemma.

Lemma 9-3

Let $\Psi(t)$ be a matrix of fundamental solutions for a dynamical system. Then

$$\dot{\Psi}(t)^{-1} = -\Psi(t)^{-1}F(t)$$

Proof

Starting with

$$1 = \Psi(t)\Psi(t)^{-1} \tag{9-95}$$

and differentiating both sides yields

$$0 = \dot{1} = \left(\Psi(t)\Psi(t)^{-1}\right)^{\cdot} = \Psi(t)\dot{\Psi}(t)^{-1} + \dot{\Psi}(t)\Psi(t)^{-1} \tag{9-96}$$

Multiplying through by $\Psi(t)^{-1}$ now yields

$$\begin{aligned}\dot{\Psi}(t)^{-1} &= -\Psi(t)^{-1}\dot{\Psi}(t)\Psi(t)^{-1} \\ &= -\Psi(t)^{-1}\big(F(t)\Psi(t)\big)\Psi(t)^{-1} \\ &= -\Psi(t)^{-1}F(t)\end{aligned} \tag{9-97}$$

which is the desired result. □

Rather than dealing with $\Psi(t)$ alone, it is convenient to deal with the transition matrix $\Phi(t,\gamma)$, which is formed from $\Psi(t)$ by the addition of an extra parameter.

Definition 9-6

Let $\Psi(t)$ be the matrix of fundamental solutions associated with a dynamical system. Then

$$\Phi(t,\gamma) = \Psi(t)\Psi(\gamma)^{-1}$$

is the transition matrix of the system. Here γ is an arbitrary parameter independent of time. □

In our previous scaler example,

$$\Psi(t) = e^{(e^{-t}-1)} \tag{9-98}$$

Hence,

$$\begin{aligned}\Phi(t,\gamma) = \Psi(t)\Psi(\gamma)^{-1} &= (e^{(e^{-t}-1)})(e^{-(e^{-\gamma}-1)}) \\ &= e^{(e^{-t}-e^{-\gamma})}\end{aligned} \tag{9-99}$$

is the required transition matrix for this case.

Since the differentiation in Lemma 9-3 and Definition 9-5 is independent of γ, the transition matrix satisfies equations similar to those satisfied by the fundamental solution matrix [207].

Lemma 9-4

Let $\Phi(t,\gamma)$ be the transition matrix of a dynamical system. Then

$$\dot{\Phi}(t,\gamma) = F(t)\Phi(t,\gamma)$$

and

$$\dot{\Phi}(t,\gamma)^{-1} = -\Phi(t,\gamma)^{-1}F(t)$$

Here the overdot denotes partial differentiation with respect to t.

Note that

$$\Phi(t,t) = \Phi(t,t)^{-1} = 1 \qquad (9\text{-}100)$$

is an initial condition on the equations of Lemma 9-4, induced by the definition of the transition matrix from the fundamental solution matrix.

With these preliminaries completed, one may solve

$$\dot{X} = F(t)X + G(t)u \qquad (9\text{-}101)$$

for X, the result of which may be substituted into

$$y = H(t)'X \qquad (9\text{-}102)$$

to obtain y and the impulse response of the finite network represented by the system.

Theorem 9-3

Let

$$\dot{X} = F(t)X + G(t)u$$

$$y = H(t)'X$$

be a dynamical system with transition matrix $\Phi(t,\gamma)$ satisfying

$$\dot{\Phi}(t,\gamma) = F(t)\Phi(t,\gamma)$$

Then

$$X(t) = \int_{-\infty}^{\infty} \Phi(t,q)G(q)U(t-q)u(q)dq$$

where q is a dummy variable of integration and $U(t-q)$ is the unit step function.

Proof

Starting with

$$\dot{X} = F(t)X + G(t)u \qquad (9\text{-}103)$$

and multiplying both sides by the inverse of the transition matrix yields

$$\Phi(t,\gamma)^{-1}\dot{X} - \Phi(t,\gamma)^{-1}F(t)X = \Phi(t,\gamma)^{-1}G(t)u \qquad (9\text{-}104)$$

Now, by Lemma 9-4,

$$-\Phi(t,\gamma)^{-1}F(t) = \dot{\Phi}(t,\gamma)^{-1} \qquad (9\text{-}105)$$

Hence, Equation (9-104) is equivalent to

$$\Phi(t,\gamma)^{-1}\dot{X} + \dot{\Phi}(t,\gamma)^{-1}X$$
$$= (\dot{\Phi}(t,\gamma)^{-1}X) = \Phi(t,\gamma)^{-1}G(t)u \qquad (9\text{-}106)$$

Upon integration of both sides of Equation (9-106) with respect to the time variable, this becomes

$$\int_\gamma^t \dot{\Phi}(q,\gamma)^{-1} X \, dq = \int_\gamma^t \Phi(q,\gamma)^{-1} G(q) u(q) dq \qquad (9\text{-}107)$$

where q is a dummy variable of integration. The left side of Equation (9-107) may be expanded as

$$\Phi(q,\gamma)^{-1} X \Big|_\gamma^t = \Phi(t,\gamma)^{-1} X(t) - \Phi(\gamma,\gamma)^{-1} X(\gamma)$$
$$= \int_\gamma^t \Phi(q,\gamma)^{-1} G(q) u(q) dq \qquad (9\text{-}108)$$

Multiplying by $\Phi(t,\gamma)$ and observing that

$$\Phi(\gamma,\gamma)^{-1} = 1 \qquad (9\text{-}109)$$

then yields

$$X(t) - \Phi(t,\gamma) X(\gamma) = \int_\gamma^t \Phi(t,\gamma) \Phi(q,\gamma)^{-1} G(q) u(q) dq$$
$$= \int_\gamma^t \Phi(t,q) G(q) u(q) dq \qquad (9\text{-}110)$$

where the last equality results from the fact that

$$\Phi(t,\gamma) \Phi(q,\gamma)^{-1} = \left(\Psi(t)\Psi(\gamma)^{-1}\right)\left(\Psi(\gamma)\Psi(q)^{-1}\right)$$
$$= \Psi(t)\Psi(q)^{-1} = \Phi(t,q) \qquad (9\text{-}111)$$

Since X is in $D_+{}^m$, it is zero for sufficiently negative (but finite) γ; hence,

$$\lim_{\gamma \to -\infty} X(\gamma) = 0 \qquad (9\text{-}112)$$

and

$$X(t) = \lim_{\gamma \to -\infty} X(t)$$
$$= \lim_{\gamma \to -\infty} \Phi(t,\gamma) X(\gamma) + \int_\gamma^t \Phi(t,q) G(q) u(q) dq$$
$$= \int_{-\infty}^t \Phi(t,q) G(q) u(q) dq \qquad (9\text{-}114)$$

Finally, if $U(t - q)$ is inserted into the integrand, one may raise the upper limit of integration to infinity (since the new integrand will be zero for q greater than t). This then yields the required result that

$$X(t) = \int_{-\infty}^\infty \Phi(t,q) G(q) U(t - q) u(q) dq \qquad (9\text{-}115)$$

□

Substituting the result of the theorem into

$$y = H(t)'X \tag{9-116}$$

yields the following corollary.

Corollary 9-2

Under the hypothesis of Theorem 9-3,

$$y(t) = \int_{-\infty}^{\infty} H(t)'\Phi(t,q)G(q)U(t-q)u(q)dq \qquad \square$$

Given this corollary, our main result, the impulse response of a finite time-variable n-port (via its dynamical system decomposition), may be obtained by adding the appropriate $J(t)$ terms to the impulse response of the strictly proper system given in the corollary.

Corollary 9-3

A finite n-port characterized by the dynamical system

$$\dot{X} = F(t)X + G(t)u$$

$$y = H(t)'X + J(t)u + \cdots + J_\gamma(t)u^{(\gamma)}$$

has impulse response

$$H(t,q) = J_\gamma(t)\delta^{(\gamma)}(t-q) + \cdots + J(t)\delta(t-q)$$
$$+ H(t)'\Phi(t,q)G(q)U(t-q) \qquad \square$$

Following our previously adopted notation for differential operators and dynamical systems, we say that an impulse response of a finite network is strictly proper if none of the $J(t)$ terms appear, proper if only the first $J(t)$ term appears, and improper otherwise.

In dealing with the impulse response of a finite network, it is usually convenient to let

$$\theta(t) = H(t)'\Psi(t) \tag{9-117}$$

and

$$\phi(t) = G(t)'\Psi(t)'^{-1} \tag{9-118}$$

Now, since the transition matrix decomposes as

$$\Phi(t,q) = \Psi(t)\Psi(q)^{-1} \tag{9-119}$$

the finite impulse response of Corollary 9-3 may be written in the separable form

$$H(t,q) = \sum_{i=0}^{\gamma} J_i(t)\delta^{(i)}(t-q) + \theta(t)\phi(q)'U(t-q) \tag{9-120}$$

Definition 9-7

An impulse response having the form of Equation (9-120) is said to be *separable*. □

For the dynamical system

$$\dot{X} = [-e^{-t}]x + [0 \ \sin(t)]\begin{bmatrix} a_1 \\ a_2 \end{bmatrix} \quad (9\text{-}121)$$

$$\begin{bmatrix} b_1 \\ b_2 \end{bmatrix} = \begin{bmatrix} 1 \\ 1 \end{bmatrix} x + \begin{bmatrix} 1 & e^{-t} \\ 1 & 0 \end{bmatrix}\begin{bmatrix} a_1 \\ a_2 \end{bmatrix} \quad (9\text{-}122)$$

constraining the scattering variables of a 2-port, we have already calculated the transition matrix as

$$\Phi(t,\gamma) = e^{(e^{-t}-e^{-\gamma})} \quad (9\text{-}123)$$

Carrying out the calculation indicated by Corollary 9-3, the scattering matrix for the finite 2-port is found to be

$$\begin{aligned} S(t,q) &= J(t)\delta(t-q) + H(t)\Phi(t,q)G(q)U(t-q) \\ &= \begin{bmatrix} 1 & e^{-t} \\ 1 & 0 \end{bmatrix}\delta(t-q) + \begin{bmatrix} 1 \\ 1 \end{bmatrix}[e^{(e^{-t}-e^{-q})}][0 \ \sin(q)]U(t-q) \\ &= \begin{bmatrix} 1 & e^{-t} \\ 1 & 0 \end{bmatrix}\delta(t-q) + \begin{bmatrix} 0 & e^{(e^{-t}-e^{-q})}\sin(q) \\ 0 & e^{(e^{-t}-e^{-q})}\sin(q) \end{bmatrix}U(t-q) \end{aligned} \quad (9\text{-}124)$$

Alternatively, we may let

$$\theta(t) = \begin{bmatrix} e^{e^{-t}} \\ e^{e^{-t}} \end{bmatrix} \quad (9\text{-}125)$$

and

$$\phi(q) = \begin{bmatrix} 0 \\ e^{-e^{-q}}\sin(q) \end{bmatrix} \quad (9\text{-}126)$$

in which case the scattering matrix may be written in its separable form as

$$\begin{aligned} S(t,q) &= J(t)\delta(t-q) + \theta(t)\phi(q)'U(t-q) \\ &= \begin{bmatrix} 1 & e^{-t} \\ 1 & 0 \end{bmatrix}\delta(t-q) + \begin{bmatrix} e^{e^{-t}} \\ e^{e^{-t}} \end{bmatrix}[0 \ e^{-e^{-q}}\sin(q)]U(t-q) \end{aligned} \quad (9\text{-}127)$$

Separability is not only a necessary condition on the impulse response of a finite network, but is also sufficient [207].

Theorem 9-4

A necessary and sufficient condition for $H(t,q)$ to be the impulse response of a time-variable RLC n-port with a dynamical system representation is that it be separable.

Proof

The necessity of the separability condition has already been established. For the sufficiency it suffices to exhibit at least one (there are many) dynamical system which posesses a given separable

$$\sum_{i=0}^{\gamma} J_i(t)\delta^{(i)}(t-q) + \theta(t)\phi(q)'U(t-q) \tag{9-128}$$

as its impulse response (since time-variable RLC n-ports can always be constructed by analog computer techniques from the dynamical system). Given $H(t,q)$ as per Equation (9-128), consider the dynamical system

$$\dot{X} = 0X + \phi(t)'u \tag{9-129}$$

$$y = \theta(t)X + J(t)u + \cdots + J_\gamma(t)u^{(\gamma)} \tag{9-130}$$

Here,

$$F(t) = 0 \tag{9-131}$$

Hence, if

$$\Psi(t) = 1 \tag{9-132}$$

$$\dot{\Psi}(t) = \dot{1} = 0 = 0\Psi(t) = F(t)\Psi(t) \tag{9-133}$$

Thus, 1 is a set of fundamental solutions for the system. Corollary 9-3 now shows that the impulse response associated with this system is

$$\sum_{i=0}^{\gamma} J_i(t)\delta^{(i)}(t-q) + \theta(t)\cdot 1 \cdot 1 \cdot \phi(q)'U(t-q) = H(t,q) \tag{9-134}$$

which was to be shown. □

It should be noted that the dynamical system constructed in the theorem is by no means a "good" realization. In particular, it is unstable and time variable, even for (externally) time-invariant networks [158], [193]. In the time-invariant case, the construction of Proposition 9-7 via the multivariable decomposition does yield a time-invariant dynamical system which is stable if the n-port is stable and "generally" well behaved [114]. This is, however, constructed from the system function rather than the impulse response, in which case an analogous construction to obtain "good" realizations is not available [193].

Applying the theorem to the scattering matrix of Equation (9-127), the dynamical system

$$\dot{X} = [0]X + [0 \quad e^{-e^{-t}}\sin(t)]\begin{bmatrix}a_1\\a_2\end{bmatrix} \tag{9-135}$$

$$\begin{bmatrix}b_1\\b_2\end{bmatrix} = \begin{bmatrix}e^{e^{-t}}\\e^{e^{-t}}\end{bmatrix}X + \begin{bmatrix}1 & e^{-t}\\1 & 0\end{bmatrix}\begin{bmatrix}a_1\\a_2\end{bmatrix} \tag{9-136}$$

is obtained. Now, this is clearly not the same dynamical system with which we began in Equations (9-121) and (9-122) in calculating the scattering matrix

$S(t,q)$. It is, however, equivalent, as one may verify by applying the transformation

$$T(t) = e^{e^{-t}} \tag{9-137}$$

to the state vector for that system, and applying Theorem 9-2.

The separable form of $H(t,q)$ naturally leads to a definition for the degree of a finite network.

Definition 9-8

The *degree* of a finite network with proper impulse response is the minimal number of columns in $\theta(t)$ [and $\phi(q)$] in any decomposition of the network as a proper separable impulse response. □

The impulse response

$$\begin{bmatrix} 1 & 0 \\ 0 & 0 \end{bmatrix} \delta(t-q) + \begin{bmatrix} tq & tq^{-1} \\ 0 & 0 \end{bmatrix} U(t-q) \tag{9-138}$$

may be decomposed as

$$\begin{bmatrix} 1 & 0 \\ 0 & 0 \end{bmatrix} \delta(t-q) + \begin{bmatrix} t & 0 \\ 0 & e^{-t} \end{bmatrix} \begin{bmatrix} q & q^{-1} \\ 0 & 0 \end{bmatrix} U(t-q) \tag{9-139}$$

or, alternatively, as

$$\begin{bmatrix} 1 & 0 \\ 1 & 0 \end{bmatrix} \delta(t-q) + \begin{bmatrix} t \\ 0 \end{bmatrix} [q \quad q^{-1}] U(t-q) \tag{9-140}$$

Hence, according to the definition, the degree of the matrix is one rather than two, since the second decomposition has only one column in $\theta(t)$, which is clearly the minimum possible. It is by no means clear that this definition of degree coincides with that used for the RLC networks (for those finite networks which are also RLC networks). This, however, can be established [86] by the following.

*Lemma 9-5

The degree of a finite network is the minimal dimension of X, the state vector, in any dynamical system decomposition for the network. □

*Proposition 9-8

Let an RLC network have a system function representation and a separable impulse response representation. Then

$$\delta H(p) = \delta H(t,q)$$

where $\delta H(p)$ is the degree of the RLC network in the MacMillan sense (as used in Chapter 6) and $\delta H(t,q)$ is the finite network degree of Definition 9-8. □

9-5 Synthesis Via Reactance Extraction

One approach to the problem of synthesizing a time-variable RLC n-port is to reverse the "analysis by reactance extraction" technique of Proposition 9-5. That is, given a dynamical system decomposing a given network matrix, one uses the system to construct a memoryless coupling network which realizes the given network under reactor loading [11], [19], [160]. Since the theory underlying such syntheses is most nearly complete in the impedance formalism, we will consider that case first. Although the impedance matrix of a finite network may, in fact, be improper, the higher order derivative terms can be removed from the matrix by an extraction technique which is somewhat similar to the removal of the poles at infinity from an RLC impedance [2]; hence, we will restrict ourselves to proper impedances. Let us, therefore, consider a proper separable impedance matrix

$$Z(t,q) \tag{9-141}$$

with associated dynamical system

$$\dot{X} = F(t)X + G(t)i \tag{9-142}$$

$$v = H(t)'X + J(t)i \tag{9-143}$$

(Note that the assumption that a dynamical system for the given impedance matrix is known does not restrict the generality of the process, since one can readily identify the dynamical system of Theorem 9-4 with any given separable impedance matrix.) Now, define a memoryless impedance matrix by

$$Z_m(t,q) = \begin{bmatrix} J(t) & H(t)' \\ \hline -G(t) & -F(t) \end{bmatrix} \delta(t - q) \tag{9-144}$$

which, by Proposition 9-5 (applied backwards), realizes $Z(t,q)$ as its "1" ports when its "2" ports are loaded in unit inductors [2].

Consider the impedance matrix

$$Z(t,q) = e^t \delta(t - q) + e^{t+1} e^{q-1} U(t - q) \tag{9-145}$$

Now, by Theorem 9-4, this may be decomposed by the dynamical system

$$\dot{X} = [0]X + [e^{t-1}]i \tag{9-146}$$

$$v = [e^{t+1}]X + [e^t]i \tag{9-147}$$

Hence, the impedance may be realized by loading the last port of the memoryless impedance matrix

$$Z_m(t,q) = \begin{bmatrix} e^t & -e^{t+1} \\ \hline -e^{t-1} & 0 \end{bmatrix} \delta(t - q) \tag{9-148}$$

with unit reactors. Writing this impedance matrix as the sum of its hermitian and skew hermitian parts, and synthesizing as per Proposition 5-6, now yields the realization of Figure 9-11. Summarizing the above yields the following proposition.

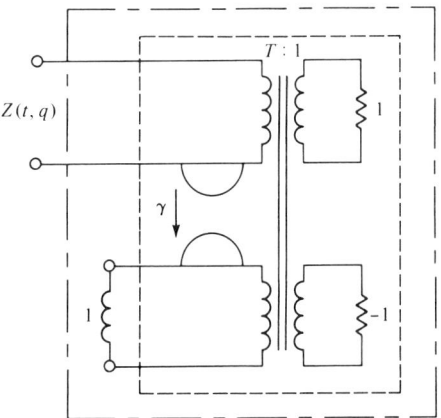

Figure 9-11 Realization of impedance by reactance extraction.

$$\gamma = \frac{-e^{t+1} + e^{t-1}}{2} \qquad T = \begin{bmatrix} e^{t/2} & 0 \\ \dfrac{e^{t/2+1} - e^{t/2-1}}{2} & \dfrac{e^{t/2-1} - e^{t/2+1}}{2} \end{bmatrix}$$

Proposition 9-9

Any proper separable impedance matrix with a dynamical system decomposition

$$R(t) = \{F(t), G(t), H(t)', J(t)\}$$

can be synthesized by loading a realization of the memoryless impedance matrix

$$Z_m(t,q) = \begin{bmatrix} J(t) & H(t)' \\ \hline -G(t) & -F(t) \end{bmatrix} \delta(t-q)$$

with unit inductors. □

The synthesis procedure of Proposition 9-9 is clearly very easily implemented and, moreover, yields realizations in a single step for active as well as passive and lossless impedances. Its disadvantage is that, in general, it yields an active realization even for passive impedance matrices. This is due to the fact that an arbitrary dynamical system decomposition is used as the starting point for the synthesis. If, rather, one started with a "good" decomposition, a passive or lossless synthesis might be expected as the result. Consider, once again, our previous example of Equation (9-145). Rather than employing the dynamical system decomposition

$$R(t)\{0, e^{t-1}, e^{t+1}, e^t\} \tag{9-149}$$

of Equations (9-145) and (9-146), as before, let us form a new dynamical system via the state vector transformation

$$\mathbf{X} = -e\mathbf{X} \tag{9-150}$$

in which case Theorem 9-2 yields the decomposition

$$R(t) = \{0, -e^t, -e^t, e^t\} \tag{9-151}$$

Applying Proposition 9-9 with this decomposition now results in the memoryless coupling network

$$Z_m(t,q) = \begin{bmatrix} e^t & -e^t \\ \hline e^t & 0 \end{bmatrix} \delta(t-q) \tag{9-152}$$

and the passive realization of Figure 9-12.

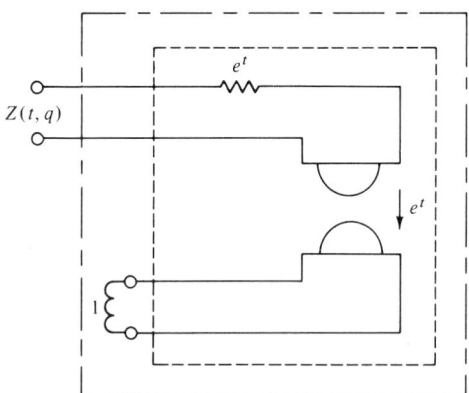

Figure 9-12 Passive realization of impedance by reactance extraction.

Clearly, for the given 1-port the realization of Figure 9-12 is far superior to that of Figure 9-11. Hence, one would desire to find a systematic procedure for finding "good" decompositions to which Proposition 9-9 may be applied. Although no such procedure is known which is assured to work for all passive impedances, a rather powerful sufficiency condition, due to Anderson, is possible. In fact, this condition is necessary in the time-invariant case, as will be demonstrated in the following section. As before, we begin with an arbitrary decomposition

$$R(t) = \{F(t), G(t), H(t)'J(t)\} \tag{9-153}$$

of a specified impedance and, moreover, we assume that the set of four simultaneous matrix equations, in the unknowns $P(t)$, $L(t)$, $W(t)$, and $T(t)$,

$$\dot{P}(t) + P(t)F(t) + F(t)'P(t) = -L(t)L(t)' \quad (9\text{-}154)$$

$$P(t)G(t) = H(t) - L(t)W(t) \quad (9\text{-}155)$$

$$W(t)'W(t) = J(t) + J(t)' \quad (9\text{-}156)$$

and

$$T(t)T(t) = P(t) = P(t)' \quad (9\text{-}157)$$

has a solution. Now, initially, these nonlinear matrix differential equations may appear forboding. In fact, given the existence of certain inverses, they can be converted into a single matrix Ricatti equation [11], [19] for which solution procedures are well known. Now, assuming that a solution to these equations is known, let us use the matrix $T(t)$ (which we assume to be invertible) as a transformation with which to construct (by Theorem 9-2) the equivalent dynamical system decomposition

$$\begin{aligned}\boldsymbol{R}(t) &= \{\boldsymbol{F}(t), \boldsymbol{G}(t), \boldsymbol{H}(t)', \boldsymbol{J}(t)\} \\ &= \{T(t)F(t)T(t)^{-1} + \dot{T}(t)T^{-1}(t), T(t)G(t), H(t)'T(t)^{-1}, J(t)\}\end{aligned} \quad (9\text{-}158)$$

It is our contention now that if one uses this new dynamical system $R(t)$ in Proposition 9-9, a passive realization is obtained. To this end, it must be shown that the coupling network

$$Z_m(t,q) = \begin{bmatrix} J(t) & | & H(t)' \\ \hline -G(t) & | & -F(t) \end{bmatrix} \delta(t-q) \quad (9\text{-}159)$$

is passive. Writing this as the sum of its hermitian and skew hermitian parts,

$$\begin{aligned}Z_m(t,q) &= \tfrac{1}{2}[Z_m(t,q) + Z_m(t,q)'] \\ &\quad + \tfrac{1}{2}[Z_m(t,q) - Z_m(t,q)']\end{aligned} \quad (9\text{-}160)$$

it suffices to show that both terms are passive. Now, being memoryless, the skew hermitian part is lossless (as are all memoryless skew hermitian impedance matrices). Writing the hermitian part of $Z_m(t,q)$ in terms of the $R(t)$ decomposition, it becomes

$$\tfrac{1}{2}[Z_m(t,q) + Z_m(t,q)'] = \tfrac{1}{2}\begin{bmatrix} J(t) + J(t)' & | & H(t)' - G(t)' \\ \hline H(t) - G(t) & | & -F(t) - F(t)' \end{bmatrix} \delta(t-q) \quad (9\text{-}161)$$

which may be factored as

$$\begin{aligned}\tfrac{1}{2}[Z_m(t,q) + Z_m(t,q)'] &= \begin{bmatrix} W(t)' \\ \hline (L(t)'T^{-1}(t))' \end{bmatrix}[W(t) \mid L(t)'T^{-1}(t)]\tfrac{1}{2}\delta(t-q) \\ &= \begin{bmatrix} W(t)' \\ \hline (L(t)'T^{-1}(t))' \end{bmatrix} \circ \tfrac{1}{2}\delta(t-q) \circ [W(q) \mid (L(q)'T^{-1}(q))]\end{aligned} \quad (9\text{-}162)$$

upon substitution of the equalities of Equations (9-154) through (9-158) into Equation (9-161) [11]. The reader is invited to verify this as an exercise (note that Equation (9-157) assures that $T(t)$ is symmetric). Observing that the factorization of Equation (9-162) is precisely the form of the impedance matrix obtained by loading the ideal transformer with turns ratio matrix

$$[W(t) \mid (L(t)'T^{-1}(t))'] \qquad (9\text{-}163)$$

with $\frac{1}{2}$-ohm resistors, we find, as required, that the hermitian part of $Z_m(t,q)$ is indeed passive. In fact, we have exhibited a passive realization for the network as a resistor-loaded transformer. Since both parts of $Z_m(t,q)$ are passive, we may conclude that application of Proposition 9-9 with the dynamical system $R(t)$ yields a passive realization. In fact, if one can solve Equations (9-154) through (9-157) with

$$L(t) = W(t) = 0 \qquad (9\text{-}164)$$

a lossless realization is obtained [since in that case the hermitian part of $Z_m(t,q)$ is zero, by Equations (9-154) through (9-158)]. In summary, we have the following theorem.

Theorem 9-5

Let $Z(t,q)$ be a proper finite impedance matrix with dynamical system decomposition

$$R(t) = \{F(t), G(t), H(t)', J(t)\}$$

and let the equations

$$\dot{P}(t) + P(t)F(t) + F(t)'P(t) = -L(t)L(t)'$$

$$P(t)G(t) = H(t) - L(t)W(t)$$

$$W'(t)W(t) = J(t) + J(t)'$$

and

$$T(t)T(t) = P(t) = P(t)'$$

be satisfied for matrices $P(t)$, $T(t)$, $L(t)$, and $W(t)$. Then application of Proposition 9-9 with the equivalent decomposition

$$R(t) = \{\boldsymbol{F}(t), \boldsymbol{G}(t), \boldsymbol{H}(t)', \boldsymbol{J}(t)\}$$
$$= \{T(t)F(t)T(t)^{-1} + \dot{T}(t)T(t)^{-1}, T(t)G(t), H(t)'T(t)^{-1}, J(t)\}$$

yields a passive realization of $Z(t,q)$. That is, the impedance matrix

$$Z_m(t,q) = \begin{bmatrix} J(t) & H(t)' \\ \hline -G(t) & -F(t) \end{bmatrix} \delta(t-q)$$

is passive and realizes $Z(t,q)$ at its "1" ports when its "2" ports are loaded in unit inductors. Moreover, if the equations have a solution with

$$W(t) = L(t) = 0$$

the realization is lossless. □

As stated, the theorem is only a sufficient condition for obtaining a passive realization, its applicability being dependent on the existence of a solution of the four equations. Now, clearly, a necessary condition for these equations to have a solution is that $Z(t,q)$ be passive (for, otherwise, one would be able to construct a passive realization of an active n-port which is, of course, impossible). In general, no necessary condition is known without some *a priori* restrictions on the given decomposition $R(t)$ [19], [160]. In the time-invariant case, however, the statement of Theorem 9-5 is, in fact, also necessary, this result being obtained in the next section, along with the corresponding results for time-invariant scattering matrices.

Applying the theorem to the dynamical system

$$R(t) = \{0, e^{t-1}, e^{t+1}, e^t\} \tag{9-165}$$

one finds that the four equations are satisfied by

$$P(t) = e^2 \tag{9-166}$$

$$T(t) = -e \tag{9-167}$$

$$W(t) = 2e^{t/2} \tag{9-168}$$

$$L(t) = 0 \tag{9-169}$$

Hence, the transformation

$$X = -eX \tag{9-170}$$

of Equation (9-151) used to obtain the equivalent system of Equation (9-152) and the realization of Figure 9-12 does, indeed, result from the systematic process of the theorem.

Consider the impedance matrix

$$Z(t,q) = \frac{t^2}{2\delta(t-q)} + [t \quad 1]\begin{bmatrix} q \\ 1 \end{bmatrix} U(t-q) \tag{9-171}$$

Now, following Theorem 9-4, a dynamical system for this 1-port is

$$\begin{bmatrix} x_1 \\ x_2 \end{bmatrix} = \begin{bmatrix} 0 & 0 \\ 0 & 0 \end{bmatrix} \begin{bmatrix} x_1 \\ x_2 \end{bmatrix} + \begin{bmatrix} t \\ 1 \end{bmatrix} i \tag{9-172}$$

$$v = [t \quad 1]\begin{bmatrix} x_1 \\ x_2 \end{bmatrix} + \begin{bmatrix} t^2 \\ 2 \end{bmatrix} i \tag{9-173}$$

Applying the theorem in this case with

$$F(t) = 0 \tag{9-174}$$

the four equations become

$$\dot{P}(t) = -L(t)L(t)' \tag{9-175}$$

$$P(t)\begin{bmatrix} t \\ 1 \end{bmatrix} = \begin{bmatrix} t \\ 1 \end{bmatrix} - [L(t)W(t)] \tag{9-176}$$

$$W(t)'W(t) = [t^2] \tag{9-177}$$

and

$$T(t)T(t) = P(t) = P(t)' \tag{9-178}$$

Clearly, one set of solutions is

$$W(t) = [t] \tag{9-179}$$

$$L(t) = \begin{bmatrix} 0 \\ 0 \end{bmatrix} \tag{9-180}$$

$$P(t) = T(t) = \begin{bmatrix} 1 & 0 \\ 0 & 1 \end{bmatrix} \tag{9-181}$$

Since a unit $T(t)$ transformation is obtained from the four equations, the dynamical system of Equations (9-172) and (9-173) must give a passive realization without transformation. This is indeed the case, as may be discerned by writing the memoryless coupling network impedance matrix as

$$Z_m(t,q) = \begin{bmatrix} \dfrac{t^2}{2} & t & 1 \\ -t & 0 & 0 \\ -1 & 0 & 0 \end{bmatrix} \delta(t-q)$$

$$= \begin{bmatrix} \dfrac{t^2}{2} & 0 & 0 \\ 0 & 0 & 0 \\ 0 & 0 & 0 \end{bmatrix} \delta(t-q) + \begin{bmatrix} 0 & t & 0 \\ -t & 0 & 0 \\ 0 & 0 & 0 \end{bmatrix} \delta(t-q) + \begin{bmatrix} 0 & 0 & 1 \\ 0 & 0 & 0 \\ -1 & 0 & 0 \end{bmatrix} \delta(t-q) \tag{9-182}$$

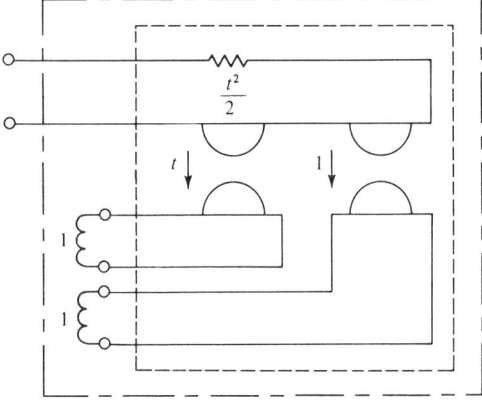

Figure 9-13 Realization of a second-order impedance matrix.

SYNTHESIS VIA REACTANCE EXTRACTION   339

realizing each of these three terms separately (as a resistor and two gyrators), and loading its last two ports with unit inductors. The resultant realization for $Z(t,q)$ is then as shown in Figure 9-13.

Although we have only considered synthesis by reactance extraction thus far, an extension of Theorem 9-5 will allow a synthesis wherein both inductors and resistors are extracted [11], [160], yielding a realization of the form shown in Figure 9-14, where the memoryless coupling network is lossless. As a pre-

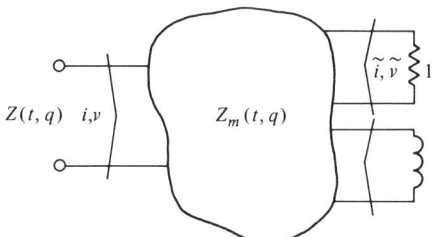

Figure 9-14 Impedance realization by extraction of resistors and reactors.

liminary to the formulation of this result, we need the following algebraic identities, which may be obtained directly from Equations (9-154) through (9-158).

*Lemma 9-6

Under the hypothesis of Theorem 9-5,

$$\boldsymbol{F}(t) + \boldsymbol{F}(t)' = \boldsymbol{L}(t)\boldsymbol{L}(t)'$$
$$\boldsymbol{G}(t) = \boldsymbol{H}(t) - \boldsymbol{L}(t)W(t)$$

where

$$\boldsymbol{L}(t) = T(t)^{-1}L(t) \qquad \square$$

The proof [160] can be obtained by applying Theorem 9-5 twice, the second time with the solution

$$P(t) = T(t) = 1 \qquad \text{(9-183)}$$

With this preliminary completed, we may now consider the problem of synthesis with both reactor and resistor extraction. Consider the dynamical system (for which the explicit dependence of the matrices on t is deleted)

$$\dot{X} = [\tfrac{1}{2}F - F']X + \left[G + \tfrac{1}{2}LW \;\Big|\; -\tfrac{1}{\sqrt{2}}L \right]\begin{bmatrix} i \\ \tilde{i} \end{bmatrix} \qquad \text{(9-184)}$$

$$\begin{bmatrix} v \\ \tilde{v} \end{bmatrix} = \begin{bmatrix} G' + \tfrac{1}{2}W'L' \\ \text{------} \\ -\tfrac{1}{\sqrt{2}}L' \end{bmatrix} X + \begin{bmatrix} \tfrac{1}{2}J + J' & \Big|\; \tfrac{1}{\sqrt{2}}W' \\ \text{------} & \text{------} \\ -\tfrac{1}{\sqrt{2}}W & \Big|\; 0 \end{bmatrix}\begin{bmatrix} i \\ \tilde{i} \end{bmatrix} \qquad \text{(9-185)}$$

where the input-output variables i and v represent the usual ports at which $Z(t,q)$ is to be realized, and the extra variables $\tilde{\imath}$ and $\tilde{v}$ represent the additional ports where the resistive load is to be applied. The effect of loading these extra ports with 1-ohm resistors is to induce the constraint

$$\tilde{\imath} = -\tilde{v} \tag{9-186}$$

into the system. Now, if one combines this constraint with the dynamical system equations and the equalities of Lemma 9-6, it may be shown [160] that the original dynamical system is realized at the v-i ports. Upon observing that the memoryless coupling network associated with this system is skew symmetric and, hence, lossless, the desired realization of Figure 9-14 is found.

*Proposition 9-10

Under the hypothesis of Theorem 9-5 $Z(t,q)$ can be realized at the "1" ports of the lossless memoryless coupling network characterized by the impedance matrix

$$Z_m(t,q) = \begin{bmatrix} \frac{1}{2}J - J' & \vline & \frac{1}{\sqrt{2}}W' & \vline & -G' - \frac{1}{2}W'L' \\ \hline -\frac{1}{\sqrt{2}}W & \vline & 0 & \vline & \frac{1}{\sqrt{2}}L' \\ \hline G + \frac{1}{2}LW & \vline & -\frac{1}{\sqrt{2}}L & \vline & -\frac{1}{2}F - F' \end{bmatrix}$$

when the "2" ports are loaded with 1-ohm resistors and the "3" ports with unit inductors. □

Applying the technique of Proposition 9-10 to the dynamical system of Equations (9-172) and (9-173), for which

$$\boldsymbol{F}(t) = \begin{bmatrix} 0 & 0 \\ 0 & 0 \end{bmatrix} \tag{9-187}$$

$$\boldsymbol{J}(t) = \begin{bmatrix} t^2 \\ 2 \end{bmatrix} \tag{9-188}$$

$$\boldsymbol{G}(t) = \boldsymbol{H}(t) = \begin{bmatrix} t \\ 1 \end{bmatrix} \tag{9-189}$$

$$\boldsymbol{W}(t) = \begin{bmatrix} t \\ 2 \end{bmatrix} \tag{9-190}$$

and

$$\boldsymbol{L}(t) = \begin{bmatrix} 0 \\ 0 \end{bmatrix} \tag{9-191}$$

yields the memoryless coupling network

$$Z_m(t,q) = \begin{bmatrix} 0 & \dfrac{t}{\sqrt{2}} & -t & -1 \\ \dfrac{-t}{\sqrt{2}} & 0 & 0 & 0 \\ t & 0 & 0 & 0 \\ 1 & 0 & 0 & 0 \end{bmatrix} \delta(t-q) \qquad \textbf{(9-192)}$$

and the realization for the given 2-port [of Equation (9-171)] shown in Figure 9-15.

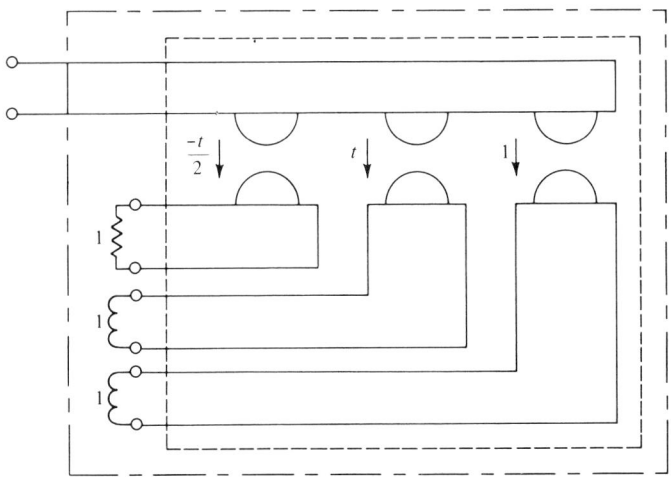

Figure 9-15 Alternate impedance realization.

Clearly, the syntheses of Theorem 9-5 and Proposition 9-10 are powerful when applicable, since they allow one to realize a network entirely by manipulating the memoryless portion, this rendering the result operationally sound, since one does not have to deal (directly) with either distributions or complex functions. Moreover, a rather slight modification of the technique will allow one to obtain a realization which simultaneously minimizes the number of reactors and resistors used. The disadvantage of the technique is that the equations of Theorem 9-5 are not assured to have solutions for all passive impedance matrices. In particular, the technique is not applicable to those networks which have irregular differential operators and, hence, no dynamical system representation. However, even given regularity, and hence the existence of at least one dynamical system representation, the equations of Theorem 9-5 may still not be solvable for all passive n-ports. Under such circumstances, one would like to approach the necessity proof assuming the existence of a passive RLC realization, extracting the reactors and constructing the passive dynamical system decomposition by analyzing the memoryless coupling network, as per Proposition 9-5. Unfortunately, there is no assurance that the memoryless coupling network left after the reactance extraction will have an impedance matrix (even if $Z(t,q)$ is known to exist). Hence, necessity cannot be derived without some additional assumptions. A natural means of circumventing this difficulty would be to deal with scattering rather than impedance matrices (in which case, the passive, linear, normal, memoryless coupling network is assured to possess a scattering matrix). In this case, however, the scattering analog of the Ricatti equation is not (as yet) known. Alternatively, an admittance approach can be taken wherein one uses capacitive loading rather than inductive loading. This follows readily from an argument dual to that of Theorem 9-5 and/or Proposition 9-10 (which the reader is invited to derive for himself), but suffers from the same difficulties as the impedance synthesis.

In the active case of Proposition 9-9, either an impedance or scattering approach may be successfully undertaken. In this case, one simply interprets the dynamical system

$$\dot{X} = F(t)X + G(t)u \qquad (9\text{-}193)$$

$$y = H(t)' + J(t)u \qquad (9\text{-}194)$$

as the hybrid matrix

$$H(t,q) = \begin{bmatrix} J(t) & H(t)' \\ \hline -G(t) & -F(t) \end{bmatrix} \delta(t-q) \qquad (9\text{-}195)$$

of the coupling network with the "1" ports characterized by the y and u variables (say, scattering or admittance) and the "2" ports by the impedance variables.

9-6 State-Space Passivity Conditions

In this section the dynamical system techniques of this chapter are restricted to the time-invariant case, thereby obtaining the so-called state-space theory of networks [17], [117], [182]. Our main results are algebraic conditions, such as were used in Theorem 9-5 of the preceding section, for the passivity of a time-invariant system function in terms of the properties of a minimal state decomposition. In particular, we show that in the time-invariant case, the existence of solutions for the four equations of Theorem 9-5 is necessary, as well as sufficient, condition for the passivity of an n-port. In addition, the corresponding result for time-invariant n-ports characterized by their scattering variables is formulated.

To verify the necessity of the conditions of the preceding section, we will work with a passive impedance (or admittance) having the decomposition

$$R = \{F,G,H',J\} \qquad (9\text{-}196)$$

Now, since it is passive, the even part of $Z(p)$ has the factorization (Theorem 6-9)

$$Z(p) + Z(p)_* = W(p)_*W(p) \qquad (9\text{-}197)$$

where $W(p)$ is analytic in the RHP and has a right inverse which is also analytic in the RHP. Following Anderson's formulation [13], we will attempt to construct a state decomposition of $W(p)$,

$$S = \{A,B,L',W\} \qquad (9\text{-}198)$$

for which L and W are the required solutions to the four equations of Theorem 9-5 (P and T following canonically once L and W are obtained). Of course, this is not true of any decomposition of $W(p)$; but we will show that $W(p)$ has a decomposition for which this is the case. In fact, the required decomposition has the form

$$S = \{F,G,L',W\} \qquad (9\text{-}199)$$

where F and G are the same matrices used in the decomposition of $Z(p)$ [12], [85].

We will make the simplifying assumption that $Z(p)$ has no $j\omega$ axis poles. Since any such poles can be extracted from $Z(p)$ by the methods of Chapter 6, before beginning synthesis this may be done without loss of generality [17]. It does, however, justify the following lemma, which is of technical value in the derivation.

Lemma 9-7

Let $Z(p)$ be PR and have no $j\omega$ axis poles, and let

$$R = \{F,G,H',J\}$$

be a minimal decomposition of $Z(p)$. Then the eigenvalues of F are in the LHP.

Proof

Since R is a decomposition of $Z(p)$,

$$Z(p) = J + H'(p1 - F)^{-1}G \qquad (9\text{-}200)$$

Hence, every pole of $Z(p)$ is an eigenvalue of F [36], [132]. Now, by Lemma 9-5, the dimension of F, and hence the number of its eigenvalues, is equal to the degree of $Z(p)$, showing that every eigenvalue must also be a pole. The eigenvalues of F and poles of $Z(p)$ thus coincide, and since the latter are in the LHP (since $Z(p)$ is PR and has no $j\omega$ axis poles), so are the former. □

Continuing with a series of lemmas preliminary to our main development, we have the following.

Lemma 9-8

Let

$$S = \{A, B, L', W\}$$

be a minimal decomposition of $W(p)$. Then the equation

$$PA + A'P = -LL'$$

has a unique positive definite symmetric solution P.

Proof

Let

$$\psi(t) = e^{At} \qquad (9\text{-}201)$$

be the fundamental solution matrix for A, and pre- and post-multiple both sides of the equation by $\psi(t)'$ and $\psi(t)$, respectively, yielding

$$\psi(t)'PA\psi(t) + \psi(t)'A'P\psi(t) = -\psi(t)'LL'\psi(t) \qquad (9\text{-}202)$$

Now, by the definition of $\psi(t)$,

$$\dot\psi(t) = A\psi(t) \qquad (9\text{-}203)$$

the left side of Equation (9-202) becomes

$$[\psi'(t)\dot P\psi(t)] = \psi'(t)P\dot\psi(t) + \dot\psi'(t)P\psi(t) \qquad (9\text{-}204)$$

and we have

$$[\psi(t)'\dot P\psi(t)] = -\psi(t)'LL'\psi(t) \qquad (9\text{-}205)$$

Finally, integrating both sides of Equation (9-205) yields

$$e^{A't}Pe^{At} - P = \psi(t)'P\psi(t) - P = \psi(q)'P\psi(q)\Big|_0^t$$

$$= -\int_0^t \psi(q)'LL'\psi(q)dq = -\int_0^t e^{A'q}LL'e^{Aq}dq \qquad (9\text{-}206)$$

which, upon taking the limit as t goes to infinity, leaves

$$P = \int_0^\infty e^{A'q}LL'e^{Aq}dq \qquad (9\text{-}207)$$

Here the $e^{A't}Pe^{At}$ term drops out in the limiting process, since the eigenvalues of A are in the LHP [36]. P is clearly positive and symmetric. The fact that it is also positive, definite, and unique results from the minimality of the decomposition. The verification of these facts [87] is left to the reader as an exercise. □

An immediate corollary of Lemma 9-8 is a "matrix partial fraction expansion," as follows.

Corollary 9-4

Let
$$S = \{A, B, L', W\}$$
be a minimal decomposition of $W(p)$. Then
$$W(p)_*W(p) = W'W + [W'L' + B'P](p1 - A)^{-1}B$$
$$- B'(p1 + A')^{-1}[LW + PB]$$
where P is a solution to the equation
$$PA + A'P = -LL'$$

Proof

Since S is a decomposition of $W(p)$,
$$W(p) = W + L'(p1 - A)^{-1}B \qquad (9\text{-}208)$$
while
$$W(p)_* = W(-p)' = W' - B'(p1 + A')^{-1}L \qquad (9\text{-}209)$$
which, upon multiplication, yields
$$W(p)_*W(p) = W'W + W'L'(p1 - A)^{-1}B - B'(p1 + A')^{-1}LW$$
$$- B'(1p + A')^{-1}LL'(p1 - A)^{-1}B \qquad (9\text{-}210)$$
Now, if
$$PA + A'P = -LL' \qquad (9\text{-}211)$$
$$(1p + A')^{-1}P - P(1p - A)^{-1}$$
$$= (1p + A')^{-1}P(1p - A) - (1p + A')P(1p - A)^{-1}$$
$$= (1p + A')^{-1}[-PA - A'P](1p - A)^{-1} \qquad (9\text{-}212)$$
$$= (1p + A')^{-1}LL'(1p - A)^{-1}$$
which, upon substitution into Equation (9-210) and rearrangement, yields the required equality. □

Finally, we have still another lemma before proceeding to our main theorems.

Lemma 9-9

If
$$T\begin{bmatrix} F & 0 \\ \hline 0 & -F' \end{bmatrix} = \begin{bmatrix} F & 0 \\ \hline 0 & -F' \end{bmatrix}T$$
then T has the form
$$T = \begin{bmatrix} T_1 & 0 \\ \hline 0 & T_2 \end{bmatrix}$$

Proof

Assume T is arbitrary, having the form

$$T = \begin{bmatrix} T_1 & | & T_{12} \\ \hline T_{21} & | & T_2 \end{bmatrix} \tag{9-213}$$

Then the equality of the hypothesis yields the four equations

$$FT_1 = T_1 F \tag{9-214}$$

$$F'T_2 = T_2 F' \tag{9-215}$$

$$FT_{12} + T_{12}F' = 0 \tag{9-216}$$

$$T_{21}F + F'T_{21} = 0 \tag{9-217}$$

Now, Equations (9-216) and (9-217) are in the form of the equality of Lemma 9-8, with $L = 0$; hence, from the proof of that lemma, it follows that their unique solutions T_{12} and T_{21} are zero, verifying the lemma. □

With these preliminaries, we are now ready to formulate our first theorem of this section, to the effect that if $Z(p)$ has a state decomposition

$$R = \{F, G, H', J\} \tag{9-218}$$

then $W(p)$ has a state decomposition

$$S = \{F, G, L'W\} \tag{9-219}$$

This is done in two steps, the first of which is used to show that $W(p)$ has a decomposition of the form

$$\bar{S} = \{F, \bar{B}, \bar{L}', W\} \tag{9-220}$$

and the second that it has a decomposition of the form of Equation (9-219).

Writing $Z(p)$ in terms of the decomposition of Equation (9-218) yields

$$Z(p) = J + H'(1p - F)^{-1}G \tag{9-221}$$

and

$$Z(p)_* + Z(p) = J' + J + G'(1p + F')^{-1}H + H'(1p - F)^{-1}G \tag{9-222}$$

from which the decomposition

$$U = \left\{ \begin{bmatrix} F & | & 0 \\ \hline 0 & | & -F' \end{bmatrix}, \begin{bmatrix} G \\ H \end{bmatrix} [H' \mid G'], J' + J \right\} \tag{9-223}$$

for $Z(p)_* + Z(p)$ is obtained. Similarly, from Corollary 9-4 it follows that

$$V = \left\{ \begin{bmatrix} A & | & 0 \\ \hline 0 & | & -A' \end{bmatrix}, \begin{bmatrix} B \\ LW + BP \end{bmatrix}, [W'L' + PB' \mid -B'], W'W \right\} \tag{9-224}$$

is a decomposition of $W(p)$. Now, since by definition

$$Z(p)_* + Z(p) = W(p)_* W(p) \tag{9-225}$$

V and U are decompositions of the same matrix. Moreover, since F is minimal and has its eigenvalues in the LHP, U and V are also minimal. By the equivalence theorem for decompositions (Theorem 8-2), we are therefore assured of the existence of (constant) matrix M such that

$$M \begin{bmatrix} A & 0 \\ \hline 0 & -A' \end{bmatrix} M^{-1} = \begin{bmatrix} F & 0 \\ \hline 0 & -F' \end{bmatrix} \tag{9-226}$$

Hence, M induces a matrix M_1 such that

$$M_1 A M_1^{-1} = F \tag{9-227}$$

Now, if this M_1 is used to transform the state vector of the decomposition S, an equivalent decomposition

$$\tilde{S} = \{F, \bar{B}, \bar{L}, W\} \tag{9-228}$$

of $W(p)$ is obtained. Applying the same argument once again, but using the decomposition $\tilde{S}$ rather than S to characterize $W(p)$, yields the decomposition

$$\bar{V} = \left\{ \begin{bmatrix} F & 0 \\ \hline 0 & -F' \end{bmatrix}, \begin{bmatrix} \bar{B} \\ \hline \bar{L}W + \overline{PB} \end{bmatrix}, [W'\bar{L} + \overline{PB}' \mid \bar{B}'], W'W \right\} \tag{9-229}$$

for $W(p)_*W(p)$. Once again, U and $\bar{V}$ are equivalent minimal decompositions of the same matrix; hence, there exists a matrix T such that

$$T \begin{bmatrix} F & 0 \\ \hline 0 & -F' \end{bmatrix} T^{-1} = \begin{bmatrix} F & 0 \\ \hline 0 & -F' \end{bmatrix} \tag{9-230}$$

$$T \begin{bmatrix} B \\ \hline \bar{L}W + \overline{BP} \end{bmatrix} = \begin{bmatrix} G \\ \hline H \end{bmatrix} \tag{9-231}$$

and

$$[W'\bar{L} + \overline{PB}' \mid \bar{B}']T^{-1} [H' \mid G'] \tag{9-232}$$

Now, by Lemma 9-9, together with Equation (9-230), T has the form

$$T = \begin{bmatrix} T_1 & 0 \\ \hline 0 & T_2 \end{bmatrix} \tag{9-233}$$

where, via Equation (9-230),

$$T_1 F T_1^{-1} = F \tag{9-234}$$

and, via Equation (9-231),

$$T_1 \bar{B} = G \tag{9-235}$$

If one now uses the matrix T_1 to transform the state vector of $\tilde{S}$, our final equivalent decomposition

$$\mathbf{S} = \{F, G, L'W\} \tag{9-236}$$

is obtained. We have therefore proven the following theorem.

Theorem 9-6

Let $Z(p)$ be a PR matrix with no $j\omega$ axis poles and state decomposition

$$R = \{F,G,H',J\}$$

Then

$$Z(p)_* + Z(p) = W(p)_*W(p)$$

where $W(p)$ has state decomposition

$$S = \{F,G,L',W\} \qquad \square$$

Note that in the theorem we have dropped the boldface italic notation used in the derivation.

Finally, we have the main result of this section, the "PR lemma" [13], [85].

Theorem 9-7

Let $Z(p)$ be a real rational matrix with all poles in the LHP and minimal state decomposition

$$R = \{F,G,H',J\}$$

Then $Z(p)$ is PR if and only if there exist matrices P, L, W, and T such that the equations

$$PF + F'P = -LL'$$

$$PG = H - LW'$$

$$W'W = J' + J$$

and

$$TT = P' = P$$

are satisfied. Moreover, if these matrices exist, $Z(p)$ may be synthesized by the methods of Theorem 9-5.

Proof

As far as the sufficiency is concerned, the theorem is just a special case of Theorem 9-5, where the $\dot{P}$ term drops out since we are dealing with constant matrices. To verify the necessity of the theorem, let

$$Z(p)_* + Z(p) = W(p)_*W(p) \qquad (9\text{-}237)$$

and let

$$S = \{F,G,L',W\} \qquad (9\text{-}238)$$

be the decomposition of $W(p)$ of Theorem 9-6, and let P be the unique positive definite symmetric solution of the equation

$$PF + F'P = LL' \qquad (9\text{-}239)$$

which exists by Lemma 9-8. We will now show that the equalities of the theorem are satisfied with this choice of L, W, and P, which are well defined when $Z(p)$ is PR.

Since R and S are decompositions of $Z(p)$ and $W(p)$, respectively, we may write

$$Z(p) = J + H'(1p - F)^{-1}G \tag{9-240}$$

and

$$W(p) = W + L'(1p - F)^{-1}G \tag{9-241}$$

which, upon substitution into Equation (9-237) and invoking Corollary 9-4, yields the equality

$$J' + J - G'(1p + F')^{-1}H + H'(1p - F)^{-1}G$$
$$= W'W + [W'L' + G'P](p1 - F)^{-1}G - G'(1p + F')^{-1}[LW + PG] \tag{9-242}$$

where P is defined by Equation (9-239). Now, evaluating Equation (9-242) as p goes to infinity yields the equality

$$J'' + J = W'W \tag{9-243}$$

while equating coefficients of the $(1p - F)^{-1}$ or $(1p + F')^{-1}$ terms yields the equality

$$PG = H - LW \tag{9-244}$$

Finally, if one lets T be the positive symmetric square root of P [132], the equality

$$TT = P = P' \tag{9-245}$$

is obtained. Thus, for PR $Z(p)$, there do indeed exist matrices P, L, W, and T satisfying the four equations, as required. □

Consider the impedance matrix

$$Z(p) = \begin{bmatrix} \dfrac{k}{p+1} & -1 \\ 1 & 0 \end{bmatrix} \tag{9-246}$$

where k is an arbitrary parameter. Now, $Z(p)$ has the decomposition

$$R = \left\{ [-1], [1 \ \ 0], \begin{bmatrix} k \\ 0 \end{bmatrix}, \begin{bmatrix} 0 & -1 \\ 1 & 0 \end{bmatrix} \right\} \tag{9-247}$$

For this realization the four equations of Theorem 9-7 are

$$-2P = -LL' \tag{9-248}$$

$$P[1 \ \ 0] = [k \ \ 0] - LW \tag{9-249}$$

$$W'W = \begin{bmatrix} 0 & 0 \\ 0 & 0 \end{bmatrix} \tag{9-250}$$

and

$$TT = P = P' \tag{9-251}$$

Hence,
$$W = [0 \ \ 0] \quad (9\text{-}252)$$
$$P = k \quad (9\text{-}253)$$
$$T = \sqrt{k} \quad (9\text{-}254)$$
and
$$L = \sqrt{2k} \quad (9\text{-}255)$$

is a solution if k is positive, while the equations have no solution for k strictly negative. $Z(p)$ is, therefore, PR if and only if k is positive.

Unlike the time-variable case, where no scattering analog of Theorem 9-5 is known, in the time-invariant case we may characterize the BR rational matrices as follows.

*Theorem 9-8

Let $S(p)$ be a real rational matrix with minimal state decomposition
$$R = \{F, G, H', J\}$$
Then $S(p)$ is BR if and only if there exists matrices P, L, W, and T such that the equations
$$PF + F'P = -HH' - LL'$$
$$PG = HJ + LW$$
$$1 - J'J = W'W$$
and
$$TT = P = P'$$
are satisfied. $\square$

The necessity proof [15], [182] for this theorem results from an argument essentially parallel to that for the PR theorem, and is left to the reader as an exercise. The sufficiency proof results from the construction of Theorem 9-10.

An additional application of the techniques of this section is to construct the spectral factorization of a matrix $R(p)$ in terms of "its" decomposition. Theorem 9-9 follows readily upon invoking Theorem 9-6 [20].

*Theorem 9-9

Let $R(p)$ be a real rational matrix with no $j\omega$ axis poles, such that
$$R(p)_* = R(p)$$
and
$$R(j\omega) \geq 0 \quad \text{for all } j\omega$$
Let
$$Z(p)_* + Z(p) = R(p)$$

where $Z(p)$ corresponds to the LHP poles of $R(p)$ and $Z(p)_*$ corresponds to the RHP poles of $R(p)$, and let

$$R = \{F, G, H', J\}$$

be a decomposition of $Z(p)$. Then

$$R(p) = W(p)_* W(p)$$

where

$$W(p) = W + L'(1p - F)^{-1}G$$

with W and L the matrices of Theorem 9-6. □

A proof of this theorem is given in reference [20], where it is also shown that $W(p)$ has the appropriate analyticity properties, and the effects of $j\omega$ axis poles are considered.

Finally, we sketch a sufficiency proof for Theorem 9-8 by constructing a passive realization for $S(p)$ whose decomposition satisfies the condition of the theorem. As in the multivariable case, one cannot carry out a scattering synthesis via reactance extraction directly in the p domain, since the extracted components would not correspond to inductors. Rather, we make a change of variable,

$$\tilde{p} = \frac{p+1}{p-1} \quad (9\text{-}256)$$

thereby converting the scattering matrix $S(p)$ to

$$\tilde{S}(\tilde{p}) = S(p)\Big|_{p=\frac{\tilde{p}+1}{\tilde{p}-1}} \quad (9\text{-}257)$$

whereupon extraction of the $1/\tilde{p}$ devices is equivalent to the extraction of p domain inductors. Of course, in the $\tilde{p}$ domain the BR conditions are translated into:

1. $\tilde{S}(\tilde{p})$ is real rational and analytic in the region $|\tilde{p}| \geq 1$
2. $1 - \tilde{S}(1/\tilde{p})\tilde{S}(\tilde{p}) \geq 0$ in the region $|\tilde{p}| \geq 1$.

Hence, theorem 9-8 is not directly applicable to a decomposition of $\tilde{S}(\tilde{p})$. We can, however, translate the condition of Theorem 9-8 to an equivalent condition on a decomposition of $\tilde{S}(\tilde{p})$. To this end, we let

$$\tilde{R} = \{\tilde{F}, \tilde{G}, \tilde{H}', \tilde{J}\} \quad (9\text{-}258)$$

be a decomposition of $\tilde{S}(\tilde{p})$ such that

$$\tilde{S}(\tilde{p}) = \tilde{J} + \tilde{H}'(1\tilde{p} - \tilde{F})^{-1}\tilde{G} \quad (9\text{-}259)$$

Now, substitution of Equation (9-256) into Equation (9-259) yields

$$S(p) = \tilde{S}\left(\frac{p+1}{p-1}\right) = \tilde{J} + \tilde{H}'\left[1\left(\frac{p+1}{p-1}\right) - \tilde{F}\right]^{-1}\tilde{G} \quad (9\text{-}260)$$

$$= J + H'(1p - F)^{-1}G$$

where
$$F = (\tilde{F} - 1)^{-1}(\tilde{F} + 1) \tag{9-261}$$
$$G = -2(\tilde{F} - 1)^{-2}\tilde{G} \tag{9-262}$$
$$H = \tilde{H} \tag{9-263}$$
$$J = \tilde{J} - \tilde{H}'(\tilde{F} - 1)^{-1}\tilde{G} \tag{9-264}$$

The details of this calculation appear in reference [182], and will not be given here. If $S(p)$ is BR, the four equalities of Theorem 9-8 hold for
$$R = \{F,G,H',J\} \tag{9-265}$$
Hence, substitution of Equations (9-261) through (9-264) into these equalities and simplifying [182] yields the four equalities of the following theorem. Here, if $S(p)$ is BR, $\tilde{S}(\tilde{p})$ is analytic on the unit circle, which in turn assures that $(\tilde{F} - 1)$ is invertible, as required for the derivation [132].

Theorem 9-10
Let
$$\tilde{S}(\tilde{p}) = S(p)\bigg|_{p=\frac{\tilde{p}+1}{\tilde{p}-1}}$$
have the decomposition
$$\tilde{R} = \{\tilde{F},\tilde{G},\tilde{H}',\tilde{J}\}$$
Then $S(p)$ is BR if and only if there exists matrices Q, L, W, and T such that the equations
$$\tilde{F}'Q\tilde{F} - Q = -\tilde{H}\tilde{H}' - LL'$$
$$-\tilde{F}'Q\tilde{G} = \tilde{H}\tilde{J} + LW$$
$$1 - \tilde{J}'\tilde{J} = W'W + \tilde{G}'Q\tilde{G}$$
$$TT = Q = Q'$$
are satisfied.

Proof

The necessity of the theorem follows from Theorem 9-8 and the argument indicated above [182]. For sufficiency, we use T to transform the state vector to the equivalent decomposition
$$\tilde{R} = \{TFT^{-1}, TG, H'T^{-1}, J\} = \{\tilde{F},\tilde{G},\tilde{H}',\tilde{J}\} \tag{9-266}$$
Now, defining a memoryless coupling network via the scattering matrix
$$\Sigma = \begin{bmatrix} \tilde{J} & \tilde{H}' \\ \hline \tilde{G} & \tilde{F} \end{bmatrix} \tag{9-267}$$

one may realize $S(p)$ as a realization of Σ loaded in inductors at its "2" ports, as shown in Figure 9-16. To verify that Σ is indeed passive, we may compute

$$1 - \Sigma'\Sigma = \begin{bmatrix} W' \\ -- \\ L \end{bmatrix} [L \mid W'] \tag{9-268}$$

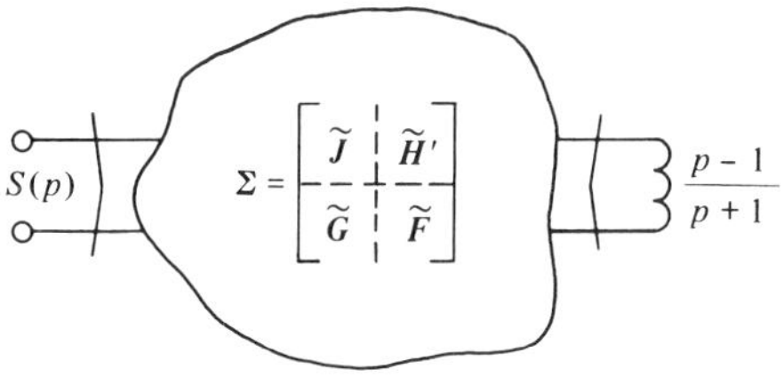

Figure 9-16 Realization of a scattering matrix via reactance extraction.

where the last equality results from the substitution of the equalities of the theorem [182]. Since this is positive, being the product of a matrix and its transpose, Σ is indeed passive, and the proof is complete. □

Note that the above synthesis is both minimal reactor and minimal resistor, the former resulting from the fact that $\tilde{F}$ is minimal, and the latter from the fact that the rank of $1 - \Sigma'\Sigma$ is the number of rows in $W(p)$, which is, in turn, known to be the minimal number of resistors with which $S(p)$ may be realized (Corollary 6-9).

It is noteworthy that the technique of Theorem 9-10 can be modified to also obtain a reciprocal synthesis of a reciprocal scattering matrix [182], [199]. This requires, essentially, some additional algebraic manipulation [182] once a solution to the equations of Theorem 9-10 has been found. The details, however, will not be considered here.

9-7 Discussion

Although the chapter began by making the seemingly modest modification of the RLC networks of allowing time-variable gyrators, the implications of this modification have been far-reaching indeed. Although a considerable theory of time-variable RLC n-ports exists, it is clearly incomplete. In particular, we have exhibited no synthesis applicable to all finite time-variable RLC n-ports. In the reactance extraction approach, it is not known whether or not the appropriate nonlinear equations have solutions. A second area where the theory falls short is the case of an n-port characterized by an irregular differential operator. For these networks, neither the dynamical system nor the separable impulse is a valid characterization; hence, the theory of this chapter is not, in general, applicable to such n-ports. In fact, virtually nothing is known about this class of "pathological" networks.

An alternative approach to the theory of finite time-variable n-ports, which has the potential of including the irregular case, is to employ the differential operator characterizing the n-port directly, invoking neither the dynamical system nor the impulse response. Although a number of results have been obtained by such an approach [145], [147] the corresponding theory is not (yet) as fully developed as the impulse response and dynamical system approaches taken here.

APPENDIX

A

Infinite-Dimensional Vector Spaces and Their Operators

A-1 Introduction

Among the various mathematical tools amenable to engineering application, the most powerful is possibly the vector space. These, in their commonly encountered finite-dimensional form, serve to characterize multiple input–multiple output systems over both real and binary fields, as well as yielding the foundation for matrix algebra [74]. For the purpose of this work, however, one is concerned with vector spaces of continuous functions. Since these are, in general, infinite dimensional, the convenient (though unnecessary) techniques of counting basis vectors and representing linear operators by matrices are not applicable. Fortunately, most of the basic results associated with finite-dimensional spaces also hold in the general case. In fact, a study of infinite-dimensional vector spaces is often more enlightening than one of the finite-dimensional special case. This is because many useful, but unnecessary, details of the finite case would otherwise obscure the fundamental ideas.

The development is divided into four sections, dealing with vector spaces, operators, norms and inner products, and topological vector spaces, respectively. Since it is assumed that the reader is familiar with the (finite-dimensional) vector space concept, we simply outline the main results, leaving it to the reader to fill in the details and prove the theorems.

A-2 Vector Spaces

Definition A-1

A (real) *vector space* is a set of objects, called *vectors*, satisfying the following conditions.

1. For every pair of vectors x and y their *sum*, $x + y$, is defined and satisfies:

    a. $x + y = y + x$

    b. $x + (y + z) = (x + y) + z$

    c. there exists a unique zero vector, denoted by 0 such that $x + 0 = x$ for all vectors x

    d. for each vector x there exists a unique vector $-x$ such that $x + (-x) = 0$.

2. For each (real) scaler a and vector x there exists a vector ax, the *scaler product* of a and x, which satisfies the following conditions:

    a. $a(bx) = (ab)x$

    b. $1x = x$

    c. $a(x + y) = ax + ay$

    d. $(a + b)x = ax + bx$. □

Examples of vector spaces abound. The real numbers themselves form a vector space, as do the complexes. The set of nth-order polynomials is a vector space P_n. Here, vector addition is component-wise addition of the polynomials,

$$\left(\sum_{i=0}^{n} a_i x^i\right) + \left(\sum_{i=0}^{n} b_i x^i\right) = \left(\sum_{i=0}^{n} (a_i + b_i) x^i\right) \tag{A-1}$$

while scaler multiplication is given by

$$c\left(\sum_{i=0}^{n} a_i x^i\right) = \left(\sum_{i=0}^{n} (ca_i) x^i\right) \tag{A-2}$$

More generally, the set of all polynomials is a vector space P, where n in Equations (A-1) and (A-2) is replaced by ∞, and the additional condition that all but a finite number of the coefficients a_i be zero is added. This latter condition is needed to assure that the summations are polynomials rather than infinite series, for which the existence of the sum is not assured. The condition that the coefficients be zero "almost everywhere" is, in fact, a rather common condition in infinite-dimensional spaces, which distinguishes them from finite spaces, resulting from the fact that the definition of the vector space only assures the existence of finite sums. If infinite sums are to be allowed, one must add some additional structure, such as a norm or inner product, to the vector space. For the purpose of this work, the most important vector spaces are the "function spaces." These include the spaces of all continuous functions, C^1, all smooth functions, C^∞, smooth functions with support bounded on the left, D_+, and smooth functions with bounded support, D. Here D_+ is the space of functions which are zero before some arbitrary (but finite) time, whereas D is a space of functions which are zero both before and after some finite time, these being illustrated in Figure A-1.

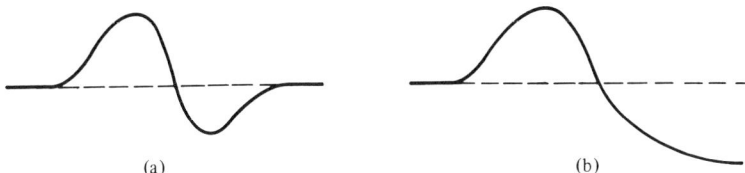

Figure A-1 Typical function from (a) D and (b) D_+.

These two classes of functions are of primary importance in the development of our theory. D_+ serves as the input-output space for the n-ports and networks, whereas D is the space underlying the theory of distributions formulated in Appendix B. As an exercise, the reader may verify that the polynomial and function spaces listed above are indeed vector spaces.

*Proposition A-1

The function spaces P_n, P, C^1, C^∞, D_+, and D are vector spaces. □

Subspaces prove to be a significant tool in the study of vector spaces.

*Definition A-2

If V is a vector space, then a subset of its vectors which is itself a vector space (under the inherited operations) is a *subspace*. □

To be a subspace, we require more than a simple subset of the vectors, for if the subset is to be a vector space in its own right, then the (sum and) difference of any two vectors in the subset must also be in the subset. Hence, P_n is a subspace of P and D is a subspace of D_+, but the positive numbers are not a subspace of the reals (since axiom 1b of Definition A-1 for the vector space is not satisfied).

In the study of finite-dimensional vector spaces, linearly independent sets of vectors and bases play a significant role. In the general case, the usefulness of these concepts is somewhat restricted by our inability to count them; however, they do remain useful and are defined as follows.

*Definition A-3

Let V be a vector space and

$$\{x_1, x_2, \cdots, x_n\}$$

be a finite set of vectors in V. Then they are *linearly independent* if whenever

$$c_1 x_1 + c_2 x_2 + \cdots + c_n x_n = 0$$

for scalers

$$c_i, \quad i = 1, 2, \cdots, n$$

then

$$c_i = 0, \quad i = 1, 2, \cdots, n \qquad \square$$

*Definition A-4

Let V be a vector space and X be a (not necessarily finite) set of vectors of V. Then X is *linearly independent* if every finite subset of X is linearly independent (as per Definition A-3). $\qquad \square$

For the polynomials P, the vectors

$$\{x^i \quad i = 0, 1, \cdots, n\}$$

are linearly independent for any finite integer n; hence,

$$\{x^i \quad i = 0, 1, 2, \cdots\}$$

is a linearly independent set of vectors in P.

A second tool needed in defining a base is the concept of the space spanned by a set of vectors.

*Proposition A-2

Let V be a vector space and X be a (not necessarily finite) set of vectors from V. Then the set of vectors which can be written as

$$\sum_{x_i \text{ in } X} (a_i x_i)$$

where $a_i = 0$ for all but a finite number of indices is a subspace of V. $\qquad \square$

*Definition A-5

The subspace of Proposition A-2 is the *subspace spanned by* X and is denoted by $S(X)$. $\qquad \square$

In Proposition A-2 we have once again encountered the significant distinction between finite- and infinite-dimensional spaces, that is, that the infinite sum may only have a finite number of nonzero terms. Of course, in a vector space an infinite sum (of nonzero terms) may not even be defined. This requirement that the coefficients be zero "almost everywhere" implies that the space spanned by

$$X = \{x^i \quad i = 0, 1, \cdots\} \qquad \text{(A-3)}$$

is the space of all polynomials P, and not the space of series (since they may have infinitely many nonzero coefficients).

*Definition A-6

Let V be a vector space and X a subset of V. Then X is a *basis* for V if X is linearly independent and $S(X) = V$ (that is, X spans V). □

The following theorem on bases, the proof of which relies on Zorn's lemma [94], a rather deep and not universally accepted mathematical axiom, assures the existence of at least one basis for every vector space.

*Theorem A-1

Every vector space has a basis. Furthermore, any two bases for the same vector space have the same cardinality (that is, number of elements). □

See Lang [94] for a proof.

Theorem A-1 justifies the definition for the dimension of a vector space.

*Definition A-7

Let V be a vector space and X be a basis for V. Then the *dimension* of V is the cardinality of X. □

Dimension is well defined, since V is assured to have at least one basis, and the definition is independent of the particular choice of basis if more than one exists.

The dimension of P_n is $n + 1$, since one of its bases is

$$X = \{x^i \quad i = 0, 1, \cdots, n\} \tag{A-4}$$

whereas the dimension of P is $\aleph_0$ (countably infinite) since

$$X = \{x^i \quad i = 0, 1, \cdots\} \tag{A-5}$$

is one of its bases. All of the function spaces thus far considered are infinite dimensional; hence, it is this class of vector spaces that should receive special attention.

A-3 Operators

In dealing with finite-dimensional spaces, the linear operators are often confused with their matrix representation. In essence, the former is a symbol for a specified correspondence between members of a vector space, whereas the latter is a means of calculating the "output" vector which is induced by a given "input." In the case of an infinite-dimensional space, it is always possible to define a correspondence between vectors, an operator in the symbolic sense; but such a correspondence may not have a representation, such as the matrix, from which one can calculate the output engendered by a given input. Of course, many infinite-dimensional spaces do have such representations, two of which, for the space D_+, do, in fact, make up the main body of Chapters 2 and 3.

With the study of these specific representations being left to these chapters, in this section we take up the study of operators in their purely symbolic sense. This approach is consistent with our desire to obtain maximum generality with a "minimum" of work, since all linear networks may be symbolically represented by an operator, while the various representations are applicable only to certain special cases.

It is possibly surprising that a great deal of network theory may be developed entirely by manipulating these operator symbols without ever resorting to an operator representation, which allows one to actually calculate the vector correspondence indicated by an operator [148], [149]. In fact, the entire theory developed in Chapters 4 and 5 is obtained by such symbol manipulation and is, for this reason, applicable to all linear networks.

In the case of finite-dimensional spaces, a linear operator may be represented as a matrix (relative to some choice of basis). In the general case, operators can be defined in an analogous manner and with similar properties, but the convenient (if unnecessary) matrix representation is not applicable.

*Definition A-8

Let V and W be vector spaces and T be a function from V to W,

$$T : V \to W$$

such that

$$T(ax + by) = aT(x) + bT(y)$$

for all x and y in V and real a and b. Then T is a *homomorphism* (linear operator) from V into W. If for each x in W there exists a y in V such that

$$T(y) = x$$

then T is an *epimorphism* (onto). If for each x and y in V with

$$x \neq y$$

$$T(x) \neq T(y)$$

then T is a *monomorphism* (one-to-one). If T is both an epimorphism and a monomorphism, then it is an *isomorphism*. □

A number of "obvious" implications of the definition are as follows.

*Proposition A-3

For every homomorphism T,

$$T(0) = 0$$

□

*Proposition A-4

A homomorphism T has an inverse (which is also a homomorphism) if and only if it is an isomorphism and, in that case, the inverse is also an isomorphism. □

In the preceding section, a subspace was defined. Its dual, a quotient space, is defined for a vector space V and a distinguished subspace W by identifying certain elements of V. We say that x is equivalent to y,

$$x \simeq y \qquad \text{(A-6)}$$

if

$$x - y \text{ is in } W \qquad \text{(A-7)}$$

"$\simeq$" partitions V into equivalence classes (that is, mutually exclusive, all inclusive subsets of equivalent vectors). W itself is one such equivalence class, since the difference between any two vectors in W is also in W. Actually, W is the equivalence class containing zero and will serve as the zero element of the quotient space. The equivalence class of x in V is denoted by $[x]$. Thus, $[0] = W$. The significant aspect of the partitioning of V defined by "$\simeq$" is that the equivalence classes may be viewed as the vectors of a new vector space, the quotient space.

*Proposition A-5

Let V be a vector space and W a subspace. Then the equivalence classes defined by

$$x \simeq y$$

if

$$x - y \text{ is in } W$$

together with the operations

$$[x] + [y] = [x + y]$$

and

$$a[x] = [ax]$$

(that is, the sum of the equivalence class of x and the equivalence class of y is the equivalence class of $x + y$ and the equivalence class of a times the equivalence of x is the equivalence class of ax) forms a vector space. □

*Definition A-9

The vector space of Proposition A-5 is the *quotient space* of V over W and is denoted by V/W. □

The quotient space is the final tool needed to define the four subspaces associated with a homorphism.

*Definition A-10

Let

$$T : V \to W$$

be a homomorphism. Then the *image of T* [im (T)] is the subspace of W given by

$$\text{im } (T) = \{w \text{ in } W, \quad w = T(x) \text{ for some } x \text{ in } V\}$$

The *kernel of* T [ker (T)] is the subspace of V given by
$$\ker (T) = \{v \text{ in } V, \quad T(v) = 0\}$$
The cokernel of T [coker (T)] is the quotient space of W given by
$$\text{coker } (T) = W/\text{im } (T)$$
The coimage of T [coim (T)] is the quotient space of V given by
$$\text{coim } (T) = V/\ker (T)$$

The validity of these definitions is, of course, contingent on the easily verified fact that im (T) and ker (T) are actually subspaces of V and W, respectively [94]. Basic properties of the subspaces associated with T are given by the following proposition.

*Proposition A-6

Let
$$T : V \to W$$
be a homorphism. Then there exists an isomorphism
$$K : \ker (T) \to \text{coker } (T)$$
and an isomorphism
$$I : \text{im } (T) \to \text{coim } (T)$$

*Proposition A-7
Let
$$T : V \to W$$
be a homomorphism. Then T is an epimorphism if and only if
$$\text{im } (T) = W$$
and T is a monomorphism if and only if
$$\ker (T) = 0$$

*Definition A-11

Let
$$T : V \to W$$
be a homomorphism. Then a *left inverse* for T is a homomorphism
$$T^{-L} : W \to V$$
such that the composite operator
$$T^{-L}T : V \to V$$

is the identity, and a *right inverse* for T is a homomorphism
$$T^{-R} : W \to V$$
such that the composite operator
$$TT^{-R} : W \to W$$
is the identity. □

*Proposition A-8

A homomorphism has a left inverse if and only if it is a monomorphism, and a right inverse if and only if it is an epimorphism. □

In Definition A-11 the intuitive fact that one can multiply (compose) operators has been employed. Before proceeding to a formal discussion of the algebraic structure of the operators, we will first show that the set of all homomorphisms between two vectors spaces is itself a vector space, the additivity structure being inherited from the range space. We define the sum of two homomorphisms by

$$(T + S)x \equiv Tx + Sx \qquad \text{(A-8)}$$

[that is, the "vector" sum of a homomorphism T and a homomorphism S is the new homomorphism $(T + S)$ which takes the vector x to the sum of the vectors (Tx) and (Sx)]. Similarly, multiplication of a homomorphism T by a scaler a yields the new homomorphism, which takes the vector x to the vector (Tx) multiplied by the scaler a; that is,

$$(aT)x \equiv a(Tx) \qquad \text{(A-9)}$$

With these operations, we have the following.

*Proposition A-9

Let V and W be vector spaces. Then the set of all homomorphisms from V into W, together with the operations of Equations (A-8) and (A-9), is a vector space. □

*Definition A-12

The vector space of homomorphisms from V into W is denoted by

$$\text{Hom } (V, W)$$

□

The vector space structure for the operators is the source of their great usefulness, for it permits the manipulation of the operators on a vector space via the same techniques employed for the manipulation of the vectors themselves. It is for this reason that the set of $n \times m$ matrices is a vector space. In fact, it is exactly Hom (R^n, R^m).

Of particular importance is the vector space of homomorphisms from V into R.

*Definition A-13

Let V be a vector space. The homomorphisms from V to R are the *linear functionals* on V and

$$\text{Hom }(V,R)$$

is the *dual space* of V denoted by V'. □

*Proposition A-10

The dimension of a vector space and its dual are the same. □

The dual space, or vector space, of linear functionals for the finite-dimensional space R^n is Hom (R^n,R), or equivalently, the space of $1 \times n$ matrices (row vectors). To see that such vectors, in fact, define functionals, one need only observe that a row vector from Hom (R^n,R) operating on a vector from R^n (via the usual matrix multiplication) yields a 1×1 matrix, a real number. For our application, a subspace of the space of linear functionals Hom (D,R) on the vector space D, of smooth functions with bounded support, is of considerable interest. In fact, as defined in Appendix B, this is the space of "distributions," the study of which is the subject of that appendix. One class of functionals on this space (though not the entire space) is given by the weighted integral

$$\int_{-\infty}^{\infty} f(u)\phi(u)du \qquad \text{(A-10)}$$

which defines a real number for each $\phi(t)$ in D and any fixed integrable $f(t)$. In Equation (A-10), it is important to note that the integral is finite for every $\phi(t)$ in D, since such a function has bounded support. A similar integral operator approach might be employed to define functionals for more general function spaces, but in such cases, care must be taken to assure that the integral is finite.

Another class of functionals which may be defined on function spaces is the "evaluation functional," which associates with a function its value at some specified time. For instance, a functional may take $\phi(t)$ in D_+ to the number $\phi(7)$.

It is possibly surprising that Hom (V,V) has more algebraic structure than V itself. In fact, it forms an "algebra."

*Definition A-14

An *algebra* is a vector space such that for any two vectors x and y there is defined a *vector product* xy satisfying:

1. $(x + y)z = xz + yz$
2. $x(y + z) = xy + xz$
3. $(ax)y = x(ay) = a(xy)$ for any scaler a. □

Unlike the vector space operations, vector multiplication is not, in general, commutative (that is, $xy \neq yx$).

*Proposition A-11

The vector space Hom (V,V), together with the vector multiplication

$$(TS)x = T(Sx)$$

[that is, the "vector" product of the homomorphisms T and S is the new homomorphism which takes a vector x to T of the vector (Sx)], is an algebra. □

In the case of the finite-dimensional space R^n, Hom (R^n,R^n) is "the matrix algebra." From the standpoint of our application, one is primarily interested in the algebra Hom $(D_+{}^n,D_+{}^n)$, which is the underlying structure for most of linear network theory.

A-4 Norms and Inner Products

In dealing with networks, to achieve maximum generality it is desirable to minimize the structure assumed for the linear operators, this being replaced by a structure on the input-output function spaces rather than the operators themselves. This takes the form of a norm and/or an inner product.

*Definition A-15

Let V be a vector space. Then a *norm* on V is a function

$$\|\cdot\| : V \to R$$

satisfying

1. $\|x\| \geq 0$
2. $\|x\| = 0$ if and only if $x = 0$
3. $\|x + y\| \leq \|x\| + \|y\|$
4. $\|ax\| = |a|\, \|x\|$. □

For the vector spaces of real or complex numbers, the norm may be taken as the absolute value, and, in the general case, the norm may be viewed as a generalized absolute value. For the vector spaces P and P_n, the l_p norm is

$$\left\|\sum_i a_i x^i\right\|_p \equiv \left(\sum_i |a_i|^p\right)^{1/p} \qquad \text{(A-11)}$$

This may also be taken as a norm for the vector space of series which satisfy

$$\sum_i |a_i|^p < \infty \qquad \text{(A-12)}$$

that is, the vector space l_p. An alternate "sup norm" for these spaces is

$$\left\|\sum_i a_i x^i\right\| = \sup_i \{|a_i|\} \tag{A-13}$$

In the case of the various function spaces, analogous norms may be defined by replacing the summation of Equation (A-11) with an integral

$$\|f(t)\|_p = \left(\int_{-\infty}^{\infty} |f(t)|^p dt\right)^{1/p} \tag{A-14}$$

These are designated the L_p norms. The fact that L_p is a norm is nontrivial, resulting from the Minkowski inequality [84]. The L_p norm may be applied to D, but not to D_+, since the norm of a D_+ function may not exist. If a norm is given for V and W, it is possible to induce a norm on Hom (W,V) via the following proposition.

*Proposition A-12

Let V be a vector space with norm $\|\cdot\|_v$. Then for any vector space W, normed by $\|\cdot\|_w$, Hom (W,V) may be normed by

$$\|\cdot\|_{\text{Hom}} : \text{Hom } (W,V) \to R$$

where

$$\|T\|_{\text{Hom}} = \sup_{\|a\|_w=1} \|Ta\|_v \qquad \square$$

Note that the norm of T may not exist; hence, Hom (W,V) is not a normed space.

*Definition A-16

The norm of Proposition A-12 is the *sup norm* on Hom (W,V). $\qquad\square$
  A somewhat stronger structure than a norm is an inner product.

*Definition A-17

Let V be a (real) vector space. Then an *inner product* for V is a function

$$\langle \cdot, \cdot \rangle : V \times V \to R$$

satisfying

  1. $\langle x,y \rangle = \langle y,x \rangle$
  2. $\langle ax + by, z \rangle = a\langle x,z \rangle + b\langle y,z \rangle$
  3. $\langle x,x \rangle \geq 0$
  4. $\langle x,x \rangle = 0$ if and only if $x = 0$

for all x, y in V and a, b in R. $\qquad\square$

An inner product is actually a stronger structure than a norm. In fact, an inner product induces a norm on its space.

*Proposition A-13

Let $\langle \cdot, \cdot \rangle$ be an inner product on a vector space. Then

$$\|x\| = \langle x, x \rangle^{1/2}$$

is a norm for the vector space. □

*Definition A-18

The norm of Proposition A-13 is the *induced norm* of the inner product. □

*Definition A-19

A vector space together with a norm is a *normed space*, and a vector space together with an inner product is an *inner product space*. □

In dealing with inner product spaces, the only norm used will be the induced norm.

The most common inner products are the l_2 and L_2 inner products given by

$$\left\langle \sum_i a_i x^i, \sum_i b_i x^i \right\rangle = \sum_i a_i b_i \quad \text{(A-15)}$$

and

$$\langle f(t), g(t) \rangle = \int_{-\infty}^{\infty} f(t) g(t) dt \quad \text{(A-16)}$$

for sequences and functions, respectively. For the most part of this work, L_2 will be the only inner product used. These two inner products clearly induce the L_2 and l_2 norms.

It is possibly surprising that an inner product on V does not, in general, induce one onto Hom (V,V). The induced norm of an inner product, however, induces a norm onto Hom (V,V) (as does any norm). Though the inner product does not induce a norm on the operator space, it does serve to define the adjoint of an operator.

*Definition A-20

Let V be an inner product space and A be in Hom (V,V). Then the adjoint of A, denoted by A^*, is the unique operator in Hom (V,V) (if it exists) such that for all x, y in V

$$\langle x, Ay \rangle = \langle A^*x, y \rangle \quad □$$

The basic properties of the adjoint are given by the following proposition.

*Proposition A-14

Let V be an inner product space and A, B be in Hom (V,V). Then

1. $(aA)^* = aA^*$
2. $(A+B)^* = A^* + B^*$
3. $(AB)^* = B^*A^*$
4. $(1)^* = 1$
5. $(0)^* = 0$
6. $\|A^*\| = \|A\|$
7. $(A^{-1})^* = (A^*)^{-1}$
8. $(A^{-L})^* = (A^*)^{-L}$
9. $(A^{-R})^* = (A^*)^{-L}$ □

The adjoint operator, in turn, permits the definition of a number of new operators derived from A.

*Definition A-21

Let V be an inner product space and A be in Hom (V,V). Then the *hermitian part* of A is

$$A_h = \tfrac{1}{2}(A + A^*)$$

The *skew hermitian part* of A is

$$A_s = \tfrac{1}{2}(A - A^*)$$

An operator is hermitian if it is equal to its *hermitian* part, and *skew hermitian* if it is equal to its skew hermitian part. □

The reader may verify that hermitian operators satisfy

$$A = A^* \qquad \text{(A-17)}$$

skew hermitian operators satisfy

$$A = -A^* \qquad \text{(A-18)}$$

and all operators satisfy

$$A = A_h + A_s \qquad \text{(A-19)}$$

The hermitian part of an operator behaves much like the real part of a complex number, whereas the skew part behaves like the imaginary part of a complex number.

*Definition A-22

Let V be an inner product space and A be in Hom (V,V). Then A is positive (≥ 0) if
$$\langle Ax,x \rangle \geq 0$$
for all x, and it is *positive definite* (> 0) if
$$\langle Ax,x \rangle > 0$$
for all $x \neq 0$. □

One application of the inner product is in the definition of orthogonality.

*Definition A-23

Let V be an inner product space. Then x and y in V are *orthogonal* if
$$\langle x,y \rangle = 0$$
and two subspaces, X and Y, are *orthogonal* if for each x in X and y in Y
$$\langle x,y \rangle = 0$$
The *orthogonal compliment* of a subspace is the set of all vectors which are orthogonal to all members of the given subspace. □

A-5 Topological Vector Spaces

Possibly the most significant property of a normed vector space is that the algebraically defined norm may be employed to define a metric, or distance, function on the vector space. This, in turn, allows topological concepts, such as convergence of vectors and continuity of operators, to be studied in a vector space context. Although there is a significant theory of topological vector spaces [22], [77], [140], [179], which includes vector spaces whose topological structure does not arise from a norm, in the present context we are interested primarily in the topological properties of a normed space.

*Definition A-24

Let S be a set and d be a function taking $S \times S$ to the reals
$$d : S \times S \to R$$
such that

1. $d(x,y) \geq 0$ with equality if and only if $x = y$
2. $d(x,y) = d(y,x)$ for all x and y in S
3. $d(x,y) + d(y,z) \geq d(x,z)$ for all x, y, and z in S.

Then d is a *metric* on S. □

*Proposition A-15

Let V be a normed space. Then the function
$$d : V \times V \to R$$
defined by
$$d(x,y) = \|x - y\|$$
is a metric on V. □

Consistent with Proposition A-15, together with the fact that all of the usual topological properties encountered in analysis [140], [144] can be defined and studied in terms of a metric, a normed space is, indeed, a topological vector space [77], [179].

Possibly the most fundamental topological concept is convergence. Unlike the real and complex numbers, wherein convergence is naturally defined in a unique manner, there are a number of distinct modes of convergence which apply in a normed space. Of course, in a function space the "obvious" mode of convergence is "pointwise convergence," where a sequence of real or complex valued functions $\{f_i\}$ is said to converge (pointwise) to a function f if

$$\lim_{i \to \infty} f_i(t) = f(t) \qquad \text{for all } t \qquad \text{(A-20)}$$

Alternatively (and, in fact, in any normed space, whether or not it is a function space), a mode of "uniform convergence" may be defined in terms of the properties of the norm.

*Definition A-25

Let $\{x_i\}$ be a sequence of vectors in a normed space V. Then the sequence $\{x_i\}$ *converges* (*uniformly*) to a vector x in V if
$$\lim_{i \to \infty} \|x_i - x\| = 0$$
(where the limit is taken in the usual sense for real numbers). □

Another topological concept closely related to convergence is that of continuity. Since this is a topological concept, the continuity of an operator in Hom (W,V) can be defined in a manner completely analogous to that used for functions if V and W are normed spaces.

*Definition A-26

Let V and W be normed spaces and T be in Hom (W,V). Then T is *continuous at a vector* x in V if for any ϵ there exists a δ such that if
$$\|y - x\|_W < \delta$$
then
$$\|Ty - Tx\|_V < \epsilon$$

If T is continuous at each x in V, then T is *continuous*. □

The relationship between continuity and convergence, and also between continuity and the norm of T, is given by the following proposition.

*Proposition A-16

Let V and W be normed spaces and T be in Hom (W,V). Then the following statements are equivalent.

1. T is continuous
2. $\|T\| < \infty$
3. if a sequence of vectors $\{x_i\}$ converges (uniformly) to x, then the sequence $\{Tx_i\}$ converges (uniformly) to Tx
4. if a sequence of vectors $\{x_i\}$ converges (uniformly) to 0, then the sequency $\{Tx_i\}$ converges (uniformly) to 0. □

Consistent with the proposition, those operators in Hom (W,V) which are continuous have finite norms; hence, the subspace (a fact which the reader may verify) of continuous linear operators in Hom (W,V) is a normed space. Of particular interest in Appendix B, on distributions, are the continuous linear functionals on D, that is, the subspace of Hom (D,R) consisting of those functionals which are continuous, this being denoted by D'.

One final mode of convergence, which, in fact, proves to be of value in the theory of distributions, uses the functionals on a space to translate convergence in that space to an equivalent problem of convergence in the reals, where the usual convergence concepts may be invoked.

*Definition A-27

Let V be a normed space and $\{x_i\}$ be a sequence of vectors in V. Then $\{x_i\}$ converges (weakly) to a vector x if for every continuous linear functional $f(\cdot)$ in Hom (V,R),

$$\lim_{i \to \infty} f(x_i) = f(x)$$

□

Of course, the roles of the functional and the vectors may be interchanged in Definition A-27 to allow one to define a weak convergence concept in the space of continuous linear functionals (in addition to the uniform convergence concept, which results from the fact that the space of continuous linear functionals is normed). That is, if $\{f_i\}$ is a sequence of continuous linear functionals on a normed space V, then $\{f_i\}$ converges (weakly) to f if for every x in V

$$\lim_{i \to \infty} f_i(x) = f(x) \tag{A-21}$$

Following the result of Proposition A-16, we may define a weak mode of continuity for an operator in Hom (W,V) in terms of the weak convergence concept.

*Definition A-28

Let V and W be normed spaces and T be in Hom (W,V). Then T is (weakly) continuous if, given any sequence of vectors x_i in W which converges (weakly) to 0, then the sequence of vectors Tx_i in V converge (weakly) to 0. □

Of course, we may also define various mixed modes of continuity, where the sequence x_i is required to converge uniformly, and Tx_i is required to converge weakly, or vice versa.

A-6 Discussion

The purpose of this appendix has been to outline the techniques applicable to infinite-dimensional vector spaces and their induced operator spaces. Though finite-dimensionality is a convenient constructive tool, it is not needed to achieve the basic results. In fact, the details added by finiteness often obscure the fundamental ideas.

APPENDIX B
Distributions

B-1 Introduction

The input-output functions for the networks are required to be smooth and in D_+, thereby eliminating many of the problems induced by discontinuous and/or singular inputs. These "functions" are, however, not completely by-passed since it is desired to represent a network by the integral equation

$$b(t) = \int_{-\infty}^{\infty} K(t,q)a(q)dq \qquad \text{(B-1)}$$

If it is required that $K(t,q)$ be a function, this cannot, in general, be achieved. Consider, for example, the open circuit for which

$$b(t) = a(t) \qquad \text{(B-2)}$$

For $K(t,q)$ to represent the open circuit it must satisfy the equation

$$a(t) = \int_{-\infty}^{\infty} K(t,q)a(q)dq \qquad \text{(B-3)}$$

which is the case only if

$$K(t,q) = 0 \quad \text{if } t \neq q \qquad \text{(B-4)}$$

and

$$\int_{-\infty}^{\infty} K(t,q)dq = 1 \qquad \text{(B-5)}$$

Now, Equation (B-4) implies that $K(t,q)$ is zero almost everywhere (that is, equal to zero except possibly on a set of measure zero); hence, if it is a function, its integral must be zero [92]. Thus the open circuit cannot be represented by an integral operator of the type shown in Equation (B-1) if $K(t,q)$ is a function. Equations (B-4) and (B-5), in fact, define the Dirac delta "function" [60], a distribution. The main result of this appendix will, in fact, be to justify the contention that every network has an integral operator representation with distributional kernel.

It is necessary to generalize the functions before a valid kernels theorem can be achieved. The classical approach to such a generalization is to start with the usual functions and then add the required generalized functions, the latter taking the form of Cauchy sequences. While a sound theory can be achieved via this approach [96], the result is a patchwork of special cases which often leads to confusion. This is illustrated by the (intentionally) "bogus" derivation of the Laplace transform of the impulse function, due to Reed [138]. Taking the impulse function to be the limit of a sequence of functions $f_i(u)$, such that

$$\int_{-\infty}^{\infty} f_i(u)du = 1 \qquad \text{(B-6)}$$

and for which the support goes, in the limit, to the single point zero, the Laplace transform

$$(L\delta(t)) = \int_{0}^{\infty} [\lim f_i(t)]e^{-pt}dt \qquad \text{(B-7)}$$

is zero, since the limit of the integrand is zero almost everywhere. Of course, the transform is not zero, the erroneous solution resulting from the fact that Equation (B-7) is not a valid definition for the Laplace transform of a generalized function.

The approach to the required generalization to be developed here is due to Schwartz [156], who uses the term distribution (which will henceforth be adopted) rather than generalized function. His approach is to find a space of "objects," not necessarily functions, which has the desired properties, and then to define an isomorphism of the usual functions into this space. This approach gives the distributions a unified character since they are all in the same space. However, it suffers from the disadvantage that the distributions thus defined are never functions. Of course, since the functions are isomorphic to a subspace of the distributions, they have a well-defined distributional interpretation. Moreover, such an approach naturally associates an integral operator with each distribution, thereby paving the way for the theory of generalized integral operators to follow.

This appendix is divided into four sections. The first deals with the basic definitions of distribution theory and their relation to the usual functions [63], [77], [156], [210]. The second is devoted to the kernels theorem [156], [157], [179] and its application, and the final sections are concerned with the distributional Fourier and Laplace transforms [40], [63], [156], [210].

B-2 Definitions

In Chapter 1, the function space D_+ of smooth functions, which are zero for sufficiently large negative t, is employed. Here a subspace of D_+, smooth functions which are zero for both sufficiently large negative and positive t, plays the dominent role.

Definition B-1

The *testing functions* are real-valued smooth functions of a real variable which have compact support (that is, they are zero outside of some finite interval on the line). The vector space of all testing functions is denoted by D. □

For any (real) vector space, a linear functional is a linear function which takes the objects in the vector space to the reals. The set of all continuous linear functionals on a vector space is itself a vector space, called the dual space. The distributions are the continuous linear functionals on D. Here continuity is taken to mean that if a sequence of testing functions converges (uniformly) to zero, then the sequence of real numbers obtained by applying the functional to the sequence also converges to zero.

Definition B-2

The distributions are the continuous linear functionals on D. They are written as

$$\int_{-\infty}^{\infty} x(q)(\cdot)dq$$

where x is the distribution name. If a particular testing function ϕ is under consideration, we denote the distribution by

$$\int_{-\infty}^{\infty} x(q)\phi(q)dq$$

The space of all distributions is denoted by D'. □

It is important to note that the distributions are not integrals, nor is x, in general, a function. The notation is, however, used because a large class of distributions take the form of weighted integrals. Moreover, the properties of the distributions closely parallel those of weighted integrals (for which reason the distributions may be employed in the definition of a generalized integral operator).

The relationship between the distributions and the weighted integrals is considered in the following proposition, this relationship serving to define an isomorphism between the locally integrable functions and a subclass of the distributions.

Proposition B-1

The locally integrable functions [144] can be isomorphically embedded (mod sets of measure zero) into the distributions by

$$f \leftrightarrow \int_{-\infty}^{\infty} f(q)(\cdot)dq$$

Proof

It must be shown that the integral operator

$$\int_{-\infty}^{\infty} f(q)(\cdot)dq \tag{B-8}$$

is, in fact, a distribution, for which purpose it suffices that the integral

$$\int_{-\infty}^{\infty} f(q)\phi(q)dq \tag{B-9}$$

exist for all testing functions ϕ in D. The compact support of ϕ permits Equation (B-8) to be replaced by

$$\int_{-\infty}^{\infty} f(q)\phi(q)dq = \int_{a}^{b} f(q)\phi(q)dq \tag{B-10}$$

since any given testing function is zero outside of some interval bounded by a and b. The local integrability of f, together with the smoothness of ϕ, now assures the existence of the integral. Finally, if $f \neq g$ (on a set of nonzero measure), the distributions defined by f and g are distinct; hence, the embedding is an isomorphism. □

Since the functions in D_+ are smooth, they are locally integrable and thus may be embedded into D' via Proposition B-1. The fact that the isomorphism of Proposition B-1 is strictly into is illustrated by the delta function.

Definition B-3

The delta function is the distribution

$$\int_{-\infty}^{\infty} \delta(q)(\cdot)dq$$

such that for any testing function ϕ in D,

$$\int_{-\infty}^{\infty} \delta(q)\phi(q)dq = \phi(0) \qquad \square$$

Clearly,

$$a(t) = \int_{-\infty}^{\infty} \delta(t-q)a(q)dq \tag{B-11}$$

but this distribution is not obtainable from a function, as was shown in the introduction.

Although many distributions may be derived from locally integrable functions, via Proposition B-1, it is a mistake to confuse this isomorphism with an identity. The distributions are functionals, the properties of which may often be related to those of the functions, but they are not themselves functions. A distribution obtained from a locally integrable function f is denoted by

$$\int_{-\infty}^{\infty} f(q)(\cdot)dq$$

not f.

In the preceding we have defined the distributions as an extension of the weighted integrals. It is now desired to define operations on the distributions which naturally extend such functional operations as addition, differentiation, and so forth. In doing this one would want to define operations which are well defined for all of the distributions and, moreover, coincide with the corresponding functional operations. For instance, one would like to define distributional addition so as to assure that if one were to form distributions from two functions, f and g, as per Proposition B-1, and add them (in the distributional sense), the same distribution would be obtained as if the functions had been added before forming the distribution. That is,

$$\int_{-\infty}^{\infty} f(q)(\cdot)dq + \int_{-\infty}^{\infty} g(q)(\cdot)dq = \int_{-\infty}^{\infty} (f+g)(q)(\cdot)dq \qquad \textbf{(B-12)}$$

In essence, one requires that the distributional operations "commute" with the isomorphism of Proposition B-1. The definitions for the distributional operations are given by Definition B-4, while the proof that the distributional operations coincide with the usual functional operations (for the distributions which are derived from integrable functions) is the subject of Proposition B-2.

Definition B-4

Let

$$\int_{-\infty}^{\infty} f(q)(\cdot)dq$$

and

$$\int_{-\infty}^{\infty} g(q)(\cdot)dq$$

be arbitrary distributions, c and T real constants, and h a smooth function. Then

1. $\displaystyle\int_{-\infty}^{\infty} (f+g)(q)(\cdot)dq \equiv \int_{-\infty}^{\infty} f(q)(\cdot)dq + \int_{-\infty}^{\infty} g(q)(\cdot)dq$

2. $\displaystyle\int_{-\infty}^{\infty} (cf)(q)(\cdot)dq \equiv c\int_{-\infty}^{\infty} f(q)(\cdot)dq$

3. $\displaystyle\int_{-\infty}^{\infty} (fh)(q)(\cdot)dq \equiv \int_{-\infty}^{\infty} f(q)(h\cdot)dq$

4. $\displaystyle\int_{-\infty}^{\infty} f(q-T)(\cdot)dq \equiv \int_{-\infty}^{\infty} f(q)\sigma(\cdot)dq$

5. $\displaystyle\int_{-\infty}^{\infty} \dot{f}(q)(\cdot)dq \equiv -\int_{-\infty}^{\infty} f(q)(\dot{\cdot})dq$

where σ is the shift operator taking $\phi(q)$ to $\phi(q+T)$, and the dot is the derivative operator. □

The objects on the left of the equalities in definition B-4 are well defined by the distributions on the right, for any $\phi(q)$ from D, since the right side of the equality is a well-defined real number (which varies linearly with ϕ). Even the distributional derivative of equality 5 exists for every distribution

$$\int_{-\infty}^{\infty} f(q)(\cdot)dq$$

(including those derived from nondifferentiable functions f). This is the case because $\phi(q)$, being in D, is always differentiable. Of course, it has yet to be shown that the (well-defined) number on the right of equality 5 coincides with that which would be obtained from the weighted integral

$$\int_{-\infty}^{\infty} \dot{f}(q)(\cdot)dq \qquad \text{(B-13)}$$

if $\dot{f}$ exists. This, in fact, is the case, as is shown in the following proposition.

Proposition B-2

If in Definition B-4,

$$\int_{-\infty}^{\infty} f(q)(\cdot)dq$$

and

$$\int_{-\infty}^{\infty} g(q)(\cdot)dq$$

are distributions derived from locally integrable functions (as per Proposition B-1), then the operations of Definition B-4 coincide with the usual operations for the functions f and g (that is, the operations of Definition B-4 commute with the isomorphism of Proposition B-1).

Proof

Equalities 1 and 2 of Definition B-4 follow from the linearity of the "integral." Equality 3 holds since

$$\int_{-\infty}^{\infty} (fh)(q)\phi(q)dq = \int_{-\infty}^{\infty} f(q)(h\phi)(q)dq \qquad \text{(B-14)}$$

where $(h\phi)$ is in D since ϕ has compact support and both h and ϕ are smooth. Equality 4 results from the change of variable, $t = T + u$, and 5 follows from integration by parts.

$$\int_{-\infty}^{\infty} \dot{f}\phi dq = -\int_{-\infty}^{\infty} f\dot{\phi}dq + f\phi\bigg|_{-\infty}^{\infty} \qquad \text{(B-15)}$$

The last term of Equation (B-15) is zero since the compact support of ϕ implies that

$$\phi(-\infty) = \phi(\infty) = 0 \qquad \text{(B-16)}$$

□

Equality 3 of Definition B-4 cannot be employed to define the product if h is not smooth, for in that case the product function $h\phi$ may not be in D [156].

Definition B-4 yields a (distributional) derivative for all distributions, since the testing functions always have infinitely many derivatives. Thus, via the isomorphism, it is possible to define the derivative of any locally integrable function. Of course, if the function does not have a derivative in the usual sense, then the distributional derivative will be a distribution which is not obtainable from a locally integrable function. The argument advanced in the proof of Proposition B-2 however, assures that if the usual derivative exists, it coincides (via the isomorphism) with the distributional derivative.

From Definition B-4, the derivative of the delta function is found to be

$$\int_{-\infty}^{\infty} \dot{\delta}(q)\phi(q)dq \equiv -\int_{-\infty}^{\infty} \delta(q)\dot{\phi}(q)dq = -\dot{\phi}(0) \qquad \text{(B-17)}$$

which coincides with the intuitive interpretation of the doublet [207]. Upon repeated application of Definition B-4, the higher order derivatives of the delta function are found to be

$$\int_{-\infty}^{\infty} \delta^{(n)}(q)\phi(q)dq = (-1)^n \frac{d^n}{dt^n}\phi(0) = (-1)^n \phi(0)^{(n)}$$

In the above development, the operations on the distributions have been defined so as to assure that they coincide with the analogous operations for the weighted integrals (which form a subclass of the distributions). For this reason, in dealing with distributions, one may manipulate them via the same techniques as are commonly employed for the weighted integrals alone, these techniques being valid for all distributions whether or not they actually are weighted integrals. In fact, it is precisely this viewpoint which is taken in the main body of the text. It is, however, important to remember that the distributions are not functions and, in fact, those distributions which are not derived from integrable functions cannot be taken from under the integral sign. Thus, the symbol δ taken alone *has no meaning whatsoever*, though the symbol

$$\int_{-\infty}^{\infty} \delta(q)(\cdot)dq$$

is well defined. In practice, this causes no real difficulty, since our primary application of the distributions is to define a generalized integral operator wherein the distributional weighting function naturally appears under the integral sign.

Since we do not generally norm the space of distribution, we cannot define convergence in the space of distributions in the uniform sense, such as we do with testing functions in D. We can, however, define a mode of weak convergence via the following definition.

Definitions B-5

A sequence of distributions $\{x_i\}$ converges (weakly) to a distribution x if for every testing function ϕ in D

$$\lim_{i \to \infty} \int_{-\infty}^{\infty} x_i(q)\phi(q)dq = \int_{-\infty}^{\infty} x(q)\phi(q)dq \qquad \square$$

Here the limit on the left side of the equation is the usual real number limit. Since the function spaces D and D_+ can be naturally embedded into D' via the isomorphism of Proposition B-1, the weak convergence concept for D' may be applied to these spaces as well. A fundamental property of the space D', together with the weak mode of convergence, is that the space is complete.

*Proposition B-3

If for a sequence of distributions $\{x_i\}$ and every testing function ϕ the sequence of real numbers

$$\int_{-\infty}^{\infty} x_i(q)\phi(q)dq$$

is Cauchy [144], then there exists a unique distribution x, such that $\{x_i\}$ converges (weakly) to x. $\qquad \square$

B-3 The Kernels Theorem

In this section the existence of the integral equation representation for a linear network is considered. The basic theorem and its corollary are (by now) classical results due to Schwartz [156], [157]. Their proofs, however, rely on topological tensor products (a mathematical concept not in the repertory of the typical network theorist) and hence will not be given.

In the two cases, the theorem applies to linear continuous operators from the space of testing functions D to the space of distributions D', and from the space D_+ to D', respectively. Since neither D' nor D_+ are normed spaces, the uniform convergence definition (Definition A-26) is not applicable. We may, however, define the continuity of an operator T, via the requirement that if a sequence of vectors $\{x_i\}$ in the domain space converges, then the sequence of vectors $\{Tx_i\}$ in the range space also converges. In the following theorem this is the approach employed, with the mode of convergence in the spaces D_+ and D' taken as weak convergence in the sense of Definition B-5, and that in the space D taken as uniform convergence in the sense of Definition A-25.

Theorem B-1   (Kernels Theorem [156], [179])

Every linear continuous operator from D into D' may be represented by a unique "integral" operator

$$\int_{-\infty}^{\infty} H(t,q)(\cdot)dq$$

where $H(t,q)$ is a distribution in two variables □

See Schwartz [156] for the original proof (in French) or Treves [179] for a proof in English.

Corollary B-1

Every linear continuous operator from D_+ into D' has an "integral" operator representation as per Theorem B-1.   □

See Schwartz [157] for a proof (in French).

Although the general proof of the kernels theorem is quite complex, in the special case of a time-invariant network where the kernel is a function of only one variable (the age variable, $t - q$) a simple proof is possible [26]. The corollary has a number of additional variations [157] wherein the input and output spaces for the operator may be varied. Usually, these spaces are required to satisfy the containment condition $D \subset W \subset D'$ for W to be an admissible input or output space. Moreover, D must be dense [144] in the input space under the mode of convergence employed to define the continuity of operators on that space. As such, a variation of the kernels theorem is applicable to most of the spaces commonly encountered in functional analysis [22], [140], though often with a weaker mode of convergence than might be desired [157]. Finally, we note that other properties of an operator may be used to weaken or eliminate the continuity requirement. For instance, in Chapter 2 we show that a linear passive (scattering) operator has a kernel representation from D_+ to itself, and, in fact, it follows from the proof of that theorem (Theorem 2-1) that a causal operator from D_+ to D' whose restriction to D is continuous has a kernel representation.

In the representation of Theorem B-1 and its corollary, q is the "variable of integration," while t is a parameter which is independent of the "integral."

In essence, the representation is actually a parameterized family of distributions (in the variable q) for which

$$\int_{-\infty}^{\infty} H(t,q)\phi(q)dq \tag{B-18}$$

represents the output evaluated at time t, a real number, due to an input $\phi(q)$ in D_+. Clearly, as t varies, the output for all time, a function in D_+, is obtained.

The distribution represented by $H(t,q)$ can be heuristically given the physical interpretation of the response, measured at time t, to an impulsive input applied at time q. Of course, we do not allow impulsive inputs (whatever they are), and hence this intuitive interpretation, though useful, has no mathematical meaning.

Consider the integral operator represented by

$$\int_{-\infty}^{\infty} \delta(t-q)(\cdot)dq \tag{B-19}$$

Now, if one considers an input $\phi(t)$, the resultant output is given by

$$\int_{-\infty}^{\infty} \delta(t-q)\phi(q)dq = \int_{-\infty}^{\infty} \delta(-u)\phi(t+u)du \tag{B-20}$$

where the equality results from the change of variable

$$u = t - q \tag{B-21}$$

Now, upon invoking the definition of the δ functional, this becomes

$$\int_{-\infty}^{\infty} \delta(t-q)\phi(q)dq = \int_{-\infty}^{\infty} \delta(-u)\phi(t+u)du = \phi(t+u)\Big|_{-u=0} = \phi(t) \tag{B-22}$$

Hence the integral operator of Equation (B-20) is, as claimed in the "Introduction," a representation of the identity operator. A more interesting operator is

$$\int_{-\infty}^{\infty} \delta(t+T-q)(\cdot)dq \tag{B-23}$$

which, after an argument similar to that used above, may be shown to represent a predictor or delay (depending on whether T is positive or negative) characterized by

$$\int_{-\infty}^{\infty} \delta(t+T-q)\phi(q)dq = \phi(t+T) \tag{B-24}$$

Finally, the operator represented by

$$\int_{-\infty}^{\infty} \dot{\delta}(t-q)(\cdot)dq \tag{B-25}$$

is the derivative for

$$\int_{-\infty}^{\infty} \dot{\delta}(t-q)\phi(q)dq = \int_{-\infty}^{\infty} \delta(u)\dot{\phi}(t-u)du = \dot{\phi}(t) \tag{B-26}$$

Here the first equality results from the definition of the distributional derivative (Definition B-4), together with a change of variable, whereas the second equality follows from the definition of δ.

The basic operations on linear operators are composition and addition. Since a kernel exists for each of the operators, it would be desirable to have formulas for calculating the kernel representation of a sum or composition of two operators from the kernels for the individual operators. Since we propose to consider the composition of operators, it is necessary that the range and domain of our operators coincide; hence, these are taken as D_+ in the sequel.

Definition B-6

Let $K(t,q)$ and $L(t,q)$ be distributions in two variables. Then

$$K(t,q) \text{ o } L(t,q) = \int_{-\infty}^{\infty} K(t,u)L(u,q)du$$

is the *product* of $K(t,q)$ and $L(t,q)$ and

$$K(t,q) + L(t,q)$$

is the *sum* (the usual distributional sum) of $K(t,q)$ and $L(t,q)$. □

*Proposition B-4

The sum and product of Definition B-6 always exist (even though the usual product of $K(t,q)$ and $L(t,q)$ may not be defined). □

The fact that the operations of Definition B-6 do indeed correspond to operator addition and composition is the subject of the following proposition. The reader may prove this for himself by writing the two operators, L and K, in their "integral" form as

$$\int_{-\infty}^{\infty} L(t,q)(\cdot)dq \tag{B-27}$$

and

$$\int_{-\infty}^{\infty} K(t,q)(\cdot)dq \tag{B-28}$$

and carrying out the substitutions indicated by the addition and composition operations. Formally, we have the following.

*Proposition B-5

Let L and K be linear operators from D_+ to D_+ which are represented by

$$L(t,q)$$

and

$$K(t,q)$$

respectively. Then the composite operator KL is represented by

$$K(t,q) \circ L(t,q) = \int_{-\infty}^{\infty} K(t,u)L(u,q)du$$

and the sum operator $K + L$ is represented by

$$K(t,q) + L(t,q) \qquad \square$$

An additional operation on the linear operators is the adjoint, for which the kernel interpretation is given by the following proposition.

*Proposition B-6

Let an operator H from D_+ to D_+ be represented by the kernel

$$H(t,q)$$

Then the adjoint of H, H^*, is represented by the kernel

$$H(q,t)' \qquad \square$$

As with Proposition B-5, the reader may prove this result by writing the defining equality for the adjoint

$$\langle x, Hy \rangle = \langle H^*x, y \rangle \qquad \text{(B-29)}$$

in its integral form and substituting the "integral" representations of H and H^*, as per the proposition.

Finally, if H has an inverse, H^{-1}, we denote its kernel, if it exists, by $H(t,q)^{-1}$. Clearly, $H(t,q)^{-1}$ is defined by the equality

$$H(t,q) \circ H(t,q)^{-1} = H(t,q)^{-1} \circ H(t,q) = \delta(t-q) \qquad \text{(B-30)}$$

since $\delta(t-q)$ is the kernel of the identity and $\circ$ corresponds to operator multiplication. Of course, similar arguments apply to left and right inverses.

B-4 The Distributional Fourier Transform

Unlike the operations defined in Section B-2, the Fourier transform is not defined for all distributions in D'. Rather, we must restrict consideration to a subclass of the distributions, termed tempered distributions. To properly define such a subclass of distributions, one requires a new space of testing functions which is larger than D (if one increases the number of testing functions each distribution will have to "pass more tests," and hence, there will be fewer distributions). For our present purpose, this will be the space S of smooth functions which go to zero faster than any polynomial. We will also allow complex-valued functions in the present context. This extension in no way effects the results thus far derived for real-valued distributions.

Definition B-7

S is the set of smooth complex-valued functions of a real variable such that

$$\sup_t \left| t^m \phi(t) \right| < \infty$$

for ϕ and all of its derivatives. □

Clearly, $D \subset S$, since functions in D are zero outside of a bounded set. In S we define a mode of weak convergence by the requirement that

$$\sup_t \left| t^m [\phi_i(t) - \phi(t)]^{(n)} \right| \qquad \text{(B-31)}$$

goes to zero as i goes to infinity for all m and n if ϕ_i is to converge to ϕ. Similarly, the continuity of an operator or functional defined on S will be taken to mean that if a sequence of inputs converges to zero in the above mode, then the sequence of outputs also converges to zero (in the mode of convergence of the range space). The space of distributions S' is thus defined as follows.

Definition B-8

The *tempered distributions* S' are the linear continuous complex-valued functionals on S. □

The reason for dealing with S' rather than D' when defining the distributional Fourier transform, is that the usual Fourier transform is especially well behaved for the functions in S and, hence, may be translated into an equally well behaved transform on S'. We recall that for a complex-valued function of a real variable, the Fourier transform is defined via

$$F(x) = \int_{-\infty}^{\infty} x(t) e^{-j\omega t} dt \qquad \text{(B-32)}$$

and its inverse is defined by

$$F^{-1}(y) = \frac{1}{2\pi} \int_{-\infty}^{\infty} y(\omega) e^{j\omega t} d\omega \qquad \text{(B-33)}$$

when the integrals exist. If x and y are in S, we then have the following proposition.

Proposition B-7

The Fourier transform and its inverse are both defined for every x in S and take S continuously onto itself. □

Proposition B-7 is a well-known result of transform theory [40], [63], [77] and, as with most of the results to follow in this section and the next, will not be proven. It does, however, justify the following definition for the Fourier transform of a tempered distribution in S'.

Definition B-9

For any tempered distribution $x(t)$ in S', its Fourier transform $F(x)$ is the tempered distribution in S' defined by

$$\int_{-\infty}^{\infty} F(x)(\omega)\phi(\omega)d\omega = \int_{-\infty}^{\infty} x(\omega)F(\phi)(\omega)d\omega$$

and similarly for F^{-1}. □

Proposition B-8

The Fourier transform and its inverse are well defined for all tempered distributions in S. Moreover, if $x(t)$ is a distribution derived from a locally integrable function which possesses a Fourier transform in the sense of Equation (B-32), then the distributional Fourier transform coincides with the usual transform.

Proof

If ϕ is in S, then by Proposition B-7, $F(\phi)$ exists and is also a function in S; hence, the right side of the defining equation for $F(x)$ is well defined [since x is defined on all functions in S including $F(\phi)$]; hence, the transform is well-defined for all tempered distributions in S'. Now, if $x(t)$ is a locally intergrable function and has Fourier transform

$$F(x)(\omega) = \int_{-\infty}^{\infty} x(t)e^{-j\omega t}dt \tag{B-34}$$

then for ϕ in S,

$$\int_{-\infty}^{\infty} F(x)(\omega)\phi(\omega)d\omega = \int_{-\infty}^{\infty}\left(\int_{-\infty}^{\infty} x(t)e^{-j\omega t}dt\right)\phi(\omega)d$$

$$= \int_{-\infty}^{\infty}\int_{-\infty}^{\infty} x(t)e^{-j\omega t}\phi(\omega)d\omega\, dt = \int_{-\infty}^{\infty} x(t)\left(\int_{-\infty}^{\infty} \phi(\omega)e^{-j\omega t}d\omega\right)dt$$

$$= \int_{-\infty}^{\infty} x(t)F(x)(t)dt = \int_{-\infty}^{\infty} x(\omega)F(x)(\omega)d\omega$$

$$\tag{B-35}$$

Hence, the distributional definition coincides with the usual definition when the latter is well defined. Of course, a similar argument applies to F^{-1}; hence, our proof is complete. □

The distributional Fourier transform has all of the usual properties associated with the Fourier transform of a function, these being summarized as follows:

*Proposition B-9

For tempered distributions x and y the following equalities hold:

1. $F(x^{(n)}) = (j\omega)^n F(x)$
2. $F(\sigma x) = e^{-jT\omega} F(x)$
3. $\langle \overline{F(x)}, F(y) \rangle = \langle x, y \rangle$ when the inner products exist.

Here σ is the shift operator taking $x(t)$ to $x(t - T)$, and the overbar denotes the complex conjugate operation. □

Of course, the Parseval relation 3 extends to the case of vectors of distributions.

B-5 The Distributional Laplace Transform

The usual Laplace transform relation is given by

$$L(x) = \int_{-\infty}^{\infty} x(t) e^{-pt} dt \quad \text{(B-36)}$$

where $p = \sigma + j\omega$ is complex. If we write

$$e^{-pt} = e^{-\sigma t} e^{-j\omega t} \quad \text{(B-37)}$$

Equation (B-36) becomes

$$L(x) = \int_{-\infty}^{\infty} x(t) e^{-\sigma t} e^{-j\omega t} dt = \int_{-\infty}^{\infty} \left(x(t) e^{-\sigma t} \right) e^{-j\omega t} dt$$

$$= F(xe^{-\sigma t}) \quad \text{(B-38)}$$

Hence, we may view the Laplace transform of x as the Fourier transform of $xe^{-\sigma t}$, whence the theory of the previous section may be invoked. For this to be well defined, $xe^{-\sigma t}$ must be in S', though x does not have to be in this space.

Definition B-10

For a distribution x in D', if $xe^{-\sigma t}$ is in S', the Laplace transform of x at $p = \sigma + j\omega$, ω real, is defined by

$$L(x) = F(xe^{-\sigma t}) \quad □$$

Note that if x is in D', so is $xe^{-\sigma t}$, via Proposition B-5, but it may not be in S', which is a subset of D'. Hence, the Laplace transform may not be well defined for all (or even any) σ. The set of real numbers σ for which $L(x)$ exists is, however, assured to be an interval (possibly infinite or semi-infinite) if it is nonvoid [40].

From an inspection of the classical definition of the Laplace transform of Equation (B-36), it might seem intuitively palatable to simply use that equation as the definition of the Laplace transform, with a distribution x replacing the function and e^{-pt} viewed as a testing function. Of course, e^{-pt} is not in D or even S; hence, this is not, in general, well defined. It is, however, a valid (and very convenient) equality for certain special classes of distributions. For our purpose, the most important such class of distributions are those which are zero for t less than zero, since these correspond to the distributional weighting functions of causal operators. The characteristics of the Laplace transforms of such distributions, as well as the validity of the distributional form of Equation (B-36) for this class of distributions, are delineated in the following fundamental theorem, the proof of which appears in reference [40], and will not be given here.

Theorem B-2

Let x be a distribution in D' such that

$$x(t) = 0 \quad t < 0$$

and

$$xe^{-\sigma t} \text{ is in } S'$$

for all real $\sigma > 0$. Then for all complex p with Re $p > 0$ (that is, p in the RHP), $L(x)(p)$ exists and is analytic. Moreover,

$$L(x) = \int_{-\infty}^{\infty} x(t)e^{-pt}dt \quad \text{Re } p > 0$$

and

$$\lim_{\sigma \to 0} L(x)(\sigma + j\omega) = F(x)(\omega) \qquad \square$$

Of course, the limiting condition of the theorem allows one to apply the Fourier transform results of Proposition B-9 to the Laplace transform after an appropriate continuation to the RHP.

With the development thus far achieved, the question of finding the Laplace transform of the delta function, considered in the introduction to this appendix, can be settled.

$$\int_{-\infty}^{\infty} \delta(q)\phi(q)dq = \phi(0) \qquad \text{(B-39)}$$

where δ satisfies the condition of Theorem B-2. Hence,

$$L(\delta) = \int_{-\infty}^{\infty} \delta(q)e^{-pq}dq = e^{-p0} = 1 \qquad \text{(B-40)}$$

In Chapter 3 it is shown that the system function associated with an "integral" operator is

$$H(p,t) = \int_{-\infty}^{\infty} H(t, t-v)e^{-pv}dv \equiv \int_{-\infty}^{\infty} \mathbf{H}(t,v)e^{-pv}dv \qquad \text{(B-41)}$$

where v is the age variable,

$$v = t - q \qquad \text{(B-42)}$$

and $H(t,q)$ is the kernel representation of a causal network. Thus, after an appropriate change of variable, the system function is the Laplace transform with respect to the age variable of the kernel,

$$H(p,t) = \int_{-\infty}^{\infty} \mathbf{H}(t,v)e^{-pv}dv \qquad \text{(B-43)}$$

B-6 Discussion

Although the intuitive aspects of distribution theory are clear, the formulation of a sound theory of distributions follows a circumferential route. In particular, the distributions are not functions.

While they are not functions, they can be employed in a manner similar to functions and have substantially the same properties so long as they remain under the "integral" sign. Additionally, when a distribution is derived (by isomorphism) from a function, the distributional properties coincide with the usual functional properties.

Since a theory of distributions can be formalized, it would be desirable to have distributional input-output signal spaces. While this is possible in certain special cases [26], [185], [210], it has not (at least as yet) been achieved for the general case. In particular, our approach fails since the kernels theorem and its corollary require the inputs to be in D and D_+, respectively, but do not allow D' inputs.

APPENDIX C
Graph Theory

C-1 Introduction

A graph can be viewed, alternatively, as a topological structure or as an algebraic structure, the "topology" being derived from the connectivity aspects of the graph, while the "algebra" is that induced by the Kirchhoff laws. In network theory, both aspects are employed to advantage, the former allowing a graph to be readily identified with a network, while the latter is amenable to the algebraic manipulations required to form the network equations.

In this appendix, both aspects of the theory are considered. The basic topological properties of the graph and its subgraphs are developed in the first section. These are then employed in the second section to define the voltage and current vectors associated with a graph, these being used to implement the transition between the algebraic and topological aspects of the theory. This is achieved via matrices which are constructed from the circuits and cocircuits of the graph, and whose respective null spaces are the admissible voltage and current vectors of the graph.

Due to the excessive number of circuits and cocircuits in a graph, a matrix of all such subgraphs would be too large for convenient application. Fortunately, smaller sets of circuits and cocircuits which suffice to define the network variables can be found. One such class, the fundamental circuits and cocircuits of a specified tree, is considered in the third section. In the fourth section, these subgraphs are further exploited with the definition of the fundamental circuit, cocircuit, and connection matrices, B_f, S_f, and F. The orthogonality of B_f and S_f is exhibited and employed to extract a single matrix C which characterizes all three graph matrices. In the final section, the B_f and S_f matrices are used to obtain two alternative forms of the Kirchhoff laws.

C-2 Graph Theory

Conceptually, a graph is a pattern of lines (the edges) which connect points (the vertices). Since one wishes to associate network variables with the edges of a graph, they are oriented as indicated by an arrowhead, this for the purpose of defining a polarity for the network variables. The graph of Figure C-1 has

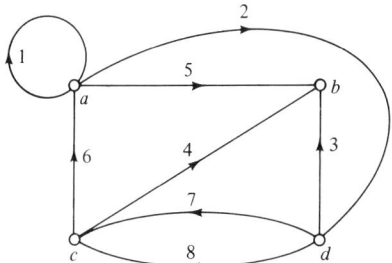

Figure C-1 Graph with eight edges and four vertices.

eight edges and four vertices, the edges being denoted by numbers and the vertices by lower case italic letters.

From the network viewpoint, the graph serves to describe the interconnection, each edge corresponding to a port. In the graph of Figure C-1, the ports represented by edges 7 and 8 are connected in parallel, whereas that represented by edge 1 is short circuited.

In formalizing the concept of a graph, each edge may be denoted by an ordered triple. For example,

$$(5,a,b)$$

denotes edge 5 whose orientation is from vertex a to vertex b. Equivalently the set of ordered triples defining the edges of a graph may be viewed as a function

$$E : J \to V \times V$$

where J is the set of edge names, V is the set of vertices, and $V \times V$ is the cartesian product of V with itself, that is, the set of all possible pairs of vertices from V.

Definition C-1

A *graph* is a pair (V,E), where V is a set, the *vertices*, and E is a function from an index set J to $V \times V$.

$$E : J \to V \times V$$

The ordered triples of which E is composed are the *edges* of the graph. □

Although the theory of graphs can be developed in the general case [23], [128] only the finite case will be considered here, since that is the only case applicable to network problems [83], [155]. In the graph of Figure C-1,

$$V = \{a,b,c,d\} \tag{C-1}$$

and

$$E = \{(1,a,a),(2,a,d),(3,d,b),(4,c,b),(5,a,b),(6,c,a),(7,d,c),(8,d,c)\} \tag{C-2}$$

Since the function E is single-valued, the edge (j,x,y) can, without inducing any ambiguity, be denoted by the index j.

The concepts of incidence, vertex degree, and subgraphs are basic to any study of graph theory.

Definition C-2

An edge $j = (j,a,b)$ of a graph is *incident* on vertices a and b. □

Definition C-3

The *degree* of a vertex is the number of edges incident to it. □

In the graph of Figure C-1, edge 5 is incident on vertices a and b, which have degrees of 5 and 3, respectively.

Definition C-4

Let (V,E) be a graph. Then a *subgraph* of (V,E) is a graph (U,F) such that U is contained in V and F is contained in E. □

The requirement that a subgraph be a graph implies that any vertices incident to edges of F are in U. This is due to the fact that the range of F is in $U \times U$, not $V \times V$. Vertices a, d, and c, together with edges 6, 7, and 1, form a subgraph of the graph shown in Figure C-1. If, however, edge 5 is added, the result is not a subgraph, since edge 5 is incident on vertex b, which is not in U.

The Kirchhoff voltage law (KVL) requires that the sum of the voltages about a closed path be zero. For the purposes of this work, it is convenient to deal with cycles, a generalization of the closed paths to which the KVL also applies.

Definition C-5

A *cycle* is a subgraph, every vertex of which is of even degree. □

In Figure C-1, typical cycles are the subgraphs composed of edges {3,5,6,8}, {4,7,3}, {4,5,6,7,8}, {1}. A circuit is the special kind of cycle which is minimal, thus it is itself a cycle, and hence, each of its vertices have even degree. Moreover, if any edge is deleted from a circuit, the resultant subgraph is no longer a cycle and, in fact, does not contain any cycle. A circuit is, therefore, a cycle which does not properly contain any other cycle.

Definition C-6

A *circuit* is a nonvoid cycle which does not properly contain any other cycle. □

In Figure C-1, {4,7,3} is a circuit, since the deletion of any edge results in a subgraph composed of two lines which clearly has no subgraph, the vertices of which all have even degree (except the subgraph with no edges, for which all vertices have degree zero, which is, however, specifically excluded by Definition C-6). An example of a cycle which is not a circuit is the subgraph composed of edges {4,5,6,7,8}. This is not minimal, since its proper subgraph containing edges {7,8} is also a cycle.

The terminology employed for the various subgraphs is anything but universal. The cycle is also called the circ, the cirk, and the Euler graph, while the term loop is commonly interchanged with circuit. A more commonly employed definition for the circuit is a connected subgraph, each vertex of which has degree two. The minimal cycle definition is employed here, so as to by-pass the concept of connectivity which does not play a significant role in the present development.

Definition C-7

A *tree* is a maximal subgraph containing no circuits. □

Several typical trees are shown in Figure C-2. Since a tree is maximal, if it is possible to join two portions of a graph without creating a circuit, such edges

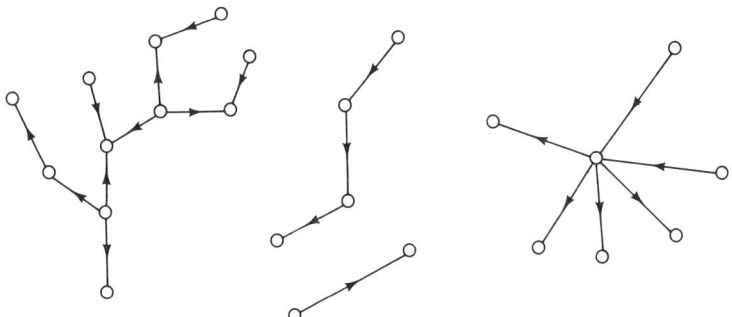

Figure C-2 Typical trees.

must be included in the tree. This subgraph therefore has the form which one might deduce from its name; that is, it is a collection of edges, none of which "circle around upon themselves," but which are connected wherever possible.

Some authors [128] reserve the term tree for a connected subgraph, while designating the subgraph of Definition C-7 a maximal forest. As with cycles and circuits, there is no standard terminology.

Fundamental to both graph theory and network theory (as well as most interesting mathematical structures), is the concept of duality. In the network context, voltage and current are dual. Correspondingly, the subgraph upon which the Kirchhoff current law (KCL) is based is dual to that upon which the KVL is founded. Subgraphs dual to the cycle, circuit, and tree are thus defined. These are denoted by the prefix "co," which is the standard (except among graph theorists) designation of duality.

Definition C-8

Let the vertices of a graph be partitioned into two classes, V^+ and V^-. Then a *cocycle* is a subgraph containing all the edges which are incident to vertices in two different classes, together with all such vertices. □

Definition C-9

A *cocircuit* is a nonvoid cocycle which does not properly contain any other cocycle. □

Definition C-10

A *cotree* is a maximal subgraph containing no cocircuits. □

The most common way to choose a cocycle, which in fact always yields a cocircuit, is to let V^+ be any one vertex and V^- be all of the remaining vertices from the graph. The edges going from class V^+ to V^- are then precisely the edges incident on the vertex in V^+ (except those which are incident twice on that vertex). In the graph of Figure C-1, if one lets V^+ be vertex b, the resultant cocycle contains edges $\{3,4,5\}$, while if a is in V^+, the resultant cocycle contains edges $\{2,5,6\}$. Edge 1 does not appear in this later cocycle, since it is incident only on vertex a and, hence, is not on vertices of different classes. It should be noted that each cocycle may be obtained from two distinct vertex partitions, for if one interchanges the sets V^+ and V^-, exactly the same edges are incident on vertices of distinct classes. This ambiguity, although on occasion inconvenient, causes no real difficulty.

Of the various subgraphs, the one with the greatest proliferation of names is the cocircuit. This is called, in various areas of the literature, a cut-set [155], simple cut-set [83], bond [180], and coboundary [24], as well as cocircuit [108].

A cotree is, in fact, the edge complement of a tree and is often defined by this property [155]. That is, if a graph with 10 edges, say $\{1,2,\cdots,10\}$, has a tree with edges {1,2,3}, then there is a cotree with edges 4 thru 10, and, conversely, if edges 1 thru 3 form a cotree, then edges 4 thru 10 form a tree. The formal proof of this statement we leave as an exercise.

*Proposition C-1

Let (V,E) be a graph. Then a subgraph (V,T) is a tree if and only if $(V, E - T)$ is a cotree. ☐

The concepts of a path in a graph and that of a connected graph are fundamental graph theoretic concepts, but of little relevance to the problems of generalized networks. For completeness, their definitions and properties are given as follows. The proofs are, however, left to the reader as exercises.

*Definition C-11

A *pseudo-path* is a subgraph, all of whose vertices have degree two, except two of them which have degree one. The vertices with degree one are the *terminal vertices*. ☐

The intuitive idea of a "path" as a sequence of edges [see, for example, Figure C-3(a)] is clearly included within Definition C-11. In this case, the two "end"

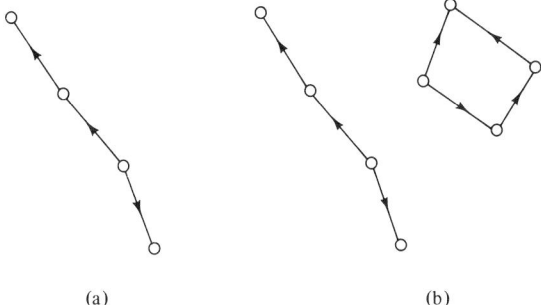

(a)   (b)

Figure C-3 (a) Path. (b) Pseudo-path formed from it by adjoining a circuit.

vertices of the path have degree one, whereas the "internal" vertices have degree two. The pseudo-paths, however, contain a larger class of subgraphs, for if one were to start with a "path" and adjoin a circuit to it, the resultant subgraph would also be a pseudo-path, as illustrated in Figure C-3. We leave it to the reader to verify that all pseudo-paths may be obtained from a "path" by this means.

Although a pseudo-path may be a "disconnected" subgraph, the two vertices of degree one, the terminal vertices, are always in the same "connected part"; hence, the pseudo-path, a "disconnected" subgraph, may be used in the formal definition of connectivity.

Definition C-12

A graph is *connected* if every pair of its vertices are the terminal vertices of some pseudo-path. □

Definition C-13

A *path* is a connected pseudo-path. □

We leave it to the reader as an exercise to verify that these definitions do indeed coincide with our intuitive view of these concepts.

While the tree may be defined without resort to connectivity, there are a number of useful characterization theorems for the trees of connected graphs.

Proposition C-2

The following are equivalent descriptions of a subgraph of a connected graph with v vertices:

1. it is a tree
2. it is a minimal connected subgraph containing all vertices
3. it is connected, has no circuits, and $v - 1$ edges. □

Proposition C-3

In a tree of a connected graph there exists a unique path between every pair of vertices. □

Any graph, connected or not, may be viewed as a collection of connected parts. For instance, the graph of Figure C-4(a) has two connected parts, one

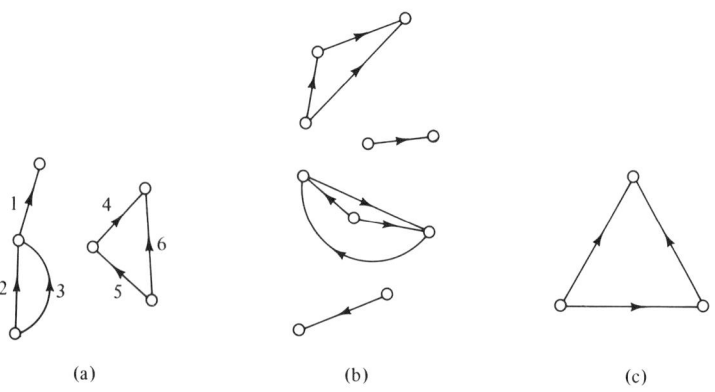

Figure C-4 Graphs with (a) 2, (b) 4, and (c) 1 part.

composed of edges {1,2,3}, and the other of edges {4,5,6}. Similarly, the graph of Figure C-4(b) has four parts, while that of Figure C-4(c) has only one part. Formally, we have the following.

*Definition C-14

A *part* of a graph is a maximal connected subgraph. □

Clearly, a connected graph has only one part, since the entire graph is connected and obviously maximal.

If p is the number of parts of a graph and v the number of its vertices, then we have the following proposition.

*Proposition C-4

Every tree of a graph has $v - p$ edges. □

*Proposition C-5

Let T be a tree of a graph, G, and P be a part of G. Then the intersection of T and P is a tree of the graph, P. □

Like the trees of a connected graph, alternate descriptions for the circuits and cocircuits are available.

*Proposition C-6

A circuit is a connected subgraph, every vertex of which has degree two. □

*Proposition C-7

A cocircuit is a minimal subgraph whose edge compliment is disconnected. □

C-3 Voltage and Current Vectors

We now consider an often neglected aspect of network theory — the definition of voltage and current. This is achieved by letting voltage be a set of functions associated with a graph which satisfy the Kirchhoff law, and similarly for current. This approach serves to "define one's problems away," since it eliminates both the consideration of physical phenomena and the proof of the Kirchhoff laws once these phenomena have been understood (if ever). An additional advantage is the generality achieved. Since no physics is required, the results hold for any structure satisfying conservation laws similar to those of electrical networks [79].

Mathematically, the voltage functions or vectors may lie in any vector space, module, or even group [94]. In fact, there are a number of applications of the theory of graphs wherein the use of an abstract mathematical structure as the setting for the voltage and current vectors is desirable [83], [108], but in the present context, the input-output space for the networks is D_+, the space of smooth functions with support bounded on the left; hence, our voltages and currents are taken to be D_+ functions.

Intuitively, one would desire to define voltage as a set of functions from D_+ such that at any instant of time their sum around a circuit was zero. As a necessary preliminary, one must first define "around." Now, if one takes the closed path viewpoint of a circuit, this may be done in the obvious manner, simply by choosing an orientation about the closed path in either of the two possible senses, as illustrated in Figure C-5(a) and (b). Precisely the same

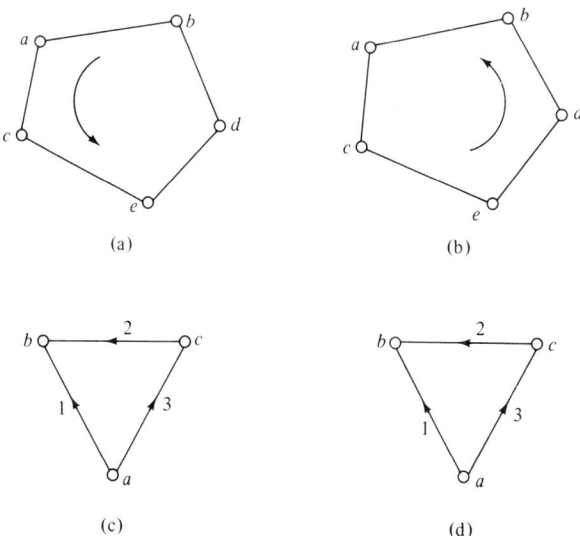

Figure C-5 Orientation of circuits.

result may be obtained by invoking the formal definition for a circuit of the preceding section. This result, which is illustrated in Figure C-5(c) and (d), is the subject of the following lemma.

Lemma C-1

Every circuit of a graph can be represented as a sequence

$$\{(1,a,b), (2,b,c),(3,c,d),\cdots, (n-1,p,q),(n,q,a)\}$$

where vertices denoted by distinct letters are distinct. Furthermore, the representation is unique except for a change of starting point and/or reversal of order. (The edge and vertex names have been arranged in sequential order only for notational convenience, but this need not be the case, in general.)

Proof

Start with any edge of the given circuit and denote it by $(1,a,b)$. Since every vertex of a cycle (hence a circuit) is of even degree, and edge 1 is incident on vertex b, there must be at least one additional edge incident on vertex b. Let edge $(2,b,c)$ be such an edge. By repeating the argument the process can be continued until it terminates by meeting a vertex for the second time. This termination must occur since the graph is finite. Let

$$\{(k,r,s), (k+1,s,t), \cdots, (n,v,r)\}$$

be the terminal portion of the sequence which closes on itself. Since r is the first vertex to be encountered twice, the internal vertices of the sequence are distinct and the edges of the terminal sequence form a cycle, since they all have degree two. Since a circuit is a minimal cycle, the given circuit cannot properly contain the constructed sequence, which leaves only the possibility that the constructed sequence is the entire circuit.

The existence of such a representation for a circuit implies that there was, in fact, no ambiguity in choosing the sequence (since there were only two edges incident on each vertex of the circuit). Thus the representation was unique, except for the choice of starting point and/or direction. □

In essence, the sequence obtained in the lemma serves to replace the path orientation of Figure C-5(a) and (b), since it determines an ordering of the edges in the circuit. For this purpose, the starting point for the sequence is of no importance; hence, the ambiguity of the lemma with regard to starting point causes no difficulty. Of course, the possibility of reversing the order of the sequence corresponds to a reversal of orientation around a closed path. In the graph of Figure C-5(c), the clockwise orientation corresponds to the sequence

$$(2,b,c)(3,c,a)(1,a,b)$$

whereas the counterclockwise orientation for the same graph, shown in Figure C-5(d), corresponds to the sequence

$$(1,b,a)(3,a,c)(2,c,b)$$

The sequential representation of a circuit permits the definition of a function e_C which characterizes the circuit.

Definition C-15

Let C be a circuit of a graph G which is represented by a sequence, as per Lemma C-1. Then e_C is a function from J to $\{-1,0,1\}$

$$e_C : J \to \{-1,0,1\}$$

such that

1. $e_C(j) = 0$ if edge j is not in C
2. $e_C(j) = 1$ if edge j is in C and the order of the vertices of edge j as encountered in the sequential representation of the circuit coincides with the orientation of the edge
3. $e_C(j) = -1$ otherwise. □

Once again, invoking the closed path view of a circuit, one may say that the edge function e_C is 1 if the path orientation induced by the given sequential representation coincides with the edge orientation, and it is -1 if the path orientation does not coincide with the edge orientation.

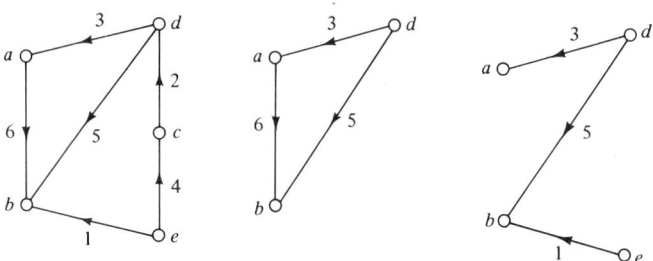

Figure C-6 Graph with circuit and cocircuit.

The graph of Figure C-6 has a circuit composed of edges 3, 5, and 6, which has six possible sequential representations, given by

$$\{(6,a,b),(5,b,d),(3,d,a)\}$$
$$\{(5,b,d),(3,d,a),(6,a,b)\} \quad \text{(C-3)}$$
$$\{(3,d,a),(6,a,b),(5,b,d)\}$$

and

$$\{(6,b,a),(3,a,d),(5,d,b)\}$$
$$\{(3,a,d),(5,d,b),(6,b,a)\} \quad \text{(C-4)}$$
$$\{(5,d,b),(6,b,a),(3,a,d)\}$$

The representations in Equation (C-3) yield an e_C function

j	1	2	3	4	5	6
$e_C(j)$	0	0	1	0	-1	1

while the representations in Equation (C-4) yield an e_C function

j	1	2	3	4	5	6
$e_C(j)$	0	0	-1	0	1	-1

Clearly, e_C is not independent of the representation employed. Fortunately, a change in representation at most negates e_C, this occurring when the order of the representation is reversed. Since our application of the function e_C is independent of a sign change, this ambiguity causes no difficulty and e_C may be considered to be representation independent.

Consistent with the pattern of duality which has thus far been established between the circuits and the cocircuits, one would like to define an orientation for the cocircuits which is dual to that formed for the circuits. The method of achieving this goal is illustrated in Figure C-7. Here one begins with a graph and an arbitrary cocircuit obtained from the vertex partition V^+, V^-, as shown in Figure C-7(a). Now, if the graph is redrawn as shown in Figure C-7(b),

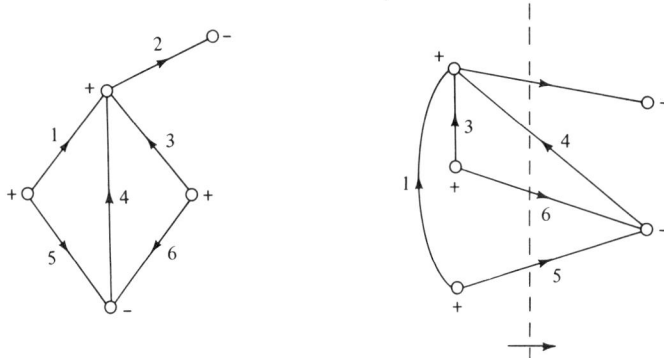

Figure C-7 Graph and cocircuit with orientation.

with all of the V^+ vertices on one side and the V^- vertices on the other, then an orientation for the cocircuit may naturally be defined by drawing an arrow from the V^+ vertices to the V^- vertices. Of course the same cocircuit would be obtained if one were to interchange V^+ and V^-, in which case the technique illustrated in Figure C-7 would yield the opposite orientation. This is similar to the ambiguity encountered in defining the circuit orientation and, as in that case, causes no difficulty. These ideas are formalized by the following definition.

Definition C-16

Let S be a cocircuit of a graph G which is defined by a vertex partition V^+ and V^-. Then e_S is a function from J to $\{-1,0,1\}$,

$$e_S : J \to \{-1,0,1\}$$

such that

1. $e_S(j) = 0$ if edge j is not in S
2. $e_S(j) = 1$ if edge j is in S with vertex a in V^+ [here edge j is represented by $j = (j,a,b)$]
3. $e_S(j) = -1$ otherwise. ☐

The cocircuit shown in Figure C-6 is defined by the vertex partition

$$V^+ = \{a,b\} ; \quad V^- = \{c,d,e\} \tag{C-5}$$

for which e_S is given by

j	1	2	3	4	5	6
$e_S(j)$	-1	0	-1	0	-1	0

The circuit and cocircuit functions play the role of the zero and ± 1 constant multipliers of the Kirchhoff laws, thus permitting the following definition.

Definition C-17

A *voltage vector* is a function associated with a graph,

$$v : J \to D_+$$

such that

$$\sum_j [v(j) e_C(j)] = 0$$

for all circuits C. □

Definition C-18

A *current vector* is a function associated with a graph,

$$i : J \to D_+$$

such that

$$\sum_j [i(i) e_S(i)] = 0$$

for all cocircuits S. □

Since functions from j into D_+ may be viewed as n vectors of functions from D_+ (where n is the number of edges in the graph under consideration), the term vector has been used in the preceding definitions and vector notation will generally be employed for v and i.

C-4 Fundamental Circuits and Cocircuits

In the preceding sections the voltage and current vectors were defined by requiring that they satisfy the Kirchhoff laws for all circuits and cocircuits. While it is possible to formulate an analysis technique which directly employs the characteristics of all circuits and cocircuits [155], this is neither desirable nor necessary. Rather, the fundamental circuits and cocircuits of a specified tree are employed, the properties of these completely characterizing the graph.

Lemma C-2

For any connected graph with a distinguished cotree T there exists a unique collection of circuits, each containing one specified edge of the cotree and no other cotree edges.

Proof

Let an edge $j = (j,a,b)$ be in T. Now, by Proposition C-3, there exists a unique path in the tree (which is the compliment of T) between vertices a and b. This path, together with the specified edge j, is the required circuit, which is unique since the path is unique. □

In the graph of Figure C-8, the tree composed of edges 1, 3, and 4 defines two circuits, one containing cotree edge 2 with tree edges 1 and 3, and the

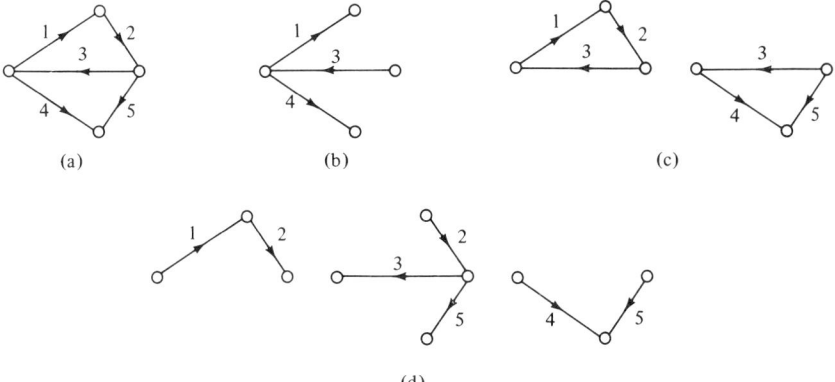

Figure C-8 (a) Graph with (b) tree, (c) f circuits, and (d) f cocircuits.

other containing cotree edge 5 with tree edges 3 and 4. Lemma C-2 may be extended to the general case of a disconnected graph by applying it separately to each connected part of the graph.

Proposition C-8

For any graph with a distinguished cotree T, there exists a unique collection of circuits, each containing one specified edge of the cotree and no other cotree edges.

Proof

By the lemma, the proposition holds for each part of the graph with the cotree

$$T^i = T \cap P^i \qquad \text{(C-6)}$$

used in the ith part. Now, the set of circuits for each part is the appropriate set of circuits for the entire graph. □

Definition C-19

For any graph with a distinguished cotree T, the collection of circuits found in Proposition C-8 are the *fundamental circuits* (f circuits) associated with T. □

Proposition C-9 is the dual of Proposition C-8, for which a similar proof is applicable.

Proposition C-9

For any graph with a distinguished tree T, there exists a unique collection of cocircuits, each containing one specified edge of the tree and no other tree edges. ☐

For the graph of Figure C-8, the cocircuits containing tree edges 1, 3, and 4 are

$$\{1,2\}, \{2,3,5\}, \{4,5\}$$

respectively.

A procedure for construction of the cocircuits associated with a tree is illustrated in Figure C-9. Here the distinguished tree is indicated by the double edges. Now, if one desires to obtain the cocircuit containing some specified

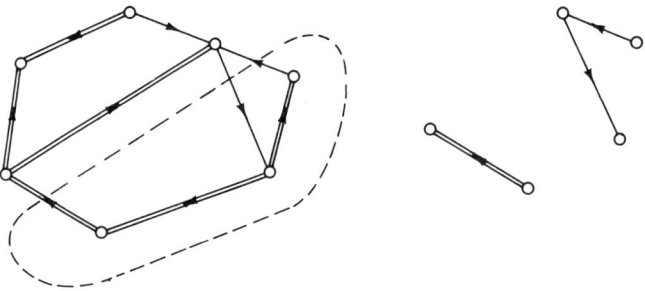

Figure C-9 Construction of fundamental cocircuits.

tree edge and no other tree edge, a closed path is drawn on the graph which crosses that tree edge and no other. Such a path clearly defines a vertex partition with V^+ taken as the vertices inside the closed path and V^- taken as those outside. Moreover, the corresponding cocircuit is just the graph composed of the edges which cross this path (an odd number of times), since such edges have one vertex in V^+ and the other in V^-. We leave it as an exercise for the reader to confirm that this cocircuit is indeed the desired one, and that it is, in fact, independent of the particular closed path used in its construction (that is, it is determined entirely by the tree and its distinguished edge).

Definition C-20

For any graph with a distinguished tree T, the collection of cocircuits found in Proposition C-9 are the *fundamental cocircuits* (f cocircuits) associated with T. ☐

In dealing with the f circuits and f cocircuits of a graph (with specified tree and cotree), it is convenient to establish special notation for their associated edge functions. With regard to a specific tree (and its cotree), its f circuits may be indexed by the (unique) cotree edge they contain, and its f cocircuits by the (unique) tree edge they contain. Thus,

$$C_k; \quad k \text{ in } \mathbf{T} \tag{C-7}$$

and

$$S_k; \quad k \text{ in } \mathbf{T} \tag{C-8}$$

are the f circuits and f cocircuits associated with T, and they have corresponding edge functions

$$e_{C_k}; \quad k \text{ in } \mathbf{T} \tag{C-9}$$

and

$$e_{S_k}; \quad k \text{ in } \mathbf{T} \tag{C-10}$$

For any circuit or cocircuit both, e_C and $-e_C$ (e_S and $-e_S$) are valid edge functions (depending on the choice of orientation). In dealing with the f circuits and f cocircuits, however, it is possible to distinguish between these two cases via the sign of the defining edge. Thus if

$$e_{C_k}(k) = 1 \tag{C-11}$$

the edge function is said to be positive and denoted by

$$e_{C_k}{}^+$$

whereas if

$$e_{C_k}(k) = -1 \tag{C-12}$$

the edge function is said to be negative and denoted by

$$e_{C_k}{}^-$$

Since the edge k is, by definition, in C_k, the third case that

$$e_{C_k}(k) = 0 \tag{C-13}$$

cannot occur. Clearly,

$$e_{C_k}{}^- = -e_{C_k}{}^+ \tag{C-14}$$

and both are valid edge functions for C_k. Of course, a similar notation can (and is assumed to) be established for the f cocircuits of a specified tree. For the f circuits of Figure C-8, the edge functions are

	1	2	3	4	5
$e_{C_2}{}^+$	1	1	1	0	0
$e_{C_2}{}^-$	-1	-1	-1	0	0
$e_{C_5}{}^+$	0	0	-1	-1	1
$e_{C_5}{}^-$	0	0	1	1	-1

406 GRAPH THEORY

The reason for giving special consideration to the fundamental circuits and cocircuits is due to the fact that these particular subsets of circuits and cocircuits are sufficient to completely characterize the voltage and current variables admissible to the graph.

*Proposition C-10

Let a graph have a distinguished cotree T. Then a function (vector in D_+^n)

$$v : J \to D_+$$

is a voltage vector if and only if

$$\sum_{j=1}^{n} [e_{C_k}^{+}(j)v(j)] = 0$$

for all C_k, k in T (that is, if the Kirchhoff voltage law is satisfied for all of the f circuits of a given cotree, then it is satisfied for all circuits of the graph). □

*Proposition C-11

Let a graph have a distinguished tree T. Then a function (vector in D_+^n)

$$i : J \to D_+$$

is a current vector if and only if

$$\sum_{j=1}^{n} [e_{S_k}^{+}(j)i(j)] = 0$$

for all S_k, k in T (that is, if the Kirchhoff current law is satisfied for all f cocircuits associated with a given tree, then it is satisfied for all cocircuits of the graph). □

The essence of the proofs for these propositions is that the edge function for any circuit or cocircuit can be written as a linear combination of those for the f circuits and f cocircuits of the specified tree. In the graph of Figure C-8, the edge function for the "outside circuit" is

j	1	2	3	4	5
$e_{(1,2,4,5)}$	1	1	0	−1	1

which could also have been written as the sum of the edge functions for the two f circuits. In this case,

$$e_{(1,2,4,5)} = e_{C_2}^{+} + e_{C_5}^{+} \tag{C-15}$$

This property, in turn, allows the Kirchhoff laws for any circuit or cocircuit to be written as a linear combination of those for the f circuits and f cocircuits. Thus, if the KVL (KCL) holds for all f circuits (f cocircuits), it must hold for all circuits (or cocircuits) their linear combinations.

C-5 The Graph Matrices

Consistent with the results of Propositions C-10 and C-11, it suffices to consider only fundamental circuits and cocircuits in defining the graph matrices.

Definition C-21

For a graph with distinguished cotree T, its *fundamental circuit matrix* (B_f) has a column for each edge of the graph and a row for each cotree edge (or, equivalently, for each f circuit of T) where

$$B_f(k,j) = e_{c_k}^+(j) \qquad \square$$

B_f is simply an array of the edge functions for all of the f circuits associated with T. For the graph of Figure C-8,

$$B_f = \begin{array}{c} \\ 2 \\ 5 \end{array} \begin{array}{cc} \begin{matrix} 1 & 2 & 3 & 4 & 5 \end{matrix} \\ \begin{bmatrix} 1 & 1 & 1 & 0 & 0 \\ 0 & 0 & -1 & -1 & 1 \end{bmatrix} \end{array} \qquad \text{(C-16)}$$

The properties of B_f are most clearly illustrated after the choice of an appropriate ordering for its rows and columns.

Definition C-22

If a B_f matrix has all columns corresponding to tree edges at the left, all columns corresponding to cotree edges at the right, and, additionally, the ordering of the rows coinciding with that of the cotree columns (by which they may be indexed), then it has the *standard ordering*. $\square$

Placing the f circuit matrix of Equation (C-16) into the standard ordering yields

$$B_f = \begin{array}{c} \\ 2 \\ 5 \end{array} \begin{array}{cc} \begin{matrix} 1 & 3 & 4 & 2 & 5 \end{matrix} \\ \begin{bmatrix} 1 & 1 & 0 & 1 & 0 \\ 0 & -1 & -1 & 0 & 1 \end{bmatrix} \end{array} \qquad \text{(C-17)}$$

The standard ordering for B_f naturally partitions B_f [as shown in Equation (C-17)] by

$$B_f = [B_f^t \mid B_f^c] \qquad \text{(C-18)}$$

where B_f^t represents the tree columns of B_f and B_f^c represents the cotree columns. Since the number of cotree columns is the same as the number of rows, the submatrix B_f^c is always square.

Proposition C-12

Let B_f be a fundamental circuit matrix with the standard ordering, partitioned as

$$B_f = [B_f^t \mid B_f^c]$$

Then
$$B_f{}^c = 1$$
is the identity matrix.

Proof

The diagonal entries in $B_f{}^c$ correspond to a cotree edge in the kth column and the f circuit defined by that cotree edge for the kth row. Thus,

$$B_f{}^c(k,k) = e_{C_k}{}^+(k) \tag{C-19}$$

which is, by the definition of $e_{C_k}{}^+$, equal to 1. The off-diagonal entries of $B_f{}^c$ correspond to the cotree edge j in the jth column and the f circuit defined by some other cotree edge in the kth row. Now,

$$B_f{}^c(k,j) = e_{C_k}{}^+(j) = 0 \tag{C-20}$$

since the jth cotree edge is not in the f circuit defined by the kth cotree edge. Thus,

$$B_f{}^c(k,j) = \begin{cases} 1 & \text{if } j = k \\ 0 & \text{if } j \neq k \end{cases} \tag{C-21}$$

Upon applying the proposition, B_f can be partitioned as

$$B_f = [B_f{}^t \mid 1] \tag{C-22}$$

The convenience of the form of Equation (C-22) is such as to justify assuming that all f circuit matrices possess the standard order.

The dual of B_f is the fundamental cocircuit matrix S_f.

Definition C-23

For a graph with distinguished tree, its *fundamental cocircuit matrix* S_f has a column for each edge of the graph and a row for each tree edge (or, equivalently, each cocircuit defined by the tree) where

$$S_f(k,j) = e_{S_k}{}^+(j) \qquad \square$$

As with B_f, S_f is an array of the edge functions e_{S_k} with k in T. For the graph of Figure C-8,

$$S_f = \begin{array}{c} \\ 1 \\ 3 \\ 4 \end{array} \begin{array}{c} \begin{array}{ccccc} 1 & 2 & 3 & 4 & 5 \end{array} \\ \begin{bmatrix} 1 & -1 & 0 & 0 & 0 \\ 0 & -1 & 1 & 0 & 1 \\ 0 & 0 & 0 & 1 & 1 \end{bmatrix} \end{array} \tag{C-23}$$

Definition C-24

If an S_f matrix has all columns corresponding to tree edges at the left, all columns corresponding to cotree edge at the right, and, additionally, the order of the rows the same as that of the tree columns (from which they are defined), then S_f has the *standard ordering*. □

The S_f matrix of Equation (C-23) is, after conversion to the standard order, given by

$$S_f = \begin{matrix} & \begin{matrix} 1 & 3 & 4 & 2 & 5 \end{matrix} \\ \begin{matrix} 1 \\ 3 \\ 4 \end{matrix} & \begin{bmatrix} 1 & 0 & 0 & -1 & 0 \\ 0 & 1 & 0 & -1 & 1 \\ 0 & 0 & 1 & 0 & 1 \end{bmatrix} \end{matrix} \quad \textbf{(C-24)}$$

which, as with B_f, may be partitioned by

$$S_f = [S_f{}^t \mid S_f{}^c] \quad \textbf{(C-25)}$$

on the basis of tree and cotree edges. As in the previous case, we have the following.

*Proposition C-13

Let S_f be a fundamental cocircuit matrix with the standard order, partitioned as

$$S_f = [S_f{}^t \mid S_f{}^c]$$

Then

$$S_f{}^t = 1$$

is the identity matrix. □

The proposition permits the partitioned form of S_f to be written as

$$S_f = [1 \mid S_f{}^c] \quad \textbf{(C-26)}$$

Hence, the standard ordering will be assumed for all S_f matrices. Additionally, when dealing simultaneously with an S_f and a B_f matrix, it will be assumed that both have identically ordered columns.

Possibly the most powerful result concerning the graph matrices is the orthogonality of S_f and B_f [83].

*Theorem C-1

Let B_f and S_f be f circuit and f cocircuit matrices for a graph based on a tree and its compliment, respectively, and assume that they have identical column order (and both have standard order). Then

$$B_f S_f' = 0$$

and

$$S_f B_f' = 0$$

The first of the many corollaries of Theorem C-1 follows. □

*Corollary C-1

Let B_f and S_f be as in Theorem C-1, and be partitioned as

$$B_f = [B_f{}^t \mid 1]$$

and

$$S_f = [1 \mid S_f{}^c]$$

respectively. Then

$$S_f{}^c = -(B_f{}^t)' \qquad \square$$

Since the submatrices $B_f{}^t$ and $S_f{}^c$ completely describe the voltage and current properties of the graph, and they are, moreover, related by Corollary C-1, a new notational scheme which shows this relationship is desirable.

*Definition C-25

Let

$$S_f = [1 \mid S_f{}^c]$$

and

$$B_f = [B_f{}^t \mid 1]$$

be as in Theorem C-1. Then

$$B_f{}^t = -(S_f{}^c)' \equiv C \qquad \square$$

For the graph of Figure C-8, the C matrix may be obtained by inspection from either the B_f matrix of Equation (C-17) or the S_f matrix of Equation (C-24) and is found to be

$$C = \begin{matrix} \\ 2 \\ 5 \end{matrix} \begin{matrix} 1 & 3 & 4 \\ \begin{bmatrix} 1 & 1 & 0 \\ 0 & -1 & -1 \end{bmatrix} \end{matrix} \qquad \text{(C-27)}$$

Substituting C into the partitioned forms of the fundamental circuit and cocircuit matrices now yields

$$S_f = [1 \mid -C'] \qquad \text{(C-28)}$$

and

$$B_f = [C \mid 1] \qquad \text{(C-29)}$$

which are the forms of S_f and B_f to be employed in the remainder of this appendix, as well as in Chapter 4. Since the tree portion of S_f and the cotree portion of B_f are always unit matrices, the submatrix C completely characterizes the graph.

Having established the fact that the circuit and cocircuit matrices are founded on the properties of a single submatrix C, it is reasonable to look for a single graph matrix which simultaneously characterizes the admissible voltage and current vectors of the graph. Such a matrix can indeed be found, and takes the form of the fundamental connection matrix F.

*Definition C-26

For any graph let C be as in Definition C-25. Then the *fundamental connection matrix* F has a row and column for each edge of the graph (assumed to have identical standard orderings) and is given by

$$F = \begin{bmatrix} -1 & | & C' \\ \hline 0 & | & 1 \end{bmatrix}$$

where the upper and left-hand row and column of the partition correspond to tree edges, and the lower and right-hand row and column correspond to cotree edges. □

For the graph of Figure C-8, F may be obtained directly from the C matrix of Equation (C-27), and is given by

$$F = \begin{bmatrix} & 1 & 3 & 4 & 2 & 5 \\ & -1 & 0 & 0 & 1 & 0 \\ & 0 & -1 & 0 & 1 & -1 \\ & 0 & 0 & -1 & 0 & -1 \\ \hline & 0 & 0 & 0 & 1 & 0 \\ & 0 & 0 & 0 & 0 & 1 \end{bmatrix} \qquad \text{(C-30)}$$

Clearly, the right-hand column of F is B'_f while the top row is $-S_f$; hence, F does indeed contain the characteristics of both matrices. Unlike the other graph, matrices F is invertible and we have the following proposition.

*Proposition C-14

F and F' are both self-inverse (that is, $F^{-1} = F$ and $(F')^{-1} = F'$). □

C-6 Matrix Characterization of the Network Variables

The graph matrices permit the voltage and current vectors of a network to be described algebraically.

Proposition C-15

A vector v in D_+^n is a voltage vector for a graph if and only if

$$B_f v = 0$$

for any B_f matrix of the graph.

Proof

The matrix equation

$$B_f v = 0 \qquad (C\text{-}31)$$

is exactly the simultaneous equations

$$\sum_{j=1}^{n} [e c_k{}^+(j) v(j)] = 0 \qquad k \text{ in } T \qquad (C\text{-}32)$$

which, by Proposition C-13, characterizes the voltage vectors of the graph. □

*Proposition C-16

A vector i in $D_+{}^n$ is a current vector for a graph if and only if

$$S_f i = 0$$

for any S_f matrix of the graph. □

Application of the partitioned form of S_f and B_f allows an alternative characterization of the voltage and current vector of a graph.

*Proposition C-17

Let

$$B_f = [C \mid \mathbf{1}]$$

be an f circuit matrix for a graph and v^t be a vector of tree voltages. Then the vector of cotree voltages is given by

$$v^c = -C v^t$$

□

*Corollary C-2

Let v^t be a vector of tree voltages for a graph and

$$S_f = [\mathbf{1} \mid -C']$$

be the f cocircuit matrix for that tree. Then the vector of all voltages is

$$v = S'_f v^t$$

□

The applications of Proposition C-17 and its corollary are two-fold. First, it allows the cocircuit matrix to be used to characterize the voltage vectors of the graph (as well as the f circuit matrix). Secondly, it exhibits the dependence of the various voltages in the network, all of which are determined by the tree voltages for any tree. The tree voltages themselves are, however, independent. The duals of Proposition C-17 and its corollary are as follows.

*Proposition C-18

Let
$$S_f = [1 \mid -C']$$
be an f cocircuit matrix for a graph and i^c be a vector of cotree currents. Then the vector of tree currents is given by
$$i^t = C'i^c \qquad \square$$

*Corollary C-3

Let i^c be a vector of cotree currents for a graph and
$$B_f = [C \mid 1]$$
be the f circuit matrix for that cotree. Then the vector of all currents is
$$i = B_f' i^c \qquad \square$$

Corollaries C-2 and C-3 can be combined to derive the law of conservation of energy from the Kirchhoff laws.

*Corollary C-4

Let v and i be any admissible voltage and current vectors for a graph (which is assumed to have no external sources). Then
$$p = v'i = 0 \qquad \square$$

The result of Corollary C-4, originally discovered by Weyl [191], is stronger than the usual conservation of energy because it does not require that the network simultaneously support the vectors v and i (in fact, the vectors v and i do not even have to be consistent with the component constraints of the network). Rather, any vectors v and i such that

$$B_f v = 0 \qquad \text{(C-33)}$$

and

$$S_f i = 0 \qquad \text{(C-34)}$$

yield the result that

$$p = v'i = 0 \qquad \text{(C-35)}$$

In particular, if $v(t_1)$ is the voltage vector admitted by a network at one time, and $i(t_2)$ is a current vector measured at a different time, one still has the equality

$$v'(t_1)i(t_2) = 0 \qquad \text{(C-36)}$$

In fact, this holds even if the components within the network vary with time. Of course, the special case wherein v and i are simultaneously supported by the network is the usual conservation of energy relationship.

C-7 Discussion

In the preceding we have defined the graph and its subgraphs and constructed minimal sets of circuits and cocircuits which suffice to describe the voltage and current variables for a graph. Based on these, voltage and current were defined and the B_f, S_f, and F matrices were constructed. Two statements for each of the Kirchhoff laws were then obtained in terms of these matrices.

It is of significance that the characteristics of B_f, S_f, and F are all completely determined by the submatrix C; hence, all three matrices contain complete information about the graph (and, in fact, can be used to reconstruct the graph from C) [83].

References

[1] B. D. O. Anderson, "Synthesis of time-varying passive networks," Ph.D dissertation, Stanford University, Stanford, Calif., 1966; also, Stanford Electronics Labs., Rept. 6560-7.

[2] ——, "Properties of time-varying n-port impedance matrices," presented at the 8th Midwest Symp. on Circuit Theory, Colorado State University, Fort Collins, 1965, paper 8.

[3] ——, "Cascade synthesis of time-varying non-dissipative n-ports," Stanford Electronics Labs., Stanford, Calif., unpublished notes, 1966.

[4] B. D. O. Anderson and R. W. Newcomb, "On reciprocity in linear time-invariant networks," Stanford Electronic Labs., Stanford, Calif., Rept. SEL-65-101, 1965.

[5] ——, "Apparently lossless time-varying networks," Stanford Electronics Labs., Stanford, Calif., Rept. SEL-65-094, 1965.

[6] ——, "Functional analysis of linear passive networks," Stanford Electronics Labs., Stanford, Calif., unpublished notes, 1966.

[7] B. D. O. Anderson, D. A. Spaulding, and R. W. Newcomb, "Useful time-variable circuit element equivalences," *Electron. Letters*, vol. 1, pp. 56–57, May 1965.

[8] ——, "The time-variable transformer," *Proc. IEEE*, vol. 53, pp. 634–635, June 1965.

[9] B. D. O. Anderson and R. W. Newcomb, "Degenerate networks," *Proc. IEEE*, vol. 53, pp. 694–695, April 1965.

[10] B. D. O. Anderson, "Passive time-variable scattering matrix synthesis," in *Notes on System Science*, vol. 3, University of Newcastle, Newcastle, Australia, Rept. EE-6902, pp. 30–38, 1969.

[11] B. D. O. Anderson and J. B. Moore, "Procedures for time-varying impedance synthesis," *Proc. 11th Midwest Symp. on Circuit Theory*, University of Notre Dame, Notre Dame, Ind., 1968, pp. 17–26.

[12] B. D. O. Anderson, R. W. Newcomb, R. E. Kalman, and D. C. Youla, "Equivalence of linear time-invariant dynamical systems," *J. Franklin Inst.*, vol. 281, pp. 271–278, May 1966.

[13] B. D. O. Anderson, "A system theory criterion for positive real matrices," *J. SIAM Control*, vol. 5, pp. 171–182, May 1967.

[14] B. D. O. Anderson and R. W. Newcomb, "Lossless n-port synthesis via state-space techniques," Stanford Electronics Labs., Stanford, Calif., Rept. SEL-6558-8, 1967.

[15] B. D. O. Anderson, "Algebraic description of bounded real matrices," *Electron. Letters*, vol. 2, pp. 466–467, December 1966.

[16] B. D. O. Anderson and R. W. Brockett, "A multiport state-space Darlington synthesis," *IEEE Trans. Circuit Theory*, vol. CT-14, pp. 336–337, September 1967.

[17] B. D. O. Anderson and R. W. Newcomb, "Impedance synthesis via state-space techniques," *Proc. IEE* (London), vol. 115, pp. 928–936, July 1968.

[18] ——, "Cascade connection for time-invariant n-ports," *Proc. IEE* (London) vol. 113, pp. 970–974, June 1966.

[19] B. D. O. Anderson and J. B. Moore, "Network realization of time-varying passive impedances," University of Newcastle, Newcastle, Australia, Rept. EE-6810, 1968.

[20] B. D. O. Anderson, "An algebraic solution to the spectral factorization problem," *IEEE Trans. Automatic Control*, vol. AC-12, pp. 410–414, August 1967.

[21] H. G. Ansell, "Networks of transmission lines and lumped reactors," *IRE Trans. Circuit Theory*, vol. CT-11, pp. 214–223, June 1964.

[22] G. Bachman and L. Narci, *Functional Analysis*. New York: Academic Press, 1968.

[23] C. Berge, *The Theory of Graphs and its Applications*. New York: Wiley, 1962.

[24] C. Berge and A. Ghouila-Houri, *Programming, Games, and Transportation Networks*. New York: Wiley, 1965.

[25] P. Beckmann, *The Scattering of Electromagnetic Waves from Rough Surfaces*. New York: Macmillan, 1963.

[26] A. V. Balakrishnan, "Foundations of state-space theory of continuous systems 1," *J. Comp. and System Sciences*, vol. 1, pp. 91–116, April 1967.

[27] V. Belevitch, "Factorization of scattering matrices with applications to passive network synthesis," *Phillips Research Repts.*, vol. 18, pp. 275–317, August 1963.

[28] ——, "Theory of $2n$-terminal networks with applications to conference telephony," *Elec. Commun.*, vol. 27, p. 223, September 1950.

[29] ——, "Synthesis des reseaux electriques passifs a n pairs de bornes de matrice de repartition prédéterminée," *Annales Telecommun.* vol. 6, pp. 302–312, November 1951.

[30] ——, "Transmission losses in $2n$-terminal networks," *J. Appl. Phys.*, vol. 19, pp. 636–638, July 1948.

[31] ——, *Classical Network Theory.* San Francisco: Holden-Day, 1968.

[32] ——, "Elementary applications of the scattering formalism to network design," *IRE Trans. Circuit Theory*, vol. CT-3, pp. 97–104, June 1956.

[33] ——, "Algebraic structure of formal realizability theory," *Revue H.F.*, vol. 4, pp. 183–194, December 1959.

[34] ——, "Summary of the history of circuit theory," *Proc. IRE*, vol. 50, pp. 848–855, May 1962.

[35] ——, "The maximum number of parameters of n-ports of various classes," *Phillips Research Repts.*, vol. 19, pp. 73–77, April 1967.

[36] R. Bellman, *Introduction to Matrix Analysis.* New York: McGraw-Hill, 1960.

[37] M. Bunge, *Causality.* Cleveland: World Publishing, 1963.

[38] R. Bott and R. J. Duffin, "Impedance synthesis without use of transformers," *J. Appl. Phys.*, vol. 20, p. 816, August 1949.

[39] O. Brune, "Synthesis of finite two-terminal networks whose driving point impedance is a prescribed function of frequency," *J. Math. and Phys.*, vol. 10, pp. 191–236, August 1931.

[40] E. J. Beltrami and M. R. Wohlers, *Distributions and Boundary Value Problems of Analytic Functions.* New York: Academic Press, 1968.

[41] R. G. Busacker and T. L. Seaty, *Finite Graphs and Networks.* New York: McGraw-Hill, 1965.

[42] M. Bayard, "Synthesis of n-terminal pair networks," *Proc. Symp. on Modern Network Synthesis*, (Polytechnic Institute of Brooklyn, Brooklyn, N.Y., 1952).

[43] ——, *Theorie des Reseaux de Kirchhoff.* Paris: Rev. d'Optique, 1954.

[44] H. J. Carlin and A. Giardano, *Network Theory.* Englewood Cliffs, N.J.: Prentice-Hall, 1964.

[45] H. J. Carlin, "General n-port synthesis with negative resistors," *Proc. IRE*, vol. 48, pp. 1174–1175, June 1960.

[46] ——, "Network theory without circuit elements," *Proc. IEEE*, vol. 55, pp. 482–497, April 1967.

[47] ——, "On the existence of the scattering matrix of passive networks," *IEEE Trans. Circuit Theory*, vol. CT-14, pp. 418–419, December 1967.

[48] ——, "Network synthesis with transmission line elements," Polytechnic Institute of Brooklyn, Brooklyn, N.Y., Rept. PIBMRI-1235-64, 1965.

[49] ——, "Singular network elements," *IEEE Trans. Circuit Theory*, vol. CT-11, pp. 67–72, March 1964.

[50] ——, "The scattering matrix in network theory," *IRE Trans. Circuit Theory*, vol. CT-3, pp. 88–97, June 1956.

[51] ——, "Cascade transmission line synthesis," Polytechnic Institute of Brooklyn, Brooklyn, N.Y., Rept. PIBMRI-889-61, 1961.

[52] H. J. Carlin and J. Rosinski, "Equivalent circuit and loss invariant for helicon mode semiconductor devices," *IEEE Trans. Circuit Theory*, vol. CT-16, pp. 365–373, September 1969.

[53] P. M. Chirilian, *Integrated and Active Network Analysis and Synthesis*. Englewood Cliffs, N.J.: Prentice-Hall, 1967.

[54] D. A. Calahan, *Modern Network Synthesis*. New York: Hayden, 1964.

[55] J. Cruz and M. Van Valkenburg, "The synthesis of models for time-varying linear systems," *Proc. Symp. on Active Networks and Feedback Systems* (Polytechnic Institute of Brooklyn, Brooklyn, N.Y., 1960).

[56] S. Darlington, "Synthesis of reactance four poles which produce prescribed insertion loss characteristics," *J. Math. and Phys.*, vol. 18, pp. 257–353, September 1939.

[57] N. DeClaris and R. Saeks, "Graph theoretic foundations of LLF networks," IEE (London), Conf. Publ. 23, pp. 163–191; also, *Proc. Conf. on Electric Networks* (University of Newcastle, Newcastle-on-Tyne, England, September 1966).

[58] ——, "Canonic and state variable characterization of networks with variable elements," *Proc. IEEE Internatl. Conv.*, pt. 5, pp. 20–31, 1967.

[59] ——, "Theoretic foundations of network and system analysis," in *Aspects of Network and System Theory*. New York: Holt, Rinehart and Winston, 1971.

[60] P. A. M. Dirac, "The physical interpretation of quantum mechanics," *Proc. Roy. Soc.* (London), Ser. A, vol. 113, pp. 621–641, 1926–1927.

[61] J. F. Delansky, "Some synthesis methods for adjustable networks using multivariable techniques," Ph.D. dissertation, Cornell University, Ithaca, N.Y., 1968.

[62] ——, "Some synthesis methods for adjustable networks using multivariable techniques," *IEEE Trans. Circuit Theory*, vol. CT-16, pp. 435–443, November 1969.

[63] W. F. Donoghue, *Distributions and Fourier Transforms*. New York: Academic Press, 1969.

[64] C. A. Desoer and P. P. Variya, "The minimal realization of a non-anticipative impulse response matrix," *J. SIAM, Appl. Math.*, vol. 15, pp. 754–764, May 1967.

[65] M. C. Davis, "Factoring the spectral matrix," *IEEE Trans. Automatic Control*, vol. AC-8, pp. 296–305, October 1963.

[66] R. J. Duffin, and D. Harzony, "The degree of a rational matrix function," *J. SIAM, Appl. Math.*, vol. 11, pp. 645–658, September 1963.

[67] R. M. Foster, "A reactance theorem," *Bell Sys. Tech. J.*, vol. 3, pp. 257–259, April 1924.

[68] A. Gersho, "Characterization of time-varying linear systems," Ph.D. dissertation, Cornell University, Ithaca, N.Y., 1963; also, Cornell University, Rept. EE 559.

[69] M. S. Ghausi and J. J. Kelly, *Introduction to Distributed Parameter Networks*. New York: Holt, Rinehart and Winston, 1968.

[70] I. Gohberg and N. Krein, "On factorization of operators in Hilbert space," *Soviet Math. Doklady*, vol. 2, pp. 1778–1782, 1962.

[71] S. C. Gupta, *Transform and State Variable Methods*. New York: Wiley, 1966.

[72] D. Harzony, *Elements of Network Synthesis*. New York: Reinhold, 1963.

[73] P. R. Halmos, *Measure Theory*. Princeton, N.J.: Van Nostrand, 1948.

[74] ——, *Finite Dimensional Vector Spaces*. Princeton, N.J.: Van Nostrand, 1942.

[75] B. L. Ho, "On effective construction of realizations from input-output descriptions," Ph.D. dissertation, Stanford University, Stanford, Calif., 1966.

[76] B. L. Ho and R. E. Kalman, "Effective construction of linear state variable models from input-output data," *Proc. 3rd Allerton Conf. on Circuit and System Theory*, pp. 449–459, 1965.

[77] J. Horvath, *Topological Vector Spaces and Distributions*. Reading, Mass.: Addison-Wesley, 1966.

[78] S. Karni, *Network Theory Analysis and Synthesis*. Boston: Allyn & Bacon, 1966.

[79] H. Koenig, Y. Tokad, and H. K. Kesavan, *Analysis of Discrete Physical Systems*. New York: McGraw-Hill, 1966.

[80] H. Koenig and A. H. Zemanian, "Necessary and sufficient conditions for a matrix distribution to have a positive real LaPlace transform," *J. SIAM, Appl. Math.*, vol. 13, pp. 1036–1040, 1965.

[81] T. Koga, "Synthesis of finite passive n-ports with prescribed two-variable reactance matrices," *IEEE Trans. Circuit Theory*, vol. CT-13, pp. 31–51, March 1966.

[82] ——, "Synthesis of finite passive n-ports with prescribed positive real matrices of several variables," *IEEE Trans. Circuit Theory*, vol. CT-15, pp. 2–24, March 1968.

[83] W. H. Kim and R. T. Chein, *Topological Analysis and Synthesis of Communication Networks.* New York: Columbia University Press, 1962.

[84] A. N. Kolmogorov and S. V. Fomin, *Functional Analysis.* Rochester, N.Y.: Graylock, 1961.

[85] R. Kalman, "On a new characterization of linear passive systems," *Proc. 1st Allerton Conf.* (University of Illinois, Urbana) pp. 456–470, 1963.

[86] ———, "Irreducible realizations and the degree of matrix rational functions," *J. SIAM, Appl. Math.*, vol. 13, pp. 520–544, June 1965.

[87] R. Kalman, M. Arbib, and P. L. Falb, *Topics in Mathematical System Theory.* New York: McGraw-Hill, 1969.

[88] B. K. Kinariwalla "Theory of cascaded transmission lines," *Proc. Symp. on Generalized Networks* (Polytechnic Institute of Brooklyn, Brooklyn, N.Y.), pp. 345–352, 1966.

[89] ———, "Theory of cascaded structures: Lossless transmission lines," *Bell Sys. Tech. J.*, vol. 45, pp. 631–649, April 1966.

[90] F. F. Kuo, *Network Analysis and Synthesis.* New York: Wiley, 1962.

[91] M. G. Krein, "Integral equations on the half line with kernels depending on differences of arguments," Am. Math. Soc., Translations Ser. 2, vol. 22, pp. 163–288, 1956.

[92] E. Kuh and R. Rohrer, *Linear Active Network Theory.* San Francisco: Holden-Day, 1966.

[93] E. S. Kuh, D. M. Layton, and J. Tow, "Network analysis and synthesis via state variables," in *Network and Switching Theory.* New York: Academic Press, 1968.

[94] S. Lang *Algebra.* Reading, Mass.: Addison-Wesley, 1965.

[95] H. Levenstein, "Theory of networks of linearly variable resistors," *Proc. IRE*, vol. 46, pp. 486–493, February 1968.

[96] M. J. Lighthill, *Introduction to Fourier Analysis and Generalized Functions.* New York: Cambridge University Press, 1960.

[97] N. Liskov, "Analytical techniques for linear time-varying systems," Ph.D. dissertation, Cornell University, Ithaca, N.Y., 1964.

[98] S. G. Loo, "A simple method for factoring spectral matrices," in *Notes on System Science*, vol. 2, University of Newcastle, Newcastle, Australia, Rept. EE-6716, pp. 32–38, 1967.

[99] S. G. Loo, J. B. Moore, and B. D. O. Anderson, "Time-varying spectral factorization, I and II," University of Newcastle, Newcastle, Australia, Repts. EE-6701, EE-6702, 1967.

[100] N. Levan, D. G. Lampard, B. D. O. Anderson, and R. W. Newcomb, "Step-up n-port networks," Stanford Electronics Labs., Stanford, Calif., Rept. 6558-17, 1967.

[101] N. Leven, D. G. Lampard, and R. W. Newcomb, "Bi-step-up n-ports," Stanford Electronics Labs., Stanford, Calif., Rept. 6558-18, 1967.

[102] N. Leven and R. W. Newcomb, "The characterization of step-up vectors of n-port networks," Stanford Electronics Labs., Stanford, Calif., Rept. 6558-1, 1965.

[103] N. Leven, D. G. Lampard, B. D. O. Anderson, and R. W. Newcomb, "Step-up n-ports in network theory," *IEEE Trans. Circuit Theory*, vol. 16, pp. 359–365, August 1969.

[104] D. M. Layton, "Equivalent realizations for linear systems with applications to n-port network synthesis," Ph.D. dissertation, University of California at Berkeley, 1966.

[105] ——, "State representation, passivity, reciprocity and n-port synthesis," *Proc. 4th Allerton Conf.*, pp. 639–646, 1966.

[106] B. McMillan, "Introduction to formal realizability theory," *Bell Sys. Tech. J.*, vol. 31, pp. 217–279, 441–600, March and May 1952.

[107] J. Mentzes, *Scattering and Defraction of Radio Waves*. London: Pergamon Press, 1955.

[108] G. J. Minty, "On the axiomatic foundations of the theories of directed linear graphs, electrical networks and network programming," *J. Math. and Mech.*, vol. 15, pp. 485–520, March 1966.

[109] C. G. Montgomery, R. H. Dicke, and E. M. Purcell, *Principles of Microwave Circuits*. Cambridge, Mass.: M. I. T. Press, 1948.

[110] R. W. Newcomb, *Linear Multiport Synthesis*. New York: McGraw-Hill, 1966.

[111] ——, "Synthesis of networks passive at p_0," Ph.D. dissertation, University of California at Berkeley, 1960.

[112] ——, "Distributional impulse theorems," *Proc. IEEE*, vol. 51, pp. 1157–1158, August 1963.

[113] ——, "The foundations of network theory," *Instn. of Engrs. (Australia), Trans. Elec. and Mech. Engrg.*, vol. EM-6, pp. 7–12, May 1964.

[114] ——, "On the realization of multivariable transfer functions," Cornell University, Ithaca, N.Y., Rept. EERL 58, December 1966.

[115] ——, "On causality, passivity and single valuedness," *IRE Trans. Circuit Theory*, vol. CT-9, pp. 87–89, March 1962.

[116] ——, "A Bayard type non-reciprocal n-port synthesis," *IEEE Trans. Circuit Theory*, vol. CT-10, pp. 85–90, March 1965.

[117] ——, *Network Theory, The State-Space Approach*. Louvain, Belgium: Library Universitaire Louvain, 1968.

[118] ——, *Concepts of Linear Systems and Controls*. Belmont, Calif.: Brooks/Cole, 1969.

[119] ——, *Active Integrated Circuit Synthesis*. Englewood Cliffs, N.J.: Prentice-Hall, 1968.

[120] R. W. Newcomb and D. A. Spaulding, "*The time-variable scattering matrix*," *Proc. IEEE*, vol. 53, pp. 651–652, June 1965.

[121] R. W. Newcomb, "Non-reciprocal transmission line n-port synthesis," *Proc. Instn. of Radio and Elec. Engrs.* (Australia), vol. 26, pp. 135–142, April 1965.

[122] R. C. Newton, *Scattering Theory of Waves and Particles.* New York: McGraw-Hill, 1966.

[123] Y. Oono and K. Yasuura, "Synthesis of finite passive $2n$-terminal network with prescribed scattering matrices," *Memoirs of the Faculty of Engineering, Kyushu University*, (Japan) vol. 14, pp. 125–127, May 1954.

[124] Y. Oono and T. Koga "Synthesis of a variable parameter 1-port," *IRE Trans. Circuit Theory*, vol. CT-10, pp. 213–227, June 1963.

[125] Y. Oono, "Synthesis of finite $2n$-terminal networks as the extension of Brune's 2-terminal network theory," *J. Inst. Elec. Commun. Engrs. Japan*, vol. 31, pp. 163–181, August 1948.

[126] ——, "Synthesis of finite $2n$-terminal networks by a group of networks each which contains only one ohmic resistance," *J. Math. and Phys.*, vol. 29, pp. 13–26, April 1950.

[127] ——, "Application of scattering matrices to the synthesis of n-ports," *IRE Trans. Circuit Theory*, vol. CT-3, pp. 111–120, June 1956.

[128] O. Ore, *Theory of Graphs.* Providence, R.I.: Am. Math. Soc., 1962.

[129] H. Ozaki and T. Kasimi "Positive real functions of several variables and their application to variable networks," *IRE Trans. Circuit Theory*, vol. CT-7, pp. 251–260, September 1960.

[130] H. Ozaki and J. Isshii, "Synthesis of transmission line networks and design of UHF filters," *IRE Trans. Circuit Theory*, vol. CT-2, p. 325, September 1955.

[131] A. Papoulis, *The Fourier Integral and its Applications.* New York: McGraw-Hill, 1962.

[132] M. C. Pease, *Methods of Matrix Algebra.* New York: Academic Press, 1965.

[133] R. Phillips and P. Lax, *Scattering Theory.* New York: Academic Press, 1966.

[134] W. A. Porter, *Modern Foundations of Systems Engineering.* New York: MacMillan, 1966.

[135] T. N. Rao and R. W. Newcomb, "Synthesis of positive real matrices," *Electron. Letters*, vol. 3, pp. 349–350, August 1967.

[136] P. I. Richards, "A special class of functions with positive real part in the half plane," *Duke Math. J.*, vol. 14, pp. 777–786, September 1947.

[137] ——, "Resistor-transmission line circuits," *Proc. IRE*, vol. 36, pp. 217–220, February 1948.

[138] M. B. Reed, private communication.

[139] ——, *Foundations of Electric Network Theory.* Englewood Cliffs, N.J.: Prentice-Hall, 1961.

[140] F. Riesz and B. Sz-Nagy, *Functional Analysis*. New York: Unger, 1955.

[141] R. A. Rohrer, "The scattering matrix — normalized to complex n-port load networks," *IEEE Trans. Circuit Theory*, vol. CT-12, pp. 223–230, March 1965.

[142] ———, "Lumped network passivity criteria," *IEEE Trans. Circuit Theory*, vol. CT-15, pp. 24–31, March 1968.

[143] G. Raisbeck, "A definition of passive linear networks in terms of time and energy," *J. Appl. Phys.*, vol. 25, pp. 1510–1514, December 1954.

[144] W. Rudin, *Real and Complex Analysis*. New York: McGraw-Hill, 1966.

[145] R. Saeks, "An algebraic time domain theory of linear time-variable networks," Ph.D. dissertation, Cornell University, Ithaca, N.Y., 1967.

[146] ———, "Synthesis of lossless time-variable networks," *2nd Princeton Conf.*, pp. 404, 1968.

[147] ———, "Synthesis of active time-variable networks," *IEEE Trans. Circuit Theory*, vol. CT-16, pp. 312–317, August 1969.

[148] ———, "Algebraic network theory — the analysis problem," *Proc. 12th Midwest Symp. on Circuit Theory* (University of Texas), Austin, Texas, paper IX-7, 1969.

[149] ———, "Synthesis of general linear networks," *J. SIAM, Appl. Math.*, vol. 16, pp. 924–930, September 1968.

[150] ———, "On finite energy networks," *IEEE Trans. Circuit Theory*, vol. CT-18, November 1970 (to be published).

[151] ———, "Analysis of multiterminal networks with indefinite scattering matrices," *Proc. 12th Midwest Symp. on Circuit Theory* (University of Texas), Austin, Texas, paper VI-6, 1969.

[152] ———, "Causality in Hilbert space," *SIAM Rev.*, vol. 12, July 1970.

[153] J. O. Scanlan, "Multivariable network functions," *IEEE Internatl. Conv. Rec.*, pt. 5, pp. 32–36, March 1966.

[154] ———, "Cascade synthesis of distributed networks," *Proc. Symp. on Generalized Networks* (Polytechnic Institute of Brooklyn, Brooklyn, N.Y.), pp. 227–256, 1966.

[155] S. Seshu and M. B. Reed, *Linear Graphs and Electric Networks*. Reading, Mass.: Addison-Wesley, 1961.

[156] L. Schwartz, *Theorie des Distributions*. Paris: Hermann, 1957.

[157] ———, "Theorie des noyaux," *Proc. Internatl. Congress of Math.* (Cambridge, Mass.), pp. 220–230, 1950.

[158] L. M. Silverman, "Representation and realization of time-variable linear systems," Ph.D. dissertation, Columbia University, New York, N.Y., 1966.

[159] L. M. Silverman, and H. E. Meadows, "Controllability and observability in time-variable linear systems," *J. SIAM, Control*, vol. 5, pp. 64–73, February 1967.

[160] L. M. Silverman, "Synthesis of impulse response matrices by internally stable and passive realizations," *IEEE Trans. Circuit Theory*, vol. CT-16, pp. 238–245, September 1968.

[161] L. M. Silverman and B. D. O. Anderson, "Controllability, observability, and stability of linear systems," *J. SIAM, Control*, vol. 6, pp. 121–130, January 1968.

[162] P. Slepian, *Mathematical Foundations of Network Analysis*. New York: Springer-Verlag, 1968

[163] M. F. Smiley, *Algebra of Matrices*. Boston: Allyn & Bacon, 1965.

[164] M. K. Sain, "On control applications of a determinant equality related to eigen value computation," *IEEE Trans. Automatic Control*, vol. AC-11, pp. 109–111, January 1966.

[165] M. Saito, "Synthesis of transmission line networks by multivariable techniques," *Proc. Symp. on Generalized Networks* (Polytechnic Institute of Brooklyn, Brooklyn, N.Y.), pp. 353–392, 1966.

[166] ——, "Characteristics of variable networks," *J. Inst. Elec. Commun. Engrs. Japan*, vol. 42, pp. 725–730, August 1959.

[167] ——, "The condition for unit element extraction in mixed-lumped and distributed networks," *IEEE Trans. Circuit Theory*, vol. CT-13, pp. 218–220, June 1966.

[168] D. A. Spaulding, "Passive time-varying networks," Ph.D. dissertation, Stanford University, Stanford, Calif., 1964.

[169] ——, "Lossless time-varying impedance synthesis," *Electron. Letters*, vol. 1, pp. 165–167, August 1965.

[170] ——, "Foster-type time-varying lossless synthesis," *Electron. Letters*, vol. 1, pp. 248–249, November 1965.

[171] T. E. Stern, *Nonlinear Networks and Systems*. Reading, Mass.: Addison-Wesley, 1965.

[172] A. R. Stubberud, *Analysis and Synthesis of Linear Time-Variable Systems*. Berkeley, Calif.: University of California Press, 1964.

[173] K. Su, *Active Synthesis*. New York: McGraw-Hill, 1966.

[174] B. D. H. Tellegen, "The gyrator, a new electric network element," *Phillips Research Repts.*, vol. 3, pp. 81–101, April 1948.

[175] ——, "On nullators and norators," *IEEE Trans. Circuit Theory*, vol. CT-13, pp. 466–469, December 1966.

[176] ——, "Synthesis of passive resistance-less 4-poles that may violate the reciprocity relation," *Phillips Research Repts.*, vol. 3, pp. 321–337, October 1948.

[177] ——, "Synthesis of passive 2-poles by means of networks containing gyrators," *Phillips Research Repts.*, vol. 4, pp. 31–37, 1949.

[178] ——, "Synthesis of $2n$-poles by networks containing the minimum number of elements," *J. Math. and Phys.*, vol. 32, pp. 1–18, April 1953.

[179] F. Treves, *Topological Vector Spaces, Distributions and Kernels.* New York: Academic Press, 1967.

[180] W. F. Tutte, *Connectivity in Graphs.* Oxford, England: Oxford University Press, 1967.

[181] H. F. Van Landingham, "Criteria for approximating solutions of linear time-varying systems," Ph.D. dissertation, Cornell University, Ithaca, N.Y., 1967.

[182] S. Vongpanitlerd and B. D. O. Anderson, "Scattering matrix synthesis via reactance extraction," University of Newcastle, Newcastle, Australia, Rept. EE-6903, 1969.

[183] L. Weinberg, *Network Analysis and Synthesis.* New York: McGraw-Hill, 1963.

[184] N. Weiner, "On the factorization of matrices," *Comment Math. Helv.*, vol. 29, pp. 97–111, 1955.

[185] M. R. Wohlers, "Complex normalization of scattering matrices and the problem of compatible impedances," *IEEE Trans. Circuit Theory*, vol. CT-12, pp. 528–539, December 1962.

[186] ——, *Lumped and Distributed Passive Networks.* New York: Academic Press, 1969.

[187] ——, "A realizability theory for smooth lossless transmission lines," *IEEE Trans. Circuit Theory*, vol. CT-13, pp. 356–363, September 1966.

[188] ——, "A realizability theory for smooth lossless transmission lines, part II," *IEEE Trans. Circuit Theory*, vol. CT-14, pp. 442–444, December 1967.

[189] M. R. Wohlers and E. J. Beltrami, "Distribution theory as a basis for generalized passive network analysis," *IEEE Trans. Circuit Theory*, vol. CT-12, pp. 164–169, June 1967.

[190] M. R. Wohlers, "Scattering matrices normalized to active n-ports at real frequencies," *IEEE Trans. Circuit Theory*, vol. CT-16, pp. 254–256, June 1969.

[191] H. Weyl, "Reparticion de corriente en una red conductora," *Rev. Matemat. Hispano-Americana*, vols. 4, 5, 1923, 1924.

[192] D. C. Youla, L. J. Castriota, and H. J. Carlin, "Bounded real scattering matrices and the foundations of linear passive network theory," *IRE Trans. Circuit Theory*, vol. CT-6, pp. 102–124, March 1959.

[193] D. C. Youla, "Synthesis of linear dynamical systems from prescribed weighting patterns," *J. SIAM, Appl. Math.*, vol. 14, pp. 527–560, May 1966.

[194] ——, "The synthesis of networks containing lumped and distributed elements," *Proc. Symp. on Generalized Networks*, (Polytechnic Institute of Brooklyn, Brooklyn, N.Y.), pp. 289–343, 1966.

[195] ——, "Cascade synthesis of passive n-ports," Rome Air Development Center, Griffiss AFB, Rome, N.Y., Rept. RADC-TDR-64-332, 1964.

[196] ——, "On scattering matrices normalized to complex port numbers," *Proc. IRE*, vol. 59, p. 1221, July 1961.

[197] ——, "An extension of the concept of a scattering matrix," *IEEE Trans. Circuit Theory*, vol. CT-11, pp. 310–312, June 1964.

[198] ——, "A new theory of braodband matching," *IEEE Trans. Circuit Theory*, vol. CT-11, pp. 30–49, March 1964.

[199] D. C. Youla and P. Tissi, "n-port synthesis via reactance extraction," *IEEE Internatl. Conv. Rec.*, pt. 5, pp. 180–203, 1966.

[200] D. C. Youla, "The synthesis of networks containing lumped and distributed elements, part 1," in *Network and Switching Theory*. New York: Academic Press, 1968, pp. 73–134.

[201] ——, "Representation theory for linear passive networks," Polytechnic Institute of Brooklyn, Brooklyn, N.Y., Rept. PIBMRI-R-655-58, 1958.

[202] ——, "A new theory of cascade synthesis," *IRE Trans. Circuit Theory*, vol. CT-9, pp. 244–260, September 1961.

[203] ——, "Physical realizability criteria," *IRE Internatl. Conv. Rec.*, pt. 2, pp. 181–192, 1960.

[204] ——, "On the factorization of rational matrices," *IRE Trans. Information Theory*, vol. IT-6, pp. 172–189, September 1961.

[205] D. C. Youla and P. Tissi, "An explicit formula for the degree of a rational matrix," Polytechnic Institute of Brooklyn, Brooklyn, N.Y., Rept. PIBMRI-1272-65, 1965.

[206] D. C. Youla, P. Tissi, and W. Kohler, "Synthesis of networks containing lumped and distributed elements, Parts I and II," Polytechnic Institute of Brooklyn, Brooklyn, N.Y., Rept. RADG-TR-66-489, 1966.

[207] L. A. Zadeh and C. A. Desoer, *Linear System Theory, The State Space Approach*. New York: McGraw-Hill, 1963.

[208] L. A. Zadeh, "Frequency analysis of variable networks," *Proc. IRE*, vol. 38, pp. 291–299, March 1950.

[209] ——, "Time-varying networks," *Proc. IRE*, vol. 49, pp. 1488–1503, October 1961.

[210] A. H. Zemanian, *Distribution Theory and Transform Analysis*. New York: McGraw-Hill, 1965.

[211] ——, "An n-port realizability theory based on the theory of distributions," *IEEE Trans. Circuit Theory*, vol. CT-2, pp. 265–274, June 1963.

[212] ——, "The Hilbert port," *J. SIAM, Appl. Math.*, vol. 18, pp. 98–138, January 1970.

Index

Index

A

A-B matrix, 87
Admittance matrix, 65. (*See also* Immittance matrix)
Age variable, 81
Algebra, definition, 364
　of kernels, 58–59, 383–384
　of operators, 55, 97–98, 106, 365
　of system functions, 97–98, 106
Analysis, of cascaded structures, 158–161, 164
　of first order n-ports, 165–170
　via immittance matrices, 155–158, 316–317
　via π-matrices, 136–140
　via scattering matrices, normalized, 149–152
　　unnormalized, 75–76, 141–149, 318
Augmented n-ports, definition, 64–65
　matrix representation of, 67

B

Bilateral n-ports, 37
Bounded reality, definition, 116
　multivariable, 268–269
　relation to passivity, 116
　of an RLC n-port, 205
　state-space criterion for, 350, 352
Bounded real-paraunitary, definition, 118
　Foster's reactance theorem for, 125
　multivariable, 269, 287–288
　relation of losslessness, 118
　of an RLC n-port, 207

C

Canonical n-port, 127–128
Capacitor, definition, 19
　losslessness of, 42, 46, 118–119, 123
　matrix representation of, 66, 67, 69, 87, 104–108
　passivity of, 40–41, 46, 118–119
Cascade, loading, 109, 130, 159–160, 164
　transmission lines, 253–254, 257–260, 296–299
Causal n-ports, definition, 38
　linear, 50
　relation to nonenergetic n-ports, 49
relation to passivity, 47
scattering matrix of, 68–85
Circulator, definition, 23
　matrix representation of, 69
　nonreciprocity of, 80
Closed n-port, 45
Circuit, 393, 403
Circuit matrix, 407, 411–412
Cocircuit, 394, 404
Cocircuit matrix, 408–409, 412–413
Companion matrix, 277, 315
Composite n-port, 31
Continuity, uniform, 3, 370–371
　weak, 372
Convergence, distributional, 380
　of normal n-ports, 92
　uniform, 370–371
　weak, 371, 385

D

Decomposition, of a dynamical system, 314–323
　multivariable, 278–283
Delta (δ) function, 373, 376, 379
Distribution, 55, 375, 378
Distributional, convergence, 38
　Fourier transform, 386
　kernel, existence, 61–63, 88, 90, 381
　　operations on, 58–59, 383–384
　　relation of dynamical system, 329
　　separable, 305, 328, 331
　Laplace transform, 387
　operations, 378
Dynamical system, analysis via, 316–317
　definition, 314
　equivalence, 320–321
　fundamental solution matrix for, 324
　proper, 316
　relation to differential operator, 315
　relation to distributional kernel, 329
　relation to multivariable decomposition, 322–323
　solution of, 326–328
　synthesis via, 332–342, 348–353
　transition matrix of, 325

429

INDEX

E

Edge, 392, 399–401
Equivalence, of decompositions, 282–283, 320–321
   of realizations, 187–189, 307–308
Energy, 17, 39–43

F

Factorization, operator, 183, 358
   rational system function, 222–223, 228, 231, 350–351
First order n-ports, 165–170
Foster's reactance theorem, 125–126
Fourier transform, 385–386
Functionals, 55, 364, 375
Fundamental connection matrix, 131, 150, 410–411

G

Graph, 13, 292
Graph matrices, circuit matrix, 131, 407, 410
   cocircuit matrix, 131, 408–410
   fundamental connection matrix, 131, 410–411
   normalized fundamental connection matrix, 150
Gyrator, definition, 20
   matrix representation of, 69
   $(n+m)$-port, 20
   nonreciprocity of, 80

H

Hankel matrix, 276
Hurwitz conjugate, 120, 202

I

Immittance matrices, analysis via, 155–158, 316–317
   augmented, 67, 85, 111, 123
   composite, 155
   definition, 54, 65
   losslessness of, 78, 122
   normalized, 85
   passivity of, 77, 122
   reciprocity of, 83
   synthesis via, general techniques, 177, 189–193
      first order n-ports, 169–170
      memoryless n-ports, 197–200
      RLC n-ports, 216–220, 227–230, 248
      time-variable RLC n-ports, 232–242
      transmission line n-ports, 253
   time-invariance of, 83
Impedance matrix, see Immittance matrices
Impulse Response, see Distributional kernel
Inductor, definition, 19
   losslessness of, 46
   matrix representation of, 69, 87, 107–108
   passivity of, 46
Inner product, 16, 366–367

K

Kernel, see Distributional kernel
Kirchhoff's laws, 13, 411–413

L

Laplace transform, 105–109, 387–388
Linear n-port, 35
Locally integrable functions, 375
Lossless n-ports, definition, 41
   immittance matrices of, 78, 122, 207–208
   instantaneous, 42
   norm of, 52, 74
   scattering matrix of, 73–75, 118, 207
   state-space characterization of, 336
   synthesis of, 213–220, 237–241, 254–255, 287–295, 337
Lossless positive reality, definition, 122
   Foster's reactance theorem for, 125–126
   relation to losslessness, 122
   of an RLC n-port, 208

M

Memoryless n-ports, definition, 36
   losslessness of, 74–75
   matrix representation of, 66, 90
   synthesis of, 184, 193–200
   system function of, 101
Metric, 369–370
Multivariable n-ports, decomposition of, 272–284
   definition, 265–267
   losslessness of, 271
   passivity of, 270–271
   synthesis of, 284–295, 297–302

N

n-port, 18
Network, 29
Network matrices, of circuit elements, 69
   definitions, 61, 65, 67, 86, 89

relation between, 93
Nollor, 91
Nonenergetic n-ports, definition, 42
  norm of, 49
  relation to causality, 49, 51, 71
  relation to normality, 111
  scattering matrix of, 71, 117, 211
Norator, 25, 46, 88
Norm, definition, 17, 365
  induced by an inner product, 367
  l_p, L_p, 365–366
  of a lossless n-port, 52
  of an n-port, 26
  of a nonenergetic n-port, 49
  of a passive n-port, 49
  of an operator, 366
Normal n-port, 92–111
Nullator, 25, 46, 88

O

Open circuit, 46
Operator, adjoint, 367–369
  algebra, 55, 97–98, 106, 362–365
  definition, 360
  derivative, 57, 101
  differential, 107, 311–315
  frequency-domain representation, 97, 107
  factorization, 183, 350
  integral, 57, 101
  representation via distributional kernels, 55–59, 381–382

P

Parseval's relation, 97, 387
Passive n-ports, definition, 39
  immittance matrices of, 77, 122, 207–208
  norm of, 49, 70
  relation to causality, 47, 51
  scattering matrix of, 71, 73, 113, 116–117, 205–206
  state-space characterization of, 335–336, 348, 350, 352
  synthesis of, 180–187, 189–193, 220–237, 256–261, 284–287, 348–353
Periodic n-ports, 44
π matrix, analysis via, 136–146
  composite, 136
  definition, 86
  existence of, 88
  nonuniqueness of, 87
  synthesis via, 176–178

π matrix, 91, 103
Positive reality, definition, 116
  relation to passivity, 116
  of an RLC n-port, 207
  state-space criterion for, 336–337, 348
Power, 17, 413
Predictor-delay, matrix representation of, 57, 72, 90, 100
  passivity of, 72

Q

Q matrix, definition, 89
  existence of, 90, 103
Q matrix, 91

R

Reality, 114–116
Realization, *see* Synthesis
Reciprocal n-ports, definition, 45
  Lorentz, 45, 83, 123–125
  matrix representation of, 78, 83
  relation to time-invariance, 84
Resistor, definition, 19
  matrix representation of, 64, 69, 87, 90, 107–108
  passivity of, 40–42, 46
Richards' theorem, 211–212, 298
Richards' transformation, definition, 242, 249, 270, 295
  for frequency dependent lines, 263
  preservation of bounded reality by, 250–251, 271
RLC n-ports, definition, 204
  losslessness of, 207–208
  passivity of, 205–207
  synthesis of, 213–242, 337, 348–353
  time-variable, 309, 331–342

S

Scattering matrix, analysis via, 75–76, 141–152, 318
  causality of, 62–63, 68–70
  composite, 142
  definition, 61
  existence of, 61–62, 103, 109–112, 153
  losslessness of, 73, 118
  nonenergetic, 71, 211
  normalized, 89, 153
  passivity of, 70–71, 73, 113, 116, 350–352
  product of, 75
  reciprocity of, 78–79

Scattering matrix, cont.
  synthesis via, first order n-ports, 165–168
    general techniques, 177, 180–187
  memoryless n-ports, 193–197
  multivariable n-ports, 284–295, 297–300
  RLC n-ports, 213–216, 220–226, 230–240, 350, 352
    transmission line n-ports, 253–261, 297–300
  time-invariance of, 80–81, 104
Short circuit, 46
Solvable n-port, 44
State-space techniques (*see* Dynamical systems)
Subgraph, 13, 392
Subgraphs, circuit, 13, 393
  cocircuit, 13–14, 394
  cocycle, 13–14, 394
  connected, 396
  cotree, 394
  cycle, 13, 393
  path, 396
  pseudo path, 395
  tree, 393
Synthesis, general techniques, active, 176–178
    first order, 167–170
    memoryless, 184, 193–200
    passive, 180–187, 189–193
    multivariable, cascade transmission line — RLC, 297–300
      lossless, 287–295
      passive, 284–287
      variable parameter, 301–302
    RLC, n-port lossless, 237–241, 337
      n-port passive, 230–237, 348–353
      1-port lossless, 213–220
      1-port passive, 220–230
      time-variable, 232–242
    transmission line, cascade, 253–261, 297–300
      n-port, 253
System function, adjoint of, 113
  analyticity of, 102
  bounded real, 116
  bounded real-paraunitary, 118
  definition, 99
  existence of, 102
  Hurwitz conjugate of, 170–171, 202
  lossless positive real, 122
  multivariable rational, bounded real, 268–269
    bounded real-paraunitary, 269
    decomposition of, 278–283, 322–323
    definition, 267
  degree, 271
  positive real, 121
  rational, bounded real, 205–206, 350–352
    bounded real-paraunitary, 207
    definition, 209
    degree, 210
    factorization of, 222–223, 228, 231, 350–351
    lossless positive real, 208
    positive real, 207–208, 348
    of an RLC n-port, 205
    of a transmission line n-port, 245–247, 250, 262–263
  reality of, 114–116
  time-invariance of, 103–104

T

Testing functions, 375, 385
Time-invariant n-ports, characterization of, 82, 322
  definition, 43
  matrix representation of, 80, 83, 103–104
Time-variable RLC n-ports, analysis via, 316–320
  definition, 309
  degree of, 331
  equivalent, 307–308
  representation of, via differential operators, 311–313
    via distributional kernels, 323–330
    via dynamical systems, 317–320
  synthesis of, 337–342
Transformer, definition, 21
  matrix representation of, 69
Transmission line, definition, 154
  frequency dependent, definition, 261
    passivity of, 262
    scattering matrix of, 262
  loaded, 248
  open-circuited, 249
  short-circuited, 249
  scattering matrix of, 154, 243
  unit element length, 244–245
Transmission line n-ports, commensurate, 244
  cascade, 253–254, 257–260, 299
  definition, 244
  synthesis of, 253–261, 297–300
  system function of, 245–250, 262
Transmission line — RLC n-port, cascade, 296–299
  definition, 295
  synthesis of, 297–300
Tree, 393, 397

V

Variable parameter n-ports, 300–302
Variables, immittance, 12–14, 402, 406
   network, 12–17, 32, 131–135, 411–413
   scattering, 15, 84
Vector space, base, 358–359
   definition, 356
   dimension, 359
   of distributions, 357, 375, 385
   dual, 364
   of homomorphisms, 363
   of network variables, 12, 357, 375
   quotient space, 361
   subspace, 357
   of testing function, 357, 375, 385

SHU-PARK CHAN